# Advances in Intelligent Systems and Computing

Volume 410

**Series editor**

Janusz Kacprzyk, Polish Academy of Sciences, Warsaw, Poland
e-mail: kacprzyk@ibspan.waw.pl

## About this Series

The series "Advances in Intelligent Systems and Computing" contains publications on theory, applications, and design methods of Intelligent Systems and Intelligent Computing. Virtually all disciplines such as engineering, natural sciences, computer and information science, ICT, economics, business, e-commerce, environment, healthcare, life science are covered. The list of topics spans all the areas of modern intelligent systems and computing.

The publications within "Advances in Intelligent Systems and Computing" are primarily textbooks and proceedings of important conferences, symposia and congresses. They cover significant recent developments in the field, both of a foundational and applicable character. An important characteristic feature of the series is the short publication time and world-wide distribution. This permits a rapid and broad dissemination of research results.

## Advisory Board

Chairman

Nikhil R. Pal, Indian Statistical Institute, Kolkata, India
e-mail: nikhil@isical.ac.in

Members

Rafael Bello, Universidad Central "Marta Abreu" de Las Villas, Santa Clara, Cuba
e-mail: rbellop@uclv.edu.cu

Emilio S. Corchado, University of Salamanca, Salamanca, Spain
e-mail: escorchado@usal.es

Hani Hagras, University of Essex, Colchester, UK
e-mail: hani@essex.ac.uk

László T. Kóczy, Széchenyi István University, Győr, Hungary
e-mail: koczy@sze.hu

Vladik Kreinovich, University of Texas at El Paso, El Paso, USA
e-mail: vladik@utep.edu

Chin-Teng Lin, National Chiao Tung University, Hsinchu, Taiwan
e-mail: ctlin@mail.nctu.edu.tw

Jie Lu, University of Technology, Sydney, Australia
e-mail: Jie.Lu@uts.edu.au

Patricia Melin, Tijuana Institute of Technology, Tijuana, Mexico
e-mail: epmelin@hafsamx.org

Nadia Nedjah, State University of Rio de Janeiro, Rio de Janeiro, Brazil
e-mail: nadia@eng.uerj.br

Ngoc Thanh Nguyen, Wroclaw University of Technology, Wroclaw, Poland
e-mail: Ngoc-Thanh.Nguyen@pwr.edu.pl

Jun Wang, The Chinese University of Hong Kong, Shatin, Hong Kong
e-mail: jwang@mae.cuhk.edu.hk

More information about this series at http://www.springer.com/series/11156

Himansu Sekhar Behera
Durga Prasad Mohapatra
Editors

# Computational Intelligence in Data Mining—Volume 1

Proceedings of the International Conference on CIDM, 5–6 December 2015

 Springer

*Editors*
Himansu Sekhar Behera
Department of Computer Science
  Engineering and Information Technology
Veer Surendra Sai University of Technology
Sambalpur, Odisha
India

Durga Prasad Mohapatra
Department of Computer Science
  and Engineering
National Institute of Technology
Rourkela, Odisha
India

ISSN 2194-5357                    ISSN 2194-5365   (electronic)
Advances in Intelligent Systems and Computing
ISBN 978-81-322-2732-8           ISBN 978-81-322-2734-2   (eBook)
DOI 10.1007/978-81-322-2734-2

Library of Congress Control Number: 2015957088

This Springer imprint is published by SpringerNature
The registered company is Springer (India) Pvt. Ltd.

# Preface

The 2nd International Conference on "Computational Intelligence in Data Mining (ICCIDM-2015)" is organized by R.I.T., Berhampur, Odisha, India on 5 and 6 December 2015. ICCIDM is an international forum for representation of research and developments in the fields of Data Mining and Computational Intelligence. More than 300 perspective authors had submitted their research papers to the conference. This time the editors have selected 96 papers after the double-blind peer review process by elegantly experienced subject expert reviewers chosen from the country and abroad. The proceedings of ICCIDM is a mix of papers from some latest findings and research of the authors. It is being a great honour for us to edit the proceedings. We have enjoyed considerably working in cooperation with the International Advisory, Program and Technical Committee to call for papers, review papers and finalize papers to be included in the proceedings.

This International Conference aims at encompassing a new breed of engineers and technologists making it a crest of global success. All the papers are focused on the thematic presentation areas of the conference and they have provided ample opportunity for presentation in different sessions. Research in data mining has its own history. But, there is no doubt about the tips and further advancements in the data mining areas will be the main focus of the conference. This year's program includes exciting collections of contributions resulting from a successful call for papers. The selected papers have been divided into thematic areas including both review and research papers and which highlight the current focus of Computational Intelligence Techniques in Data Mining. The conference aims at creating a forum for further discussion for an integrated information field incorporating a series of technical issues in the frontier analysis and design aspects of different alliances in the related field of intelligent computing and others. Therefore, the call for paper was on three major themes like Methods, Algorithms, and Models in Data Mining and Machine learning, Advance Computing and Applications. Further, the papers discussing the issues and applications related to the theme of the conference were also welcomed to ICCIDM.

The proceedings of ICCIDM are to be released to mark this great day of ICCIDM more special. We hope the author's own research and opinions add value to it. First and foremost are the authors of papers, columns and editorials whose works have made the conference a great success. We had a great time putting together this proceeding. The ICCIDM conference and proceedings are a credit to a large group of people and everyone should be proud of the outcome. We extend our deep sense of gratitude to all for their warm encouragement, inspiration and continuous support for making it possible.

Hope all of us will appreciate the good contributions made and justify our efforts.

# Acknowledgments

The 2015 edition of ICCIDM has drawn hundreds of research articles authored by numerous academicians, researchers and practitioners throughout the world. We thank all of them for sharing their knowledge and research findings on an international platform like ICCIDM and thus contributing towards producing such a comprehensive conference proceedings of ICCIDM.

The level of enthusiasm displayed by the Organizing Committee members right from day one is commendable. The extraordinary spirit and dedication shown by the Organizing Committee in every phase throughout the conference deserves sincere thanks from the core of my heart.

It has indeed been an honour for us to edit the proceedings of the conference. We have been fortunate enough to work in cooperation with a brilliant International Advisory, Program and Technical Committee consisting of eminent academicians to call for papers, review papers and finalize papers to be included in the proceedings.

We would like to express our heartfelt gratitude and obligations to the benign reviewers for sparing their valuable time and putting in effort to review the papers in a stipulated time and providing their valuable suggestions and appreciation in improvising the presentation, quality and content of this proceedings. The eminence of these papers is an accolade not only to the authors but also to the reviewers who have guided towards perfection.

Last but not the least, the editorial members of Springer Publishing deserve a special mention and our sincere thanks to them not only for making our dream come true in the shape of this proceedings, but also for its hassle-free and in-time publication in Advances in Intelligent Systems and Computing, Springer.

The ICCIDM conference and proceedings are a credit to a large group of people and everyone should be proud of the outcome.

# Conference Committee

## Patron

Er. P.K. Patra, Secretary, Roland Group of Institutions, Odisha, India

## Convenor

Dr. G. Jena, Principal, RIT, Berhampur, Odisha, India

## Organizing Secretary

Mr. Janmenjoy Nayak, DST INSPIRE Fellow, Government of India

## Honorary General Chair

Prof. Dr. P.K. Dash, Director, Multi Disciplinary Research Center, S'O'A University, Odisha, India

## General Chair

Prof. Dr. A. Abraham, Director, Machine Intelligence Research Labs (MIR Labs) Washington, USA

# Honorary Advisory Chair

Prof. Dr. S.K. Pal, Padma Shri Awarded Former Director, J.C. Bose National Fellow and INAE Chair Professor Distinguished Professor, Indian Statistical Institute, Kolkata, India

Prof. Dr. L.M. Patnaik, Ex-Vice Chancellor, DIAT, Pune; Former Professor, IISc, Bangalore, India

Prof. Dr. D. Sharma, University Distinguished Professor, Professor of Computer Science, University of Canberra, Australia

# Program Chair

Prof. Dr. B.K. Panigrahi, Ph.D., Associate Professor, Department of Electrical Engineering, IIT Delhi, India

Prof. Dr. H.S. Behera, Ph.D., Associate Professor, Department of Computer Science Engineering and Information Technology, Veer Surendra Sai University of Technology (VSSUT), Burla, Odisha, India

Prof. Dr. R.P. Panda, Ph.D., Professor, Department of ETC, VSSUT, Burla, Odisha, India

# Volume Editors

Prof. H.S. Behera, Associate Professor, Department of Computer Science Engineering and Information Technology, Veer Surendra Sai University of Technology (VSSUT), Burla, Odisha, India

Prof. D.P. Mohapatra, Associate Professor, Department of Computer Science and Engineering, NIT, Rourkela, Odisha, India

# Technical Committee

Dr. P.N. Suganthan, Ph.D., Associate Professor, School of EEE, NTU, Singapore

Dr. Istvan Erlich, Ph.D., Chair Professor, Head, Department of EE and IT, University of DUISBURG-ESSEN, Germany

Dr. Biju Issac, Ph.D., Professor, Teesside University, Middlesbrough, England, UK

Dr. N.P. Padhy, Ph.D., Professor, Department of EE, IIT, Roorkee, India

Dr. Ch. Satyanarayana, Ph.D., Professor, Department of Computer Science and Engineering, JNTU Kakinada

Dr. M. Murugappan, Ph.D, Senior Lecturer, School of Mechatronic Engineering, Universiti Malaysia Perlis, Perlis, Malaysia

Dr. G. Sahoo, Ph.D., Professor and Head, Department of IT, B.I.T, Meshra, India

Dr. R.H. Lara, Ph.D., Professor, The Electrical Company of Quito (EEQ), Ecuador

Dr. Kashif Munir, Ph.D., Professor, King Fahd University of Petroleum and Minerals, Hafr Al-Batin Campus, Kingdom of Saudi Arabia

Dr. L. Sumalatha, Ph.D., Professor, Department of Computer Science and Engineering, JNTU Kakinada

Dr. R. Boutaba, Ph.D., Professor, University of Waterloo, Canada

Dr. K.N. Rao, Ph.D., Professor, Department of Computer Science and Engineering, Andhra University, Visakhapatnam

Dr. S. Das, Ph.D., Associate Professor, Indian Statistical Institute, Kolkata, India

Dr. H.S. Behera, Ph.D., Associate Professor, Department of Computer Science Engineering and Information Technology, Veer Surendra Sai University of Technology (VSSUT), Burla, Odisha, India

Dr. J.K. Mandal, Ph.D., Professor, Department of CSE, Kalyani University, Kolkata, India

Dr. P.K. Hota, Ph.D., Professor, Department of EE, VSSUT, Burla, Odisha, India

Dr. S. Panda, Ph.D., Professor, Department of EEE, VSSUT, Burla, Odisha, India

Dr. S.C. Satpathy, Professor and Head, Department of Computer Science and Engineering, ANITS, AP, India

Dr. A.K. Turuk, Ph.D., Associate Professor and Head, Department of CSE, NIT, Rourkela, India

Dr. D.P. Mohapatra, Ph.D., Associate Professor, Department of CSE, NIT, Rourkela, India

Dr. R. Behera, Ph.D., Asst. Professor, Department of EE, IIT, Patna, India

Dr. S. Das, Ph.D., Asst. Professor, Department of Computer Science and Engineering, Galgotias University

Dr. M. Patra, Ph.D., Reader, Berhampur University, Odisha, India

Dr. S. Sahana, Ph.D., Asst. Professor, Department of CSE, BIT, Mesra, India

Dr. Asit Das, Ph.D., Associate Professor, Department of CSE, IIEST, WB, India

## International Advisory Committee

Prof. G. Panda, IIT, BBSR
Prof. Kenji Suzuki, University of Chicago
Prof. Raj Jain, W.U, USA
Prof. P. Mohapatra, University of California
Prof. S. Naik, University of Waterloo, Canada
Prof. S. Bhattacharjee, NIT, Surat
Prof. BrijeshVerma, C.Q.U, Australia
Prof. Richard Le, Latrob University, AUS
Prof. P. Mitra, IIT, KGP

Prof. Amit Das, IIEST, Kolkata
Prof. Michele Nappi, University of Salerno, Italy
Prof. P. Bhattacharya, NIT, Agaratala
Prof. G. Chakraborty, Iwate Prefectural University

## Conference Steering Committee

**Publicity Chair**
Dr. R.R. Rath, RIT, Berhampur
Prof. S.K. Acharya, RIT, Berhampur
Dr. P.D. Padhy, RIT, Berhampur
Mr. B. Naik, VSSUT, Burla

**Logistic Chair**
Dr. B.P. Padhi, RIT, Berhampur
Dr. R.N. Kar, RIT, Berhampur
Prof. S.K. Nahak, RIT, Berhampur
Mr. D.P. Kanungo, VSSUT, Burla
Mr. G.T. Chandra Sekhar, VSSUT, Burla

**Organizing Committee**
Prof. D.P. Tripathy, RIT, Berhampur
Prof. R.R. Polai, RIT, Berhampur
Prof. S.K. Sahu, RIT, Berhampur
Prof. P.M. Sahu, RIT, Berhampur
Prof. P.R. Sahu, RIT, Berhampur
Prof. Rashmta Tripathy, RIT, Berhampur
Prof. S.P. Tripathy, RIT, Berhampur
Prof. S. Kar, RIT, Berhampur
Prof. G. Prem Bihari, RIT, Berhampur
Prof. M. Surendra Prasad Babu, AU, AP
Prof. A. Yugandhara Rao, LCE, Vizianagaram
Prof. S. Halini, SSCE, Srikakulam
Prof. P. Sanyasi Naidu, GITAM University
Prof. K. Karthik, BVRIT, Hyderabad
Prof. M. Srivastava, GGU, Bilaspur
Prof. S. Ratan Kumar, ANITS, Vizag
Prof. Y. Narendra Kumar, LCE, Vizianagaram
Prof. B.G. Laxmi, SSCE, Srikakulam
Prof. P. Pradeep Kumar, SSCE, Srikakulam
Prof. Ch. Ramesh, AITAM
Prof. T. Lokesh, Miracle, Vizianagaram
Prof. A. Ajay Kumar, GVPCE, Vizag
Prof. P. Swadhin Patro, TAT, BBSR
Prof. P. Krishna Rao, AITAM

Prof. R.C. Balabantaray, IIIT, BBSR
Prof. M.M.K. Varma, GNI, Hyderabad
Prof. S.K. Mishra, RIT, Berhampur
Prof. R.K. Choudhury, RIT, Berhampur
Prof. M.C. Pattnaik, RIT, Berhampur
Prof. S.R. Dash, RIT, Berhampur
Prof. Shobhan Patra, RIT, Berhampur
Prof. J. Padhy, RIT, Berhampur
Prof. S.P. Bal, RIT, Berhampur
Prof. A.G. Acharya, RIT, Berhampur
Prof. G. Ramesh Babu, SSIT, Srikakulam
Prof. P. Ramana, GMRIT, AP
Prof. S. Pradhan, UU, BBSR
Prof. S. Chinara, NIT, RKL
Prof. Murthy Sharma, BVC, AP
Prof. D. Cheeranjeevi, SSCE, Srikakulam
Prof. B. Manmadha Kumar, AITAM, Srikakulam
Prof. B.D. Sahu, NIT, RKL
Prof. Ch. Krishna Rao, AITAM
Prof. L.V. Suresh Kumar, GMRIT, AP
Prof. S. Sethi, IGIT, Sarang
Prof. K. Vijetha, GVP College of Engineering, Vizag
Prof. K. Purna Chand, BVRIT, Hyderabad
Prof. H.K. Tripathy, KIIT, BBSR
Prof. Aditya K. Das, KIIT, BBSR
Prof. Lambodar Jena, GEC, BBSR
Prof. A. Khaskalam, GGU, Bilaspur
Prof. D.K. Behera, TAT, BBSR

# About the Conference

The International Conference on "Computational Intelligence in Data Mining" (ICCIDM) has become one of the most sought-after International conferences in India amongst researchers across the globe. ICCIDM 2015 aims to facilitate cross-cooperation across diversified regional research communities within India as well as with other International regional research programs and partners. Such active discussions and brainstorming sessions among national and international research communities are the need of the hour as new trends, challenges and applications of Computational Intelligence in the field of Science, Engineering and Technology are cropping up by each passing moment. The 2015 edition of ICCIDM is an opportune platform for researchers, academicians, scientists and practitioners to share their innovative ideas and research findings, which will go a long way in finding solutions to confronting issues in related fields.

The conference aims to:

- Provide a sneak preview into the strengths and weakness of trending applications and research findings in the field of Computational Intelligence and Data Mining.
- Enhance the exchange of ideas and achieve coherence between the various Computational Intelligence Methods.
- Enrich the relevance and exploitation experience in the field of data mining for seasoned and naïve data scientists.
- Bridge the gap between research and academics so as to create a pioneering platform for academicians and practitioners.
- Promote novel high-quality research findings and innovative solutions to the challenging problems in Intelligent Computing.
- Make a fruitful and effective contribution towards the advancements in the field of data mining.
- Provide research recommendations for future assessment reports.

By the end of the conference, we hope the participants will enrich their knowledge by new perspectives and views on current research topics from leading scientists, researchers and academicians around the globe, contribute their own ideas on important research topics like Data Mining and Computational Intelligence, as well as collaborate with their international counterparts.

# Contents

A Novel Approach for Biometric Authentication System Using Ear,
2D Face and 3D Face Modalities ................................ 1
Achyut Sarma Boggaram, Pujitha Raj Mallampalli,
Chandrasekhar Reddy Muthyala and R. Manjusha

An Optimal Partially Backlogged Policy of Deteriorating Items
with Quadratic Demand ....................................... 13
Trailokyanath Singh, Nirakar Niranjan Sethy, Ameeya Kumar Nayak
and Hadibandhu Pattanayak

A Novel Approach for Modelling of High Order Discrete Systems
Using Modified Routh Approximation Technique via W-Domain ...... 21
G.V.K.R. Sastry, G. Surya Kalyan, K. Tejeswar Rao
and K. Satyanarayana Raju

Pose Invariant Face Recognition for New Born: Machine
Learning Approach ........................................... 29
Rishav Singh and Hari Om

Implementation of Data Analytics for MongoDB
Using Trigger Utility ......................................... 39
Kalpana Dwivedi and Sanjay Kumar Dubey

Heart Disease Prediction Using k-Nearest Neighbor Classifier
Based on Handwritten Text .................................... 49
Seema Kedar, D.S. Bormane and Vaishnavi Nair

Optimizing a Higher Order Neural Network Through Teaching
Learning Based Optimization Algorithm ........................ 57
Janmenjoy Nayak, Bighnaraj Naik and H.S. Behera

Deploying a Social Web Graph Over a Semantic Web Framework .... 73
Shubhnkar Upadhyay, Avadhesh Singh, Kumar Abhishek
and M.P. Singh

**Detecting and Analyzing Invariant Groups in Complex Networks** . . . . .   85
Dulal Mahata and Chanchal Patra

**Automatic Mucosa Detection in Video Capsule Endoscopy
with Adaptive Thresholding** . . . . . . . . . . . . . . . . . . . . . . . . . . .   95
V.B. Surya Prasath and Radhakrishnan Delhibabu

**Optimization Approach for Feature Selection and Classification
with Support Vector Machine** . . . . . . . . . . . . . . . . . . . . . . . . . .   103
S. Chidambaram and K.G. Srinivasagan

**Design and Development of Controller Software
for MAS Receiver Using Socket Programming** . . . . . . . . . . . . . . . . .   113
K. Haritha, M. Jawaharlal, P.M.K. Prasad and P. Kalyan Chakravathi

**Inference of Replanting in Forest Fire Affected Land
Using Data Mining Technique** . . . . . . . . . . . . . . . . . . . . . . . . . .   121
T.L. Divya and M.N. Vijayalakshmi

**A New Approach on Color Image Encryption Using Arnold
4D Cat Map** . . . . . . . . . . . . . . . . . . . . . . . . . . . . . . . . . . . .   131
Bidyut Jyoti Saha, Kunal Kumar Kabi and Chittaranjan Pradhan

**Neutrosophic Logic Based New Methodology to Handle
Indeterminacy Data for Taking Accurate Decision** . . . . . . . . . . . . . .   139
Soumitra De and Jaydev Mishra

**Prognostication of Student's Performance: An Hierarchical
Clustering Strategy for Educational Dataset** . . . . . . . . . . . . . . . . . .   149
Parag Bhalchandra, Aniket Muley, Mahesh Joshi, Santosh Khamitkar,
Nitin Darkunde, Sakharam Lokhande and Pawan Wasnik

**Design and Implementation ACO Based Edge Detection
on the Fusion of Hue and PCA** . . . . . . . . . . . . . . . . . . . . . . . . .   159
Kavita Sharma and Vinay Chopra

**Effective Sentimental Analysis and Opinion Mining
of Web Reviews Using Rule Based Classifiers** . . . . . . . . . . . . . . . . .   171
Shoiab Ahmed and Ajit Danti

**Trigonometric Fourier Approximation of the Conjugate Series
of a Function of Generalized Lipchitz Class by Product
Summability** . . . . . . . . . . . . . . . . . . . . . . . . . . . . . . . . . . .   181
B.P. Padhy, P.K. Das, M. Misra, P. Samanta and U.K. Misra

**Optimizing Energy Efficient Path Selection Using Venus Flytrap
Optimization Algorithm in MANET** . . . . . . . . . . . . . . . . . . . . . . .   191
S. Sivabalan, R. Gowri and R. Rathipriya

**Biclustering Using Venus Flytrap Optimization Algorithm** . . . . . . . . . . 199
R. Gowri, S. Sivabalan and R. Rathipriya

**Analytical Structure of a Fuzzy Logic Controller for Software
Development Effort Estimation** . . . . . . . . . . . . . . . . . . . . . . . . . . . 209
S. Rama Sree and S.N.S.V.S.C. Ramesh

**Malayalam Spell Checker Using N-Gram Method** . . . . . . . . . . . . . . . 217
P.H. Hema and C. Sunitha

**Sample Classification Based on Gene Subset Selection** . . . . . . . . . . . . 227
Sunanda Das and Asit Kumar Das

**A Multi Criteria Document Clustering Approach Using Genetic
Algorithm** . . . . . . . . . . . . . . . . . . . . . . . . . . . . . . . . . . . . . . . . . . 237
D. Mustafi, G. Sahoo and A. Mustafi

**Positive and Negative Association Rule Mining
Using Correlation Threshold and Dual Confidence Approach** . . . . . . . . 249
Animesh Paul

**Extractive Text Summarization Using Lexical Association
and Graph Based Text Analysis** . . . . . . . . . . . . . . . . . . . . . . . . . . . 261
R.V.V. Murali Krishna and Ch. Satyananda Reddy

**Design of Linear Phase Band Stop Filter Using Fusion
Based DEPSO Algorithm** . . . . . . . . . . . . . . . . . . . . . . . . . . . . . . . . 273
Judhisthir Dash, Rajkishore Swain and Bivas Dam

**A Survey on Face Detection and Person Re-identification** . . . . . . . . . . . 283
M.K. Vidhyalakshmi and E. Poovammal

**Machine Translation of Telugu Singular Pronoun Inflections
to Sanskrit** . . . . . . . . . . . . . . . . . . . . . . . . . . . . . . . . . . . . . . . . . . 293
T. Kameswara Rao and T.V. Prasad

**Tweet Analyzer: Identifying Interesting Tweets Based
on the Polarity of Tweets** . . . . . . . . . . . . . . . . . . . . . . . . . . . . . . . . 307
M. Arun Manicka Raja and S. Swamynathan

**Face Tracking to Detect Dynamic Target in Wireless
Sensor Networks** . . . . . . . . . . . . . . . . . . . . . . . . . . . . . . . . . . . . . . 317
T.J. Reshma and Jucy Vareed

**Abnormal Gait Detection Using Lean and Ramp Angle Features** . . . . . . 325
Rumesh Krishnan, M. Sivarathinabala and S. Abirami

**Fuzzy Logic and AHP-Based Ranking of Cloud Service Providers** . . . . . 337
Rajanpreet Kaur Chahal and Sarbjeet Singh

**Classification of Tabla Strokes Using Neural Network** . . . . . . . . . . . . . 347
Subodh Deolekar and Siby Abraham

**Automatic Headline Generation for News Article** . . . . . . . . . . . . . . . . . 357
K.R. Rajalakshmy and P.C. Remya

**Identifying Natural Communities in Social Networks Using
Modularity Coupled with Self Organizing Maps** . . . . . . . . . . . . . . . . . 367
Raju Enugala, Lakshmi Rajamani, Kadampur Ali
and Sravanthi Kurapati

**A Low Complexity FLANN Architecture for Forecasting Stock
Time Series Data Training with Meta-Heuristic Firefly Algorithm** . . . . . 377
D.K. Bebarta and G. Venkatesh

**A Novel Scheduling Algorithm for Cloud Computing Environment** . . . . 387
Sagnika Saha, Souvik Pal and Prasant Kumar Pattnaik

**PSO Based Tuning of a Integral and Proportional Integral
Controller for a Closed Loop Stand Alone Multi
Wind Energy System** . . . . . . . . . . . . . . . . . . . . . . . . . . . . . . . . . . . 399
L.V. Suresh Kumar, G.V. Nagesh Kumar and D. Anusha

**Statistically Validating Intrusion Detection Framework Against
Selected DoS Attacks in Ad Hoc Networks: An NS-2
Simulation Study** . . . . . . . . . . . . . . . . . . . . . . . . . . . . . . . . . . . . . . 411
Sakharam Lokhande, Parag Bhalchandra, Santosh Khamitkar,
Nilesh Deshmukh, Satish Mekewad and Pawan Wasnik

**Fourier Features for the Recognition of Ancient Kannada Text** . . . . . . . 421
A. Soumya and G. Hemantha Kumar

**A Spelling Mistake Correction (SMC) Model for Resolving
Real-Word Error** . . . . . . . . . . . . . . . . . . . . . . . . . . . . . . . . . . . . . . 429
Swadha Gupta and Sumit Sharma

**Differential Evolution Based Tuning of Proportional Integral
Controller for Modular Multilevel Converter STATCOM** . . . . . . . . . . . 439
L.V. Suresh Kumar, G.V. Nagesh Kumar and P.S. Prasanna

**Designing of DPA Resistant Circuit Using Secure Differential
Logic Gates** . . . . . . . . . . . . . . . . . . . . . . . . . . . . . . . . . . . . . . . . . . 447
Palavlasa Manoj and Datti Venkata Ramana

**Efficient Algorithm for Web Search Query Reformulation
Using Genetic Algorithm** . . . . . . . . . . . . . . . . . . . . . . . . . . . . . . . . . 459
Vikram Singh, Siddhant Garg and Pradeep Kaur

**Throughput of a Network Shared by TCP Reno and TCP Vegas
in Static Multi-hop Wireless Network** . . . . . . . . . . . . . . . . . . . . . . . . 471
Sukant Kishoro Bisoy and Prasant Kumar Pattnaik

**A Pronunciation Rule-Based Speech Synthesis Technique for Odia
Numerals** . . . . . . . . . . . . . . . . . . . . . . . . . . . . . . . . . . . . . . . . 483
Soumya Priyadarsini Panda and Ajit Kumar Nayak

**Erratum to: A New Approach on Color Image Encryption
Using Arnold 4D Cat Map** . . . . . . . . . . . . . . . . . . . . . . . . . . . . . E1
Bidyut Jyoti Saha, Kunal Kumar Kabi and Chittaranjan Pradhan

**Author Index** . . . . . . . . . . . . . . . . . . . . . . . . . . . . . . . . . . . . . 493

# About the Editors

**Prof. Himansu Sekhar Behera** is working as an Associate Professor in the Department of Computer Science Engineering and Information Technology, Veer Surendra Sai University of Technology (VSSUT)—An Unitary Technical University, Established by Government of Odisha, Burla, Odisha. He has received M.Tech. in Computer Science and Engineering from N.I.T., Rourkela (formerly R.E.C., Rourkela) and Doctor of Philosophy in Engineering (Ph.D.) from Biju Pattnaik University of Technology (BPUT), Rourkela, Government of Odisha, respectively. He has published more than 80 research papers in various international Journals and Conferences, edited 11 books and is acting as a Member of the Editorial/Reviewer Board of various international journals. He is proficient in the field of Computer Science Engineering and served in the capacity of program chair, tutorial chair and acted as advisory member of committees of many national and international conferences. His research interests include Data Mining, Soft Computing, Evolutionary Computation, Machine Intelligence and Distributed System. He is associated with various educational and research societies like OITS, ISTE, IE, ISTD, CSI, OMS, AIAER, SMIAENG, SMCSTA, etc.

**Prof. Durga Prasad Mohapatra** received his Ph.D. from Indian Institute of Technology Kharagpur and is presently serving as Associate Professor in NIT Rourkela, Odisha. His research interests include software engineering, real-time systems, discrete mathematics and distributed computing. He has published more than 30 research papers in these fields in various international Journals and Conferences. He has received several project grants from DST and UGC, Government of India. He has received the Young Scientist Award for the year 2006 by Orissa Bigyan Academy. He has also received the Prof. K. Arumugam National Award and the Maharashtra State National Award for outstanding research work in Software Engineering for the years 2009 and 2010, respectively, from the Indian Society for Technical Education (ISTE), New Delhi. He is going to receive the Bharat Sikshya Ratan Award for significant contribution in academics awarded by the Global Society for Health and Educational Growth, Delhi.

# A Novel Approach for Biometric Authentication System Using Ear, 2D Face and 3D Face Modalities

Achyut Sarma Boggaram, Pujitha Raj Mallampalli,
Chandrasekhar Reddy Muthyala and R. Manjusha

**Abstract** Biometric system using face recognition is the frontier of the security across various applications in the fields of multimedia, medicine, civilian surveillance, robotics, etc. Differences in illumination levels, pose variations, eye-wear, facial hair, aging and disguise are some of the current challenges in face recognition. The ear, which is turning out to be a promising biometric identifier having some desirable properties such as universality, uniqueness, permanence, can also be used along with face for better performance of the system. A multi-modal biometric system combining 2D face, 3D face (depth image) and ear modalities using Microsoft Kinect and Webcam is proposed to address these challenges to some extent. Also avoiding redundancy in the extracted features for better processing speed is another challenge in designing the system. After careful survey of the existing algorithms applied to 2D face, 3D face and ear data, we focus on the well-known PCA (Principal Component Analysis) based Eigen Faces algorithm for ear and face recognition to obtain a better performance with minimal computational requirements. The resulting proposed system turns out insensitive to lighting conditions, pose variations, aging and can completely replace the current recognition systems economically and provide a better security. A total of 109 subjects participated in diversified data acquisition sessions involving multiple poses, illuminations, eyewear and persons from different age groups. The dataset is also a first attempt on the stated combination of biometrics and is a contribution to the field of Biometrics by itself for future experiments. The results are obtained separately against each biometric and final decision is obtained using all the individual results for higher accuracy. The proposed system performed at 98.165 % verification rate

A.S. Boggaram (✉) · P.R. Mallampalli · C.R. Muthyala · R. Manjusha
Computer Science and Engineering, Amrita University, Coimbatore, India
e-mail: achyutsarmaboggaram@gmail.com

P.R. Mallampalli
e-mail: poojithampalli@gmail.com

C.R. Muthyala
e-mail: mchandrasekharreddym@gmail.com

R. Manjusha
e-mail: r_manjusha@cb.amrita.edu

© Springer India 2016
H.S. Behera and D.P. Mohapatra (eds.), *Computational Intelligence
in Data Mining—Volume 1*, Advances in Intelligent Systems
and Computing 410, DOI 10.1007/978-81-322-2734-2_1

which is greater than either of the dual combinations or each of the stated modality in a statistical and significant manner.

**Keywords** Biometrics · Security system · Person identification · Modalities · Kinect · Feature extraction · Ear recognition · Face recognition · PCA (Principal component Analysis) · Eigen faces

# 1   Introduction

Identifying a unique person in this evolving day to day life plays a major role for security purposes. We come across many types of security systems like Personal Identification Number (PIN), ID cards, user name and password logins, etc., and these systems are unreliable as they have dis-advantages such as difficulty in remembrance, vulnerability to theft, etc. Biometric systems can easily over ride these type of problems because persons are identified with their own biological traits such as retina, iris, fingerprint, palm print and DNA, etc. They demand the person for authentication and not by someone else carrying cards and remembering PIN's. Biometrics are defined as the unique physical characteristics or traits of human body. The details of human body which will be used as unique biometric data to serve as ID, such as retina, iris, fingerprint, palm print and DNA. Since these biometric systems provide superior level of person identity which gives a main point about the identification is that "inalienable" which means no subject can be taken away from or given away by the possessor, so no more hackings can be done to authentication systems—they have to go through the control point personally.

Many researches has been conducted over unimodal and multimodal biometric systems. It is proved that multimodal biometric systems are more effective when compared to unimodal systems. Multimodal biometric systems are suggested for any industry where there is a requirement of high accuracy because the integration of two or more different types of biometrics results in rigid performance. When feature extraction is done appropriately 2D face, 3D face (depth image) and ear modalities can produce unique data. As observed, each of the different methods has inherent advantages and disadvantages it is not very easy to find the best method for data acquisition as.

2D face recognition is usually done by using Eigen faces based on (Principal Component Analysis) PCA to extract features vectors after achieving dimensionality reduction. As we know that biometric technology has its own merits and demerits, this system also faces difficulty in recognition when images are taken from two extremely different views and under distinct illumination conditions.

Moreover, to upgrade the system accuracy to a higher degree considering the aging, facial hair, eyewear and disguise factors, we have combined ear recognition with face recognition to endow "multimodal system" where security is guaranteed in multiple ways. In the need of upgraded security considering the insensitivity to pose and illumination variations, 3D face (depth image) verification can be used in integration with 2D face and ear verification systems where 3D facial data cab be kept as an evidence.

This paper proposes a multi-modal biometric system for authentication that uses the features of 2D face and 3D face (depth image) with that of the ear for a better authentication accuracy rate and better security. The features extracted are subjected to verification by the respective recognition modules and for higher authentication accuracy the results are fed to the system using a decision threshold strategy.

The rest of the paper is organized as follows. In this paper, Sect. 2 talks about previous work to the related area to this paper. Section 3 gives the information of proposed biometric system. Section 4 examine the obtained results. The paper is terminated by Sect. 5 conclusion elucidating the obtained results and future work.

## 2 Previous Works

Mae Alay-ay et al. [1] introduced Oto-ID. They aimed to provide solution for uniquely identifying a human ear recognition for login, implemented Ear Recognition and Shape-based Human Detection method for User Identification using PCA and Feature Extractions in their study.

Perumal and Somasundar [2] proposed a method for ear recognition using kernel based algorithm. They aimed at addressing the problems of traditional biometric systems that lack in anticipation accuracy and with the noisy images. Their ear recognition system is proposed with the following phases (a) Preprocessing of ear image using Contrast Enhancement, (b) Extraction of features from a modality using Kernel Principal Component Analysis (c) Classification using Kernel Support Vector Machine Analysis, Accuracy of the system was tested using various neural network configuration and noised images.

Gambhir et al. [3] have made a research on person recognition using multimodal biometrics and found that unimodal biometric systems often face significant limitations due to quality of data, noised image and other factors. Multimodal biometric systems solves some of these problems by individual modality recognition rate of the same identity. These type of systems help increase in which is not possible using a single-biometric system. A biometric system uses various modalities in order to increase the performance and improve high security. They developed multimodal biometric system which possesses a number of unique qualities, using PCA for unique identity with face and ear modalities. Their results showed that fusion of individual modalities can give good accuracy in the overall performance of the biometric system.

Kamencay et al. [4] a proposed methodological phenomenon for face recognition based on an information theory on face images. They proposed a 2D face 3D face-matching method using a principal component analysis (PCA) algorithm using canonical correlation analysis (CCA) to learn the mapping technique between a 2D face image and 3D face data. Their method makes it possible to match a 2D face image with enrolled 3D face data. Their proposed algorithm is based on the PCA method, which is applied to extract base features. PCA feature-level fusion from the source data with extraction of different features before features are merges. Their experimental results on the TEXAS face image based on the modified CCA-PCA method.Testing the 2D face 3D face match results gave a recognition rate for the CCA method of a quite poor result while the modified CCA method based on PCA-level fusion achieved a higher result.

## 3 Proposed System [5]

The system is developed using MATLAB completely. We used Microsoft Kinect camera for image acquisition of 3D face images (Fig. 1) and webcam for image acquisition of 2D face and ear images (Figs. 2 and 3). The system requires the use of the above mentioned cameras for training the database. The developed system can be used for user recognition and verification purposes. A user is said to be recognized if the person is identified as one registered in the training database. If a person needs to be added to the database, the images of the person are taken using mentioned cameras and saved in appropriate database folder. For conformity, in the system only right ears of the subjects are used. Normal ears and faces can be detected by the system easily but it may not have the ability to detect an ear with any deformities.

**Fig. 1** Captured 3D face images using Kinect

**Fig. 2** Captured 2D face images using webcam

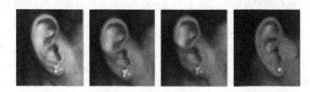

**Fig. 3** Captured ear image using webcam

## 3.1   System Architecture [6]

Initially, the user's 2D face and ear is captured by using webcam, 3D face is captured using Microsoft Kinect camera. The image which is captured for each modality is then saved in specific path in JPEG format. The images of all three modalities get automatically cropped into 180 × 200 pixels. The whole recognition process is divided into three phases: Preparation phase, training phase and testing phase. The system architecture is shown in Fig. 4.

### 3.1.1   Preparation Phase

In this phase image acquisition is done and all the images in the database undergoes preprocessing to enhance the quality of image. Then specific features are selected from the preprocessed image using PCA based Eigen Faces method which gives best recognition rate. (PCA CCA modification is PCA itself for 2D face and 3D face but PCA-Kaiser is used for ear recognition)

**Fig. 4** System architecture

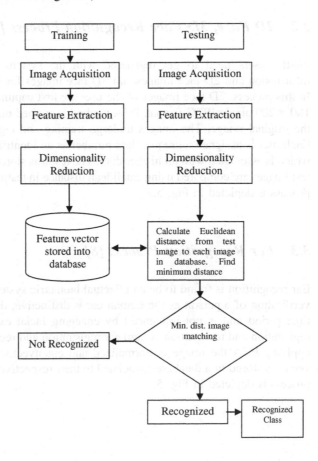

### 3.1.2 Training Phase

Feature vectors are obtained for each modality using appropriate technique (Eigen Faces for faces and Eigen Ears for ears) after PCA is applied which does dimensionality reduction. For future comparisons these feature vectors are stored in a database.

### 3.1.3 Testing Phase

Same process of image acquisition and feature selection is done for the test image, feature vector is obtained for the same. The test image feature vector is compared with the database feature vector by calculating Euclidean distance. The image in database feature vector is recognized whose Euclidean distance is minimum of all feature vectors corresponding to other images in the database. The corresponding class number is identified and is used by the integration system if recognized correctly.

## 3.2   2D Face, 3D Face Recognition Process [7]

Firstly, users need to get registered with the system by providing necessary information and 2D face images and 3D face images for later recognition process. In this process, 2D face images of the user are first captured and then cropped into 180 × 200 pixels JPEG format. PCA method is applied on the images to normalize the original image. The images undergo training and eigenvectors are generated. Each user is assigned a unique class number as an identifier index for the database which is associated with all his/her data. The result is stored in a database where the test images are compared using Euclidean distance in testing phase. The registration process is depicted in Fig. 5.

## 3.3   Ear Recognition Process [8]

Ear recognition is found to be an effectual biometric system which can be used for verification of a person as the human ear is distinctive, does not alter over a long time period and is not influenced by changing facial expressions. Same as face registration and recognition system ear registration and recognition is also done. By applying PCA the image is normalized and eigenvector is obtained. Ear feature vector is stored in a database associated to their respective classes. The registration process is depicted in Fig. 5.

**Fig. 5** Registration process

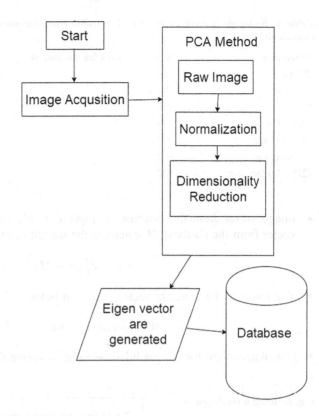

## 3.4 Integration Design [9]

We implemented an Experimental Research Method in this study of 2DFace, 3D Face and Ear biometrics. This method involves in the manipulation of illumination and viewing conditions while capturing the images. The main purpose of our research is to evaluate the system's reliability under distinct possible combinations of biometrics (refer Table 1). For 3D face recognition the variable is pose variation, for 2D face recognition the variables are lighting condition, facial expressions and the variables for ear recognition are pose variation and distance from the camera.

As shown in the Fig. 6, the respective recognized class numbers from each modality are compared with the test image class index number and the respective results are returned as 1 if they correspond to the correct associated class and 0 if not. The decision module sums up the returned recognition results into a decision variable corresponding to the test index class number. If the final decision variable value is greater than or equal to the threshold value of 2, then the user is said to be identified by the system and the associated access is provided.

Pseudo Code for Individual Recognition for 2D Face, 3D Face and Ear Modalities

**Table 1** Recognition accuracy rate for all possible combinations of modalities tried over 109 individuals' data

| Modalities | Number of samples recognized | Accuracy rate |
|---|---|---|
| 2D face | 89 | 81.651 |
| 3D face | 94 | 86.238 |
| Ear | 91 | 83.486 |
| (2D+3D) face | 99 | 90.825 |
| 2D face, ear | 94 | 86.238 |
| 3D face, ear | 102 | 93.577 |
| (2D+3D) face ear | 107 | 98.165 |

- Image vector from the database as input $U \in \Re^n$, calculate the mean image vector from the database $M$, calculate the weight of the kth Eigen face as:

$$w_k = V_k^T(U - M)$$

- The result of the a weight vector as shown below:

$$W = [w_1, w_2, \ldots, w_k, \ldots, w_n]$$

- Calculate weight for the candidate as same as above W

**Fig. 6** Integration design

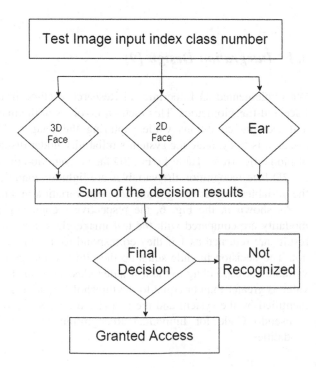

- Compare W with weight vectors $W_m$ of images in the database. Calculate the Euclidean distance with the weight vector:

$$d = ||W - W_m||^2$$

- If $d < \epsilon_1$, identified as the mth position in the database is a candidate of recognition.
- If $\in_1 < d < \epsilon_2$, then $U$ will considered as unknown face and can be stored into the database.
- If $d > \epsilon_2, U$ is considered as not an image.

### 3.4.1 Integration Pseudo Code

- Given input test image class number N
- If the recognized 2D Face image and ear image number is either 3N or 3N − 1 or 3N − 2 then recognition value is given as 1 else it is given as 0
- If the recognized 3D Face image number is either 2N or 2N − 1 or 2N − 2 then recognition value is given as 1 else it is given as 0
- Add all the three recognition values
- If the sum is ≧2 the person is said to be recognized and access is granted
- Else not recognized.

## 4　Results and Discussion

Figure 7 illustrates comparison of the results of the various tried combinations of the used modalities for person authentication over the trained database of 109 individuals (refer Table 1).

**Fig. 7** Comparison chart

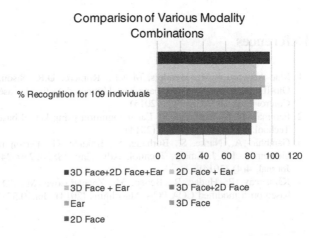

Comparision of Various Modality Combinations

% Recognition for 109 individuals

0　20　40　60　80　100　120

- 3D Face+2D Face+Ear　　2D Face + Ear
- 3D Face + Ear　　3D Face+2D Face
- Ear　　3D Face
- 2D Face

We can clearly see the increase in recognition accuracy with the increase in the usage of modalities. We can also deduce that the most effective modalities over the tried and stated variations in the input images are the 3D Face, Ear, and 2D Face in the respective order. Also we can deduce that more the modalities when used for recognition over the highly corrupt and variable illuminated multi-viewed inputs, more the effectivity of the performance.

## 5 Conclusions and Future Work

Different statistical approaches are studied in this paper to recognize persons through their ear, (2D+3D) face images (depth image). The images which are considered are corrupted by different types of noise like variant illumination and poses with distinct intensities of noise, and for the average case the recognition rate is evaluated. It was observed that 3D face and ear modalities when added to the system guarantee the finest recognition rate compared to all cases.

PCA based Eigen Faces and Eigen Ears algorithms are used to extract the features from their respective modalities. The extracted features were then used to calculate the similarity between images. This resulted in match score for each modality. The match scores obtained from the two modalities (ear and face) were used to calculate the nearest recognized image and the respective decisions from each modality are combined to recognize the individual using the stated threshold technique.

Future work will include the employment of automated 3D face, 2D face and ear segmentation from the captured images for direct application of the biometric recognition. This paper strongly supports the experimentation of various respective feature extraction techniques for each modality along with various untried combination methodologies for the system integration using the available biometrics. The use of the 3D modality for both face and ear may also lead to an increase in the robustness to both illumination and pose variations.

## References

1. Mae Alay-ay, K.M., Meralpis, M.A.L., Romero, D.R., Sison, R.F., Comendador, B.E.V.: Oto-ID: ear recognition and shape-based human detection for user identification. Int. J. Autom. Control Eng. (JOACE) 3(2) (2015)
2. Perumal, T., Somasundar, S.: Ear recognition using kernel based algorithm. Int. J. Sci. Eng. Technol. Res. (IJSETR) 3(4) (2014)
3. Gambhir, A., Narke, S., Borhade, S., Bokade, G.: Person recognition using multimodal biometrics. Int. J. Emerg. Technol. Adv. Eng. ISSN:2250–2459; ISO 9001:2008 Certified Journal, 4(4) (2014)
4. Kamencay, P., Hudec, R., Benco, M., Zachariasova, M.: 2D-3D face recognition method based on a modified CCA-PCA Algorithms (2014). doi:10.5772/58251

5. Kumar, D., Rajni: Face recognition based on PCA algorithm using simulink in matlab. Int. J. Adv. Res. Comput. Eng. Technol. (IJARCET) **3**(7) (2014)
6. Jamal, S., Muhammad, S., Mudassar, R., Aisha, A.: A survey: linear and non linear PCA based face recogonition techniques. Int. Arab J. Inform. Technol. **6** (2013)
7. Gao, Q.H.: Biometric authentication to prevent cheating. Int. J. Instruct. Technol. Dist. Learn. **9**(2), 3–13 (2012)
8. Bahgat, S.F., Ghoniemy, S., Alotaibi, M.: Proposed multimodal palm veins-face biometric authentication. Int. J. Adv. Comput. Sci. Appl. **4**(6) (2013)
9. Dhoke, P., Parsai, M.P.: Amatlab based face recognition using PCA with back propagation neural network. Int. J. Innov. Res. Comput. Commun. Eng. **2**(8) (2014)
10. Zhang, L., Ding, Z., Li, H., Shen, Y.: 3D ear identification based on sparse representation. PLoS ONE (2014)
11. Jiamin, L., Fulin, L., Hong, H., Yizhe, L.: Ear detection based on improved camshift and adaboost algorithm. J. Comput. Inform. Syst. **10**, 13 (2014)
12. Dhanda, M.: Face recognition using eigen vector from PCA. JMIET, Radaur, Yamuna Nagar, Haryana, India, March 2012
13. Bowyer, K.W., Chang, K., Flynn, P.: A survey of approaches and challenges in 3D and multimodal 3D+2D face recognition. Comput. Vision Image Understand. **101**(1), 1–15
14. Javed, A.: Face recognition based on principal component analysis. I. J. Image, Graphics Signal Process. 238–244 (2013)

5. Kumar D., Rahul: Face recognition based on PCA algorithm using simplified artificial... 5. Int. J. Res. Comput. Engg. Technol. (IJRCET) **5** (2015).
6. Jain S., Shrivastava S., Mahapatra: Alam A.: A survey: liveness and non-liveness IVA based face recognition techniques. Int. Arab. J. Inform. Technol. **6**, 1–11 (2016).
7. O.B. Boswell: A framework to prevent presentation, in: J.Internat. Technol. Dist. Learn. **9**, 1–5 (2012).
8. Batra S.: Gaurihar V.: Atombe M.: Proposed multimodal palm vein-facial feature fusion, Int. J. Adv. Comput. Sci. Appl. **4** (2015).
9. Dutta K., Pasha S.P.: A multi-facet framework using PCA with fisher propagation, Appl. Int. Comput. Intel. J. Engg. Sci. **2** (2014).
10. Zhang C., Ding Z., Li H., Shen J., Xu R.: An improved sliding-window algorithm, PLoS ONE (2015).
11. Jianfei C.: Foo L., Hong K., Yan L.: Kewei J.: An active-face recognition algorithm and feedback integral... Comput. Inform. Syst. **18**, 13 (2015).
12. Dsouza A.: Face recognition using integrative vector iteration PCA, Master's thesis, North Eastern Univ. March 2012.
13. Bowyer K.W., Chang K.: Flynn P.: A survey of approaches and challenges in 3D and multimodal 3D+2D Face Recognition, Comput. Vision Image Understand. **101**, 1–15 (2006).
14. Wei H.: Face recognition based on linear parameterized model vectors, Image, Graphics Signal Process. 235–236 (2013).

# An Optimal Partially Backlogged Policy of Deteriorating Items with Quadratic Demand

Trailokyanath Singh, Nirakar Niranjan Sethy, Ameeya Kumar Nayak
and Hadibandhu Pattanayak

**Abstract** An EOQ (Economic Order Quantity) model for a deteriorating item with quadratic demand pattern and quadratic holding cost and constant deterioration rate is considered in this paper. In addition, shortages and partial backlogging are allowed. It is assumed that the backlogging rate acts as not only a variable, but also depends on the length of the waiting time up to next replenishment during the stock out period. For this model, average total cost is derived. Finally, a numerical example for illustration is provided.

**Keywords** EOQ · Partial backlogging · Quadratic demand · Quadratic holding cost

**2000-Mathematics Subject Classification:** 90B05

T. Singh (✉)
Department of Mathematics, C. V. Raman College of Engineering, Bhubaneswar 752054, Odisha, India
e-mail: trailokyanaths108@gmail.com

N.N. Sethy
Research Scholar, Ravenshaw University, Cuttack 753003, Odisha, India
e-mail: nirakarn8@gmail.com

A.K. Nayak
Department of Mathematics, IIT Roorkee, Roorkee 247667, India
e-mail: ameeyakumar@gmail.com

H. Pattanayak
Department of Mathematics, Institute of Mathematics and Applications, Andharua, Bhubaneswar 751003, Odisha, India
e-mail: h.pattnayak@gmail.com

© Springer India 2016
H.S. Behera and D.P. Mohapatra (eds.), *Computational Intelligence in Data Mining—Volume 1*, Advances in Intelligent Systems and Computing 410, DOI 10.1007/978-81-322-2734-2_2

# 1 Introduction

In the past few decades, several mathematical techniques have been developed to control and maintain the inventories of deteriorating items. Generally, deterioration is defined as the change, damage, decay, dryness, spoilage, vaporization, etc., that cannot be used in original purposes. Researches began the inventory models by considering the time-proportional demand in their models. Bahari-Kasani [1] and Dave and Patel [3] studied their models by considering time-proportional demand. The literature survey by Raafat [10], Goyal and Giri [4] and Li et al. [8] discuss the up to date review on deteriorating inventory model. In the real market situation, the demand rate for any product cannot be constant. Many researchers have studied their models on linear and exponential time-dependent demand. A time-dependent linear demand leads to uniform change whereas time-dependent exponential demand seems unrealistic. Therefore, the more realistic approach is to consider time-dependent quadratic demand rate because it experiences the accelerated growth or decline in demand. Khanra and Chaudhuri [6], Khanra et al. [7] and Singh and Pattnayak [11] developed inventory models considering time-quadratic demand rate which is more realistic.

In real life situations, some but not all customers prefer to wait for backlogged items during the shortage period. Therefore, the longer the waiting time is, the smaller the backlogging rate will be. The opportunity cost due to lost sales should be considered in the inventory models. Researchers like Hollier and Mak [5], Chang and Dye [2] and Ouyang et al. [9] developed the inventory model considering partial backlogging. A general EOQ model for deteriorating items with time-dependent linear demand, time-proportional and partial backlogging have been developed by Singh and Pattnayak [12].

In this paper, an EOQ model for deteriorating items has been developed by considering constant deterioration rate. The demand rate and the holding cost are assumed to be quadratic function of time. The optimal shortage time, optimal cycle length, optimal order quantity and average total cost are derived. Finally, a numerical example for illustration of the model is provided.

# 2 Notations and Assumptions

The following assumptions and notations are used to develop the model.

(i)   A single item is considered over a prescribed time period.

(ii)   Shortages in the inventory system are allowed and partially backlogged.

(iii)   Replenishment rate is infinite.

(iv)   $R(t)$ : The demand rate at any time $t$ is given by $R(t) =$
$\begin{cases} a+bt+ct^2, & I(t) > 0, \\ K, & I(t) \leq 0, \end{cases}$ where $a, b, c(a \geq 0, b \neq 0, c \neq 0) \& K$ are positive constant.

(v)  $\theta$ : The constant deterioration rate where $0 < \theta < < 1$.

(vi)  The rate of backlogging during the stock out period is assumed as not only variable but also it depends on the length of the cycle time up to next replenishment. If the waiting time is longer, then the rate of backlogging will be smaller. Let $B(t) = \frac{1}{1+\delta t}$ be the negative inventory rate of backlogging where $\delta(>0)$ is the backlogging parameter in the interval $0 \leq t \leq T$.

(vii)  $h$ : The quadratic inventory holding cost nd is given by $h(t) = h + \gamma t + \beta t^2$ where $h > 0, \gamma \neq 0$ & $\beta \neq 0$.

(viii)  $C_P$ : The purchase cost of the inventory item per unit.

(ix)  $C_O$ : The ordering cost of the inventory per order.

(x)  $C_S$ : The constant shortage cost per unit time unit.

(xi)  $C_L$ : The cost of constant lost sales per unit.

(xii)  $M$ : The length of time at which the shortage period starts.

(xiii)  $T$ : The length of cycle time.

(xiv)  $W_{Max.}$ : The unit of items that arrive in the system at the starting of each cycle.

(xv)  $S_{Max.}$ : The maximum amount of demand backlogged per cycle.

(xvi)  $Q_0$ : the order quantity per cycle.

## 3  Model Development

Considering the effect of quadratic demand and constant deterioration during $[M, T]$, the inventory level at any instant during the period $[M, T]$ can be represented by the differential equation as:

$$\frac{dI_1(t)}{dt} + \theta I_1(t) = -R(t), \quad 0 \leq t \leq M \tag{1}$$

where $R(t) = a + bt + ct^2$.

Further, the inventory level reaches zero at time $t = M$. Thus, using the boundary condition $I_1(M) = 0$, the solution of Eq. (1) becomes

$$I_1(t) = \left[ \frac{a + bM + cM^2}{\theta} - \frac{b + 2cM}{\theta^2} + \frac{2c}{\theta^3} \right] e^{\theta(M-t)} \\ - \frac{a + bt + ct^2}{\theta} + \frac{b + 2ct}{\theta^2} - \frac{2c}{\theta^3}, \quad 0 \leq t \leq M. \tag{2}$$

With the help of $I_1(0) = W$, the inventory will be

$$W_{Max.} = \left[ \frac{a + bM + cM^2}{\theta} - \frac{b + 2cM}{\theta^2} + \frac{2c}{\theta^3} \right] e^{\theta M} - \frac{a}{\theta} + \frac{b}{\theta^2} - \frac{2c}{\theta^3}. \tag{3}$$

During the stock out period $[M, T]$, a fraction of demand is backlogging. Thus, the state of inventory is governed by the differential equation as:

$$\frac{dI_2(t)}{dt} = R(t)B(t), \quad M \le t \le T. \tag{4}$$

where $R(t) = K$ and $B(T - M) = \frac{1}{1+\delta(T-M)}$.

The solution of Eq. (4) with $I_2(M) = 0$ is

$$I_2(t) = \frac{K}{\delta}[\ln(1 + \delta(T - t)) - \ln(1 + \delta(T - M))], \quad M \le t \le T. \tag{5}$$

Maximum backordered units will be

$$S_{Max.} = -I(T) = \frac{K}{\delta}\ln[1 + \delta(T - M)]. \tag{6}$$

Thus, the order size during the time period $[0, T]$ will be

$$Q_0 = W_{Max.} + S_{Max.} = \left[\frac{a + bM + cM^2}{\theta} - \frac{b + 2cM}{\theta^2} + \frac{2c}{\theta^3}\right]e^{\theta M}$$
$$- \frac{a}{\theta} + \frac{b}{\theta^2} - \frac{2c}{\theta^3} + \frac{K}{\delta}\ln[1 + \delta(T - M)]. \tag{7}$$

The following cost components are needed for the calculation of per cycle average total cost:

The per cycle ordering cost is

$$CO = C_O. \tag{8}$$

The per cycle cost of carrying is

$$CC = \int_0^M h(t)I_1(t)dt = \int_0^M [(h + \gamma t + \beta t^2)I_1(t)]dt$$
$$= \left[\frac{a + bM + cM^2}{\theta} - \frac{b + 2cM}{\theta^2} + \frac{2c}{\theta^3}\right]$$
$$\times \left[\frac{he^{\theta M} - h - \gamma M - \beta M^2}{\theta} + \frac{\gamma e^{\theta M} - \gamma - 2\beta M}{\theta^2} + \frac{2\beta e^{\theta M} - 2\beta}{\theta^3}\right] \tag{9}$$
$$- \frac{h}{\theta^3}\left[\theta^2\left(aM + \frac{bM^2}{2} + \frac{cM^3}{3}\right) - \theta(bM + cM^2) + 2cM\right]$$
$$- \frac{\gamma}{\theta^3}\left[\theta^2\left(\frac{aM^2}{2} + \frac{bM^3}{3} + \frac{cM^4}{4}\right) - \theta\left(\frac{bM^2}{2} + \frac{2cM^3}{3}\right) + cM^2\right]$$
$$- \frac{\beta}{\theta^3}\left[\theta^2\left(\frac{aM^3}{3} + \frac{bM^4}{4} + \frac{cM^5}{5}\right) - \theta\left(\frac{bM^3}{3} + \frac{cM^4}{4}\right) + \frac{2cM^3}{3}\right].$$

The per cycle cost of deterioration is

$$
\begin{aligned}
CD &= C_P \left[ W_{Max.} - \int_0^M (a + bt + ct^2)\, dt \right] \\
&= C_P \left[ \left( \frac{a + bM + cM^2}{\theta} - \frac{b + 2cM}{\theta^2} + \frac{2c}{\theta^3} \right) e^{\theta M} \right] \\
&\quad - C_P \left[ \frac{a}{\theta} - \frac{b}{\theta^2} + \frac{2c}{\theta^3} + aM + \frac{bM^2}{2} + \frac{cM^3}{3} \right].
\end{aligned}
\tag{10}
$$

The per cycle cost of shortage is

$$
CS = C_S \left[ -\int_M^T I_2(t)\, dt \right] = \frac{C_S K}{\delta} \left[ (T - M) - \frac{1}{\delta} \ln(1 + \delta(T - M)) \right].
\tag{11}
$$

The per cycle cost of opportunity due to lost sales is

$$
\begin{aligned}
COLS &= C_L \left[ \int_M^T K \left( 1 - \frac{1}{1 + \delta(T - M)} \right) dt \right] \\
&= C_L K \left[ (T - M) - \frac{1}{\delta} \ln(1 + \delta(T - M)) \right].
\end{aligned}
\tag{12}
$$

Therefore, the per cycle average total cost is

$$
CO + CC + CD + CS + COLS.
\tag{13}
$$

The per cycle average total cost per unit is

$$
\begin{aligned}
ATC(M, T) &= \frac{1}{T} [CO + CC + CD + CS + COLS] \\
&= \frac{1}{T} \left[ \frac{a + bM + cM^2}{\theta} - \frac{b + 2cM}{\theta^2} + \frac{2c}{\theta^3} \right] \\
&\quad \times \left[ \frac{he^{\theta M} - h - \gamma M - \beta M^2}{\theta} + \frac{\gamma e^{\theta M} - \gamma - 2\beta M}{\theta^2} + \frac{2\beta e^{\theta M} - 2\beta}{\theta^3} \right] \\
&\quad - \frac{h}{\theta^3 T} \left[ \theta^2 \left( aM + \frac{bM^2}{2} + \frac{cM^3}{3} \right) - \theta(bM + cM^2) + 2cM \right] \\
&\quad - \frac{\gamma}{\theta^3 T} \left[ \theta^2 \left( \frac{aM^2}{2} + \frac{bM^3}{3} + \frac{cM^4}{4} \right) - \theta \left( \frac{bM^2}{2} + \frac{2cM^3}{3} \right) + cM^2 \right] \\
&\quad - \frac{\beta}{\theta^3 T} \left[ \theta^2 \left( \frac{aM^3}{3} + \frac{bM^4}{4} + \frac{cM^5}{5} \right) - \theta \left( \frac{bM^3}{3} + \frac{cM^4}{4} \right) + \frac{2cM^3}{3} \right] \\
&\quad + \frac{C_P}{T} \left[ \left( \frac{a + bM + cM^2}{\theta} - \frac{b + 2cM}{\theta^2} + \frac{2c}{\theta^3} \right) e^{\theta M} \right] \\
&\quad - \frac{C_P}{T} \left[ \frac{a}{\theta} - \frac{b}{\theta^2} + \frac{2c}{\theta^3} + aM + \frac{bM^2}{2} + \frac{cM^3}{3} \right] \\
&\quad + \frac{C_O}{T} + \frac{C_S K}{\delta T} \left[ (T - M) - \frac{1}{\delta} \ln(1 + \delta(T - M)) \right].
\end{aligned}
\tag{14}
$$

The necessary conditions for minimizing the average total cost per unit will be

$$
\begin{aligned}
\frac{\partial ATC(M,T)}{\partial M} = \frac{1}{T} &\left[ \frac{a+bM+cM^2}{\theta} - \frac{b+2cM}{\theta^2} + \frac{2c}{\theta^3} \right] \\
&\times \left[ \frac{h\theta e^{\theta M} - \gamma - 2\beta M}{\theta} + \frac{\alpha\theta e^{\theta M} - 2\beta}{\theta^2} + \frac{2\beta\theta e^{\theta M}}{\theta^3} \right] + \frac{1}{T}\left[ \frac{b+2cM}{\theta} - \frac{2c}{\theta^2} \right] \\
&\times \left[ \frac{he^{\theta M} - h - \gamma M - \beta M^2}{\theta} + \frac{\alpha e^{\theta M} - \gamma - 2\beta M}{\theta^2} + \frac{2\beta e^{\theta M} - 2\beta}{\theta^3} \right] \\
&- \frac{h+\gamma M+\beta M^2}{\theta^3 T}\left[ \theta^2\left( a+bM+cM^2 \right) - \theta(b+2cM) + 2c \right] \\
&+ \frac{C_P}{T}\left[ a+bM+cM^2 \right]\left[ e^{\theta M} - 1 \right] - \frac{K(C_S+\delta C_L)(T-M)}{T(1+\delta(T-M))} = 0.
\end{aligned}
$$

$$(15)$$

and

$$
\frac{\partial ATC(M,T)}{\partial T} = \frac{1}{T}\left[ \frac{K(C_S+\delta C_L)(T-M)}{1+\delta(T-M)} - (ATC(M,T)) \right] = 0 \qquad (16)
$$

The pair of $[M,T]$ calculated from (15) and (16) subject to the conditions $\frac{\partial^2 ATC(M,T)}{\partial M^2} > 0, \frac{\partial^2 ATC(M,T)}{\partial T^2} > 0$ & $\left( \frac{\partial^2 ATC(M,T)}{\partial M^2 \partial T} \right)^2 > 0$ will minimize $ATC(M,T)$.

## 4 Numerical Example

*Example 1* Let us consider the following parametric values of the inventory system as: $a = 12$, $b = 8$, $c = 2$, $h = 4$, $\gamma = 3$, $\beta = 2$, $C_P = 2$, $C_O = 4$, $C_S = 4$, $C_L = 3$, $K = 10$, $\theta = 0.05$ and $\delta = 3$ in appropriate units.

Solving the simultaneous Eqs. (15) and (16), the optimal shortage period and optimal cycle length are obtained as $M^* = 0.262993$ unit time and $T^* = 0.480607$ unit time respectively. Now substituting the pair $(M^*, T^*)$ in Eqs. (3), (7) and (14), we get the optimal order quantity $Q_0^* = 5.14309$ units, the optimal maximum inventory level $W_{Max.}^* = 3.4681$ units and the minimum average total cost per unit time $ATC^*(M, T) = 17.1159$ respectively.

## 5 Conclusion

For the business scenario, a large amount of capital was invested for maintaining and controlling inventory. Therefore, the demand pattern and deterioration rate are two important factors for growing the business. In this paper, a deterministic inventory model is considered with quadratic demand rate, quadratic carrying cost and constant deterioration rate. Shortages are allowed and partially backlogged in this model. The main reason of considering a quadratic demand instead of a linear demand or an exponential demand because of the accelerated growth and accelerated decline in demand. During the shortage period, the backlogging rate is inversely proportional to the waiting time for the next replenishment. Furthermore, the model has been solved with a numerical example by minimizing the total cost by simultaneously optimizing the shortage period and the length of the cycle.

There are many scopes of extending the present model as future work. Firstly, it can be extended by considering more realistic as stock dependent demand and stochastic demand. Secondly, it can be extended to variable deterioration like the generalized Weibull distribution deterioration rate.

# References

1. Bahari-Kashani.: Replenishment schedule for deteriorating items with time-proportional demand. J. Oper. Res. Soc. **40**, 75–81 (1989)
2. Chang, H.J., Dye, C.Y.: An EOQ model for deteriorating items with time varying demand and partial backlogging. J. Oper. Res. Soc. **50**, 1176–1182 (1999)
3. Dave, U., Patel, L.K.: (T, Si) policy inventory model for deteriorating items with time proportional demand. J. Oper. Res. Soc. **32**, 137–142 (1981)
4. Goyal, S.K., Giri, B.C.: Recent trends in modeling of deteriorating inventory. Eur. J. Oper. Res. **134**, 1–16 (2001)
5. Hollier, R.H., Mak, K.L.: Inventory replenishment policies for deteriorating items in a declining market. Int. J. Prod. Res. **21**, 813–826 (1983)
6. Khanra, S., Chaudhuri, K.S.: A note on an order level inventory model for a deteriorating item with time dependent quadratic demand. Comput. Oper. Res. **30**, 1901–1916 (2003)
7. Khanra, S., Ghosh, S.K., Chaudhuri, K.S.: An EOQ model for a deteriorating item with time dependent quadratic demand under permissible delay in payment. Appl. Math. Comput. **218**, 1–9 (2011)
8. Li, R., Lan, H., Mawhinney, J.: A review on deteriorating inventory study. J. Service Sci. Manage. **3**, 117–129 (2010)
9. Ouyang, L.Y., Wu, K.S., Cheng, M.C.: An inventory model for deteriorating items with exponential declining demand and partial backlogging. Yugoslav J. Oper. Res. **15**, 277–288 (2005)
10. Raafat, F.: Survey of literature on continuously deteriorating inventory model. J. Oper. Res. Soc. **42**, 27–37 (1991)
11. Singh, T., Pattnayak, H.: An EOQ model for a deteriorating item with time dependent quadratic demand and variable deterioration under permissible delay in payment. Appl. Math. Sci. **7**, 2939–2951 (2013)
12. Singh, T., Pattnayak, H.: An EOQ model for deteriorating items with linear demand, variable deterioration and partial backlogging. J. Service Sci. Manage. **6**, 186–190 (2013)

# A Novel Approach for Modelling of High Order Discrete Systems Using Modified Routh Approximation Technique via W-Domain

G.V.K.R. Sastry, G. Surya Kalyan, K. Tejeswar Rao
and K. Satyanarayana Raju

**Abstract** A novel Procedure is presented for modeling of higher order discrete systems based on matching the time responses of the original and reduced order systems. The flexibility of method is shown through familiar example.

**Keywords** Discrete systems · Order reduction · Large scale systems

## 1 Introduction

It is Observed that even through familiar methods are available for the Modelling of continuous time systems [1–16], the methods for discrete systems are seldom available like [17, 18]. Some of the familiar methods available in the literature for the reduction of high order discrete-time systems are like C.F. method, Stability equation method, Error minimization technique and methods given in reference [17]. The Routh Approximation Method (RAM) is considered as one of the most attractive methods of model reduction as it generates stable models for stable systems. Later RAM is extended to discrete time systems as Bilinear Routh Approximation Method, which tends to generate poor models as it approximates only dominant poles around z = 1. The Improved Bilinear Routh Approximation Method (IBRAM) [18] was proposed as an extension to BRAM. This method does

G.V.K.R.Sastry (✉) · K. Tejeswar Rao · K. Satyanarayana Raju
EEE Department, GIT, GITAM University, Visakhapatnam, India
e-mail: profsastrygvkr@yahoo.com

K. Tejeswar Rao
e-mail: tejeshk222@yahoo.com

K. Satyanarayana Raju
e-mail: ksatyanarayanaraju2011@yahoo.com

G. Surya Kalyan
EEE Department, Chaitanya Engineering College, JNT University, Kakinada, India
e-mail: suryakalyangarimella@yahoo.com

© Springer India 2016                                                                                          21
H.S. Behera and D.P. Mohapatra (eds.), *Computational Intelligence
in Data Mining—Volume 1*, Advances in Intelligent Systems
and Computing 410, DOI 10.1007/978-81-322-2734-2_3

not possess any optimal properties since only the total impulse energy is considered. In a very recent paper, Younseok Choo suggested a model reduction method known as Suboptimal Bilinear Routh Approximation Method [17] which modifies IBRAM so that the reduced order model gives minimum Integral Square Error.

## 2 Proposed Reduction Method

The procedure of the present method is as follows.
    Consider the original stable nth order discrete time system defined as

$$G_n(z) = \frac{N(z)}{D(z)} = \frac{B_0 + B_1 z + B_2 z^2 + \ldots + B_{n-1} z^{n-1}}{A_0 + A_1 z + A_2 z^2 + \ldots + A_n z^n}$$
$$= \frac{\sum_{i=0}^{n-1} B_i z^i}{\sum_{j=0}^{n} A_j z^j} ; \tag{1}$$

where $A_j$, $B_i$ are scalar constants.
    The kth order model is defined as

$$R_k(z) = \frac{b_0 + b_1 z + b_2 z^2 + \ldots + b_{k-1} z^{k-1}}{a_0 + a_1 z + a_2 z^2 + \ldots + a_k z^k}$$
$$= \frac{\sum_{i=0}^{k-1} b_i z^i}{\sum_{j=0}^{k} a_j z^j} ; \tag{2}$$

where $a_j$, $b_i$ are scalar constants and k = 1, 2, ..., n
    To obtain the unknown coefficients of the kth order model, i.e., $a_0$, $a_1$, ..., $a_k$ and $b_0$, $b_1$, ..., $b_{k-1}$ the following procedure steps are proposed.
    Consider the original nth order transfer function, $G_n(z)$ and apply the Bilinear Transformation with $z = \frac{1+w}{1-w}$ and hence

$$G_n(w) = \frac{N(w)}{D(w)} = \frac{d_0 + d_1 w + d_2 w^2 + \ldots + d_{n-1} w^{n-1}}{e_0 + e_1 w + e_2 w^2 + \ldots + e_n w^n} \tag{3}$$

Then the model in w-domain is

$$R_k(w) = \frac{N_k(w)}{D_k(w)} = \frac{D_0 + D_1 w + D_2 w^2 + \ldots + D_{k-1} w^{k-1}}{E_0 + E_1 w + E_2 w^2 + \ldots + E_k w^k} \tag{4}$$

The coefficients of $D_k(w)$ and $N_k(w)$ are determined as follows.
Reduced Denominator $D_k(w)$:
    The Routh table is obtained using the denominator $D(w)$ of Eq. (3) as given below (Table 1):

**Table 1** Routh table

$$
\begin{array}{llll}
e_n & e_{n-2} & \cdots\cdots & e_0 \\
e_{n-1} & e_{n-3} & \cdots e_1 \\
\vdots \\
e_0
\end{array}
$$

The algorithms for $D_k(w)$ are:

$$\text{for } k = 1; \quad D_1(w) = E_0 + E_1 w$$
$$\text{for } k = 2; \quad D_2(w) = E_0 + E_1 w + E_2 w^2$$
$$\vdots$$
$$\text{and in general, } D_k(w) = E_0 + E_1 w + \ldots + E_k w^k$$

where $E_0 = e_n$, $E_1 = e_{n-1}$ and so on from the first column of the Routh table.

Proposed Numerator $N_k(w)$: The unknown coefficients of $N_k(w)$ are obtained either by matching initial time moments and Markov parameters or retaining only initial time moments of the original system in its low order model with $k = t + m$.

Let 't' be the no. of initial time moments to be retained and 'm' be the number of Markov parameters to be retained.

$$R_k(w) = \frac{D_0 + D_1 w + D_2 w^2 + \ldots + D_{k-1} w^{k-1}}{E_0 + E_1 w + E_2 w^2 + \ldots + E_k w^k}; \text{ with } k = t + m; t = k; m = 0,$$

where,

$$D_0 = E_0 \times \left(\frac{d_0}{e_0}\right); \quad D_1 = \frac{\left(\left[\left(\frac{e_0 d_1 - d_0 e_1}{e_0^2}\right) \times E_0^2\right] + D_0 E_1\right)}{E_0};$$

$$\vdots$$

$$D_k = \frac{\left(\left[\left(\frac{e_{n-1} d_n - d_{n-1} e_n}{e_{n-1}^2}\right) \times E_{k-1}^2\right] + D_{k-1} E_k\right)}{E_{k-1}}$$

with $k = t + m; t = k - 1; m = 1,$

$$D_0 = E_0 \times \left(\frac{d_0}{e_0}\right); D_1 = E_2 \times \left(\frac{d_3}{e_4}\right)\ldots\ldots D_{k-1} = E_k \times \left(\frac{d_{n-1}}{e_n}\right)$$

For $w = \frac{z-1}{z+1}$, $R_k(w)$ can be written as

$$R_k(z) = \frac{b_0 + b_1 z + b_2 z^2 + \ldots + b_k z^k}{a_0 + a_1 z + a_2 z^2 + \ldots + a_k z^k}$$

# 3 Illustrative Example

The original system is defined as, [2]

$$G(z) = \frac{0.3124 z^3 - 0.5743 z^2 + 0.3879 z - 0.0889}{z^4 - 3.233 z^3 + 3.9869 z^2 - 2.2209 z + 0.4723}$$

Application of proposed method to obtain a 2nd order reduced model:
For $z = \frac{1+w}{1-w}$,

$$G(w) = \frac{-1.3635 w^4 + 0.5066 w^3 + 0.6152 w^2 + 0.2046 w + 0.0371}{10.9131 w^4 + 4.135 w^3 + 0.86 w^2 + 0.0866 w + 0.0053} = \frac{N(w)}{D(w)} \quad (5)$$

Thus the 2nd order model in w-domain will be defined as

$$R_2(w) = \frac{D_0 + D_1 w}{E_0 + E_1 w + E_2 w^2} = \frac{N_2(w)}{D_2(w)} \quad (6)$$

Routh table to obtain $E_0$, $E_1$, ... is given below (Table 2).
Using the proposed reduction method,

$$D_2(w) = 0.00873 + 0.1426w + w^2 \quad (7)$$

The 2nd order model in w-domain with t = 1; m = 1; is obtained as

$$R_2(w) = \frac{0.0464w + 0.0611}{0.00873 + 0.1426w + w^2} \quad (8)$$

**Table 2** Implementation of Routh table for above example

| | | | |
|---|---|---|---|
| $\omega^4$ | 0.0053 | 0.86 | 10.931 |
| $\omega^3$ | 0.0866 | 4.135 | |
| $\omega^2$ | 0.6069 | 10.931 | |
| $\omega^1$ | 2.5777 | | |
| $\omega^0$ | 10.931 | | |

For $w = \frac{z-1}{z+1}$,

$$R_2(z) = \frac{0.0933z^2 + 0.1061z + 0.0127}{z^2 - 1.722z + 0.7523}; t = 1, m = 1.$$

The 2nd order model in w-domain with t = 2; m = 0; using the proposed method is obtained as

$$R_2^1(w) = \frac{0.3365w + 0.0611}{0.00873 + 0.1426w + w^2} \tag{9}$$

For $w = \frac{z-1}{z+1}$,

$$R_2^1(z) = \frac{0.3453z^2 + 0.1061z - 0.2392}{z^2 - 1.722z + 0.7523}; t = 2; m = 0 \tag{10}$$

The responses of $G_n(z)$ and $R_2(z)$ and $R_2^1(z)$ are compared in Fig. 1.

The second order reduced model is obtained using the method mentioned in reference [2] as:

$$R_2''(z) = \frac{0.2766z - 0.1076}{z^2 - 1.776z + 0.8028}$$

The responses of $G_n(z)$ and $R_2(z)$ and $R_2^1(z)$ and $R_2''(z)$ are compared in Fig. 2.

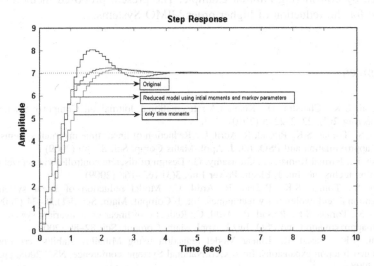

**Fig. 1** Comparison of step responses of original system and reduced order models obtained by proposed method

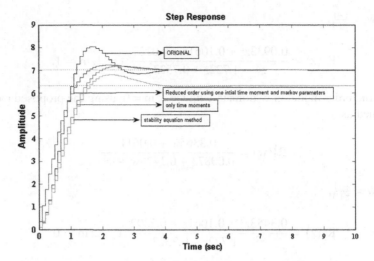

**Fig. 2** Comparison of step responses of original system and reduced order models obtained by proposed method and Stability Equation method

# 4   Conclusions

A new, computationally very simple procedure for order reduction of higher order discrete system is proposed using modified Routh Approximation technique with initial Time moments and Markov parameters matching technique. The method is illustrated by considering familiar example. The present proposed method can be extended for the reduction of higher order MIMO Systems.

# References

1. Agrawal, S.K., Chandra, D., Khan, I.A.: International Journal on Computer Science and Engineering **2**(7), 2242–2245 (2010)
2. Panda, S., Tomar, S.K., Prasad, R. Adril, C.: Reduction of linear time invariant systems using routh approximation and PSO. Int. J. Appl. Maths Comp. Sci. 82–89 (2009)
3. Rmesh, K., Nirmal Kumar, A., Gurusamy, G.: Design of discrete controller via a novel model reduction technique. Int. J. Electr. Power Eng. **3**(3):163–168 (2009)
4. Panda, S., Tomar, S.K., Prasad, R. Ardil, C.: Model reduction of linear systems by conventional and evolutionary techniques. Int. J. Comput. Math. Sci. **3**(1):28–34 (2009)
5. Panda, S., Tomar, S.K., Prasad, R., Ardil, C.: Reductio of linear time invariant systems using Routh-approximation and PSO. Int. J. Appl. Math. Comput. Sci. 82–89 (2009)
6. Tomar, S.K., Prasad, R.: Linear Model reduction using Mihailov stability criterion and continued fraction expansions. In: XXXII National Systems conference, NSC 2008, pp. 603–605 (2008)

7. Parmar, G., Bhandari, M.: Reduced order modelling of linear dynamic systems using eigen spectrum analysis and Modified cauer continued fraction, In: XXXII National Systems Conference, NSC 2008, pp. 497–602 (2008)
8. Parmar, G., Prasad, R., Mukherjee, S.: Order reduction by least squares methods about general point 'a'. Int. J. Math. Phys. Eng. Sci. 1(1), 26–29 (2008)
9. Vishwakarma C.B., Prasad, R.: Order reduction using the advantages of differentiation method and factor division method. IJEMS 15(6), (2008)
10. Parmar, G., Mukherjee, S., Prasad, R.: Reduced order modelling of linear dynamic system using particle swarm optimized eigen spectrum analysis. Int. J. Comput. Math. Sci. 1(1), 45–49 (2007)
11. Parmar, G., Mukherjee, S., Prasad, R.: Relative mapping errors of linear time invariant systems caused by particle swamp optimized reduced order model. Int. J. Comput. Inform. Syst. Sci. Eng. 1(2), 83–89 (2007)
12. Sastry, G.V.K.R., Mallikarjuna Rao, P.: A new method for modelling of large scale interval systems. IETE J. Res. 49(6), 423–430 (2003)
13. Sastry, G.V.K.R., Chakrapani, K.V.R.: A simplified approach for biased model reduction of linear systems in special canonical form. IETE J. Res. 42 (1996)
14. Sastry, G.V.K.R., Chittamuri, B.S.: An improved approach for biased model reduction using impulse energy approximation technique. IETE J. Res. 40 (1995)
15. Sastry, G.V.K.R., Raja Rao, G.: A simplified CFE method for large-scale systems modelling about s = 0 and s = a. IETE J. Res. 47(6), 327–332 (2001)
16. Sastry, G.V.K.R., Raja Rao, G., Mallikarjuna Rao, P.: Large scale interval system Modelling using Routh approximants. Electron. Lett. 36(8) (2000)
17. Choo, Y.: Sub-optimal Bilinear Routh approximation for discrete-time systems. ASME, J. Dyn. Syst. Meas, Control 125,130–134 (2006)
18. Choo, Y.: Improved bilinear Routh approximation for discrete-time systems. ASME, J. Dyn. Syst. Meas. Control 125–127 (2001)

# Pose Invariant Face Recognition for New Born: Machine Learning Approach

Rishav Singh and Hari Om

**Abstract** Pose is a natural and important covariate in case of newborn and face recognition across pose can troubleshoot the approaches dealing with uncooperative subjects like newborn, in which the full power of face recognition being a passive biometric technique requires to be implemented and utilized. To handle the large pose variation in newborn, we propose a pose-adaptive similarity method that uses pose-specific classifiers to deal with different combinatorial poses. A texture based face recognition method, Speed Up Robust Feature (SURF) transform, is used to compare the descriptor of testing (probe) face with given training (gallery) face descriptor. Probes executed on the face template data of newborn described here, offer comparative benefits towards affinity for pose variations and the proposed algorithm verdicts the rank 1 accuracy of 92.1 %, which demonstrates the strength of self learning even with single training face image of newborn.

**Keywords** SURF · Face recognition · Machine learning · Classification · Regression

## 1 Introduction

Swapping, abduction and missing of newborns is a great challenge to neonatalogical societies and It is a grave need of the time to work out the solution. Image processing research may solve this problem substantially. According to survey 80–90 million newborn are added every year and the instant global population of newborn and young children below 5 years remains around 400–500 million [1].

R. Singh (✉) · H. Om
Department of Computer Science & Engineering, Indian School
of Mines, Dhanbad, Jharkhand, India
e-mail: rishavsingh559@gmail.com

H. Om
e-mail: hariom4india@gmail.com

© Springer India 2016
H.S. Behera and D.P. Mohapatra (eds.), *Computational Intelligence
in Data Mining—Volume 1*, Advances in Intelligent Systems
and Computing 410, DOI 10.1007/978-81-322-2734-2_4

A study conducted by Gray et al. [2] infers that out of 34 newborns admitted to a neonatal intensive care unit everyday in USA, it is about 50 % probability of incorrect recognition. Further approximately 100,000–500,000 newborn babies reported switched mistakenly in USA [3], this is a technical issue. With the aid of authentic and efficient image recognition method these could be controlled or at least be reduced. The objective of this research is to propose semi supervised algorithm for face recognition of newborns while evaluating the performance of existing methods systematically. We propose the method primarily be user friendly, hygienic and economical solution for morphological recognition of newborns. The proposed method also evades the dilemma of least training dataset (works on single training image) apart from pose covariate, and provides better accuracy compared to existing methods.

Recent bracelet identifier technique remains inefficient, which requires to tag newborn hands/legs just after birth, to check mixing and switching of newborn. The recommended foot and fingerprint of child and mother by the FBI too even fails to correctly identify. Researchers from both medical and computer science background have surveyed the plausibility of using footprints for newborn recognition and concluded in uncertainty for the footprint usage [4–8]. Furthermore there have been explorations towards applicability of other biometric modalities like fingerprints, palmprints and ear for verification of newborn's identity [9–11].

The work done on recognition of newborn by Tiwari et al reports the performance of recognition on face, ear, headprint and soft biometric data on newborn database [12–21]. But there is no study on face recognition of newborn across pose or recognition of newborn with single training image. After inspection of newborn data base it was found that proper frontal face photograph of some subjects are not there and in some cases only single training image is present in the database. The neonates are naturally non cooperative towards biometric snaps. It is difficult to prepare the face data in reality for training image. This motivates to develop a technique which can handle this important issue. In this research paper we have implemented self learning method using SURF descriptor which is considered to be a pose invariant descriptor.

## 2  Proposed Methodology

Earlier people stored only important data because of limited and expensive storage space. However, as the time passed most of the practitioners and researchers felt the need for extra labelled data. They required it for building mathematical models to analyse complex problems. The data would help them build and test these models.

In the real world only a small amount of labelled face image of newborn is available due to which most of the time the unlabelled faces are misclassified. To overcome this challenge we used semi-supervised parameters to classify the newborn face image and reduce the number of misclassification. Thus semi-supervised learning helps in training the database more effectively [22–24].

Feature extraction methods extract the distinct features from the images like edges, corners, etc. [25]. SURF is one of the fast feature extraction techniques. The method uses advanced schemes for detection and description of images. It generates a compact representation of the image. The result is obtained based on information gathered from the intensity contained within required points. These points are found using scale space. The scale space is divided into octaves, and there are 4 scale levels in each octave. Each octave represents a series of filter. This helps in localizing the required interesting points using basic second order Hessian matrix approximation. Once the interesting points are detected they are described using Haar wavelet response in x and y direction. The descriptor is calculated for a square region around the interesting point along the orientation determined by the Haar method [25].

Below are the steps used to calculate the descriptors for newborn face image.

- A square region is sampled around the interest point along the origin of the interest point.
- The square region is further divided into 4 × 4 sub-regions.
- Haar wavelet responses for 5 × 5 evenly sampled points are calculated for each sub-region.
- $\sum dx$ and $\sum dy$ are computed by summing up the responses of wavelets in x–y direction
- Information about the polarity of changes in the intensities values are incorporated on the descriptors, by calculating the absolute value of the response $(\sum |dx|$ and $\sum |dy|)$.
- $S = [\sum dx, \sum |dx|, \sum dy, \sum |dy|]$, where S a vector calculated at each sub-region.
- For the intensity structure, an $m \times m \times 4$ descriptor is generate (generally m = 4).

In first stage the training is done with labelled face and classifies the unlabelled face, which contains small amount of classification information. The process is repeated until the training set reaches up to minimum required amount depending on a particular threshold value. In the next stage classification is performed on the unlabelled set which is lacking their classification information. While implementing the algorithm the face database of newborn is divided into two sets: Training and Testing.

- Let c(i) where i = 1, 2, 3, ..., n; be the subjects in the newborn database.
  P—Set of newborn face image which have c (i) and is classified, 1 image per class.
  G—Set of newborn face image which satisfy semi-supervised parameter and get classified.
- Classification: Newborns are selected for classification on the basis of threshold value (TV).

**Fig. 1** Methodology for face
recognition of newborn

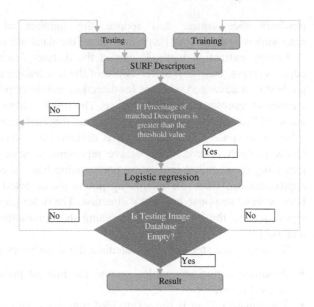

As shown in Fig. 1 the training and testing face image set is given as an input to
the SURF for feature extraction. If the total number of the matching descriptors of a
testing exceeds a given threshold value (TV), then the image of newborn is selected
for classification or else it remains in testing. Logistic regression classifies the
image of newborn on the basis of best fit and adds the image to the training [26, 27].
The steps are repeated until all the face image of newborn present in the testing
database are classified.

## 3 Experimental Setup

In this experiment 130 individuals (class) data is selected out of 280 subjects from
IIT (BHU) newborn face data base with neutral expression in different pose with
same illumination conditions, each class containing 10 face image of newborn. The
newborn database ($nD$) is further divided into two sections newborn training ($nG$)
and newborn testing ($nP$). Initially $nG$ contains 1 data per class and rest of the data
is treated as $nP$.

The experiment involves multiple steps and multiple iterations. Before splitting
up the $nD$ into $nG$ and $nP$, $nD$ is pre-processed by projecting the RGB image to
gray scale plane, followed by the process of normalization. SURF is used to identify
and match the descriptors. The face image of a newborn present in $nP$ is input for
classification, only if it exceeds a given threshold value. After classifying and
labelling the images of a newborn, the images are included in $nG$. The process is
repeated till all the images of $nP$ are classified.

# 4 Experimental Results and Analysis

As shown in Table 1 the descriptor match column the below image represents the instance of *nG* database, newborn face of the same subject is shown. The descriptors of the *nP* is matched with those of the *nG*. The result is described as below:

**Table 1** Represents the descriptors identified by the SURF

| Sl. No. | Descriptors | Descriptors Match | Result |
|---|---|---|---|
| 1 | | | Object Descriptors: 8 reaching here in surf Image Descriptors: 11 Extraction time = 29.2202ms no of pairs 5 |
| 2 | | | Object Descriptors: 20 reaching here in surf Image Descriptors: 11 Extraction time = 17.7736ms no of pairs 9 |
| 3 | | | Object Descriptors: 15 reaching here in surf Image Descriptors: 11 Extraction time = 13.5892ms no of pairs 4 |
| 4 | | | Object Descriptors: 8 reaching here in surf Image Descriptors: 11 Extraction time = 17.1084ms no of pairs 5 |
| 5 | | | Object Descriptors: 20 reaching here in surf Image Descriptors: 11 Extraction time = 20.3202ms no of pairs 9 |
| 6 | | | Object Descriptors: 15 reaching here in surf Image Descriptors: 11 Extraction time = 26.5778ms no of pairs 4 |
| 7 | | | Object Descriptors: 11 reaching here in surf Image Descriptors: 11 Extraction time = 20.5037ms no of pairs 5 |
| 8 | | | Object Descriptors: 11 reaching here in surf Image Descriptors: 11 Extraction time = 18.7654ms no of pairs 11 |
| 9 | | | Object Descriptors: 14 reaching here in surf Image Descriptors: 11 Extraction time = 33.9419ms no of pairs 6 |

- Object Descriptors: Number of identified descriptors in the $nP$.
- Image Descriptors: Number of identified descriptors in the $nG$.
- Extraction Time: Time elapsed for identifying and matching the descriptors of $nP$ with those of $nG$.
- No of pairs: Total count of matched descriptors.

In the first stage the best identified matches on the basis of descriptors are 2, 5, 7 and 8 in serial number. The other data also belong to the same class shown in Table 1 but they did not have sufficient amount of classification on the basis of a single training image (Table 2).

The values of the different parameters of the model in the first iteration are shown below.

Bottom asymptote: 5.156364
Top asymptote: 35.60644
Inflexion point at (x, y): 1.320568 20.3814
Goodness of fit: 0.8764276
Standard error: 0.4331482

After adding the classified data to the training set, some more classification information's are now provided to the classifier where training data contains 46 % of the entire database, which helps it to reduce the percentage of misclassification.

In the second iteration the model uses this information for classifying the remaining $nP$ database on the basis of new $nG$.

**Table 2** Represents the result of iteration 2

| Descriptor of Testing Data | Descriptors Match | Results |
| --- | --- | --- |
| 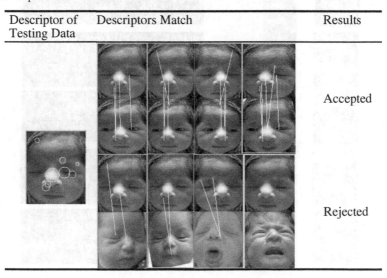 | | Accepted |
| | | Rejected |

The percentage of rejection in the second iteration has been reduced, because $nG$ in the second iteration provides some more descriptor information by using the newly classified images in the previous iteration. After each iteration few instances of $nP$ database is added to $nG$ database, depending upon the classification information and in return the new classified images helps to identify the remaining instances of the newborn.

Table 3 shows the results of 4 existing algorithm and 1 proposed algorithm for the face recognition of newborn in different pose (Fig. 2).

The Performance of the proposed algorithm is evaluated on single Newborn image per class for training and 9 newborn images per class for testing. Whereas the existing algorithms training set is 6 images per class and testing is 4 per class. Among the appearance based algorithm ICA provides the better accuracy of 82.5 % at Rank1 among the holistic based techniques. The performance of the proposed method is computed 5 times random cross validation. The average rank 1 identification accuracy of the proposed method is observed to be 91.2 %.

**Table 3** Identification accuracy of newborn

| Rank | PCA | LDA | ICA | SURF | Proposed |
|------|-------|-------|-------|-------|----------|
| 1 | 75.01 | 81.14 | 82.5 | 83.17 | 91.2 |
| 2 | 81.35 | 82.46 | 83.31 | 86.82 | 93.14 |
| 3 | 83.33 | 83.67 | 85.12 | 87.67 | 94.96 |
| 4 | 84.16 | 86.01 | 86.79 | 89.91 | 96.13 |
| 5 | 85.33 | 87.39 | 88.27 | 90.14 | 96.86 |

**Fig. 2** Comparison of Proposed methods with the existing ones

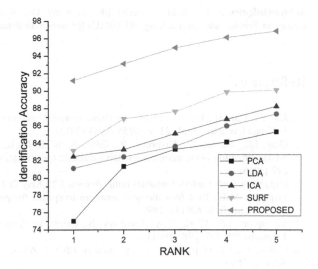

# 5 Conclusion

The problem of face recognition, pose variation has gathered a prime attention in Image processing and pattern recognition communities. The primary goal of this paper is to demonstrate the researchers towards role of self learning in improving the image similarity search performance in newborn face recognition problem. This was further constrained for overcoming the modalities of pose and the limitation of training images. The Performance of our proposed algorithm is evaluated on single newborn image per class for training and 9 newborn images per class for testing. The average rank 1 identification accuracy of the proposed method is 91.2 %.

We describe our approach to be relatively positive and encouraging lead in solving the face recognition problem of newborns. This is a prima facie level proof of concept stage, however more research need to be done to establish this on par industrial standard. We suggest that covariate based algorithms may effectively be deployed here, as the values of covariates evaluated among the couple of images (being compared) leads us to differentiate pose anomalies. This study report can observe changes due o covariates certainly. Incidentally, we may not try to explicate the cause of the outcome in detail right now, may further insights be revealed. Perhaps, answering the root cause of the effects shall assist to design more robust but critical face recognition algorithms in above said domain. Alternatively for fine graining, we suggest that the weighting of an algorithm response might change based on same estimated covariates. Finally by evaluating the boundaries of many face recognition algorithms with various covariates, we arrive on conclusion that we yet are at the beginning of the exploration of a confident neonatal face recognition system.

**Acknowledgments** The authors would like to thank Dr. Shrikant Tiwari (Department of Computer Science and Engineering, IIT (BHU) for providing database of newborn.

# References

1. Jia, W., Cai, H., Gui, J., et al.: Newborn footprint recognition using orientation feature. J. Neural Comput. Appl. **21**(8), 1855–1863 (2012)
2. Gray, J.E., Suresh, G., Ursprung, R., Edwards, W.H., Nickerson, J., Shinno, P.H.: Patient misidentification in the neonatal intensive careunit: quantification of risk. Pediatrics **117**, e46–e47 (2006)
3. http://www.amfor.net/stolenbabies.html. Accessed 25 May 2011
4. Stapleton, M.E.: Best foot forward: newborn footprints for personalidentification. FBI Law Enforcement Bull. **63**(11), 1999
5. Thompson, J.E., Clark, D.A., Salisbury, B., Cahill, J.: Footprinting the: not cost-effective. J. Pediatr. **99**, 797–798 (1981)
6. Galton, F.: Finger prints of young children (British Association for the Advancement of Science, 1899)
7. Shepard, K.S., Erickson, T., Fromm, H.: Limitations of footprinting as a means of newborn identification. Pediatrics **37**(1), 107–108 (1966)

8. Pela, N.T.R., Mamede, M.V., Tavares, M.S.G.: Analise crtica de impressoes plantares de recem-nascidos. Revista Brasileira de Enfermagem **29**, 100–105 (1975)
9. Weingaertner, D., Bellon, O.R.P., Cat, M.N.L., Silva, L.: Newborn's biometric identification: can it be done? Int. Joint Conference on Computer Vision, Imaging and Computer Graphics Theory and Applications, 2008
10. Fields, C., Hugh, C.F., Warren, C.P., Zimberoff, M.: The ear of the newborn as an identification constant. J. Obstet. Gynecol. **16**, 98–101 (1960)
11. Lemes, R.P., Bellon, O.R.P., Silva, L., Jain, A.K.: Biometric recognition of newborns: identification using palmprints, pp. 11–13. Int. Joint Conf. Biometrics, Washington, DC, USA (2011)
12. Tiwari, S., Singh, S.K.: Face recognition for newborns. IET Biometrics. **1**(4), 200–208 (2012)
13. Tiwari, S., Singh, S.K.: Newborn verification using headprint. J. Inform. Technol. Res. (JITR), **5**(2), 15–30, April–June 2012, ISSN: 1938-7857, doi:10.4018/jitr.2012040102. (IGI Global)
14. Tiwari, S., Singh, A., Singh, S.K.: Fusion of ear and soft-biometrics for recognition of newborn. Sign. Image Process.: Int. J. **3**(3), 103–116, ISSN: 0976-710X (2012)
15. Tiwari, S., Singh, Aruni, Singh, S.K.: Intelligent method for face recognition of infant. Int. J. Comput. Appl. **52**(4), 36–50 (2012)
16. Tiwari, Shrikant, Singh, Aruni, Singh, Sanjay Kumar: Integrating Faces and Soft-biometrics for Newborn Recognition. Int. J. Adv. Comput. Eng. Archit. **2**(2), 201–209 (2012)
17. Tiwari, S., Singh, A., Singh, S.K.: Can Face and Soft-biometric Traits Assist in Recognition of Newborn? In: Proceedings of IEEE, International Conference on Recent Advanced in Information Technology, pp. 74–79, March 2012
18. Tiwari, S., Singh, A., Singh, S.K.: Can Ear and Soft-biometric Traits Assist in Recognition of Newborn? In: Proceedings of International Conference on Computer Science, Engineering and Applications, doi:10.1007/978-3-642-30157-5, ISBN: 978-3-642-30157-5, pp. 179–192. Springer, Berlin (May 2012)
19. Tiwari, S., Singh, S.K., Multimodal Biometric Recognition for Newborn. In: Srivastava, R., Singh, S.K., Shukla, K.K. (eds.) Research Developments in Biometrics and Video Processing Techniques (IGI Global Publishing). Chapter 2. ISSN: 1948-9730, ISBN: 978-1-4666-4870-8
20. Tiwari, S., Singh, A., Singh, S.K.: Multimodal database of newborns for biometric recognition. Int. J. Bio-Sci. Bio-Technol. **5**(2), 89–99 (2013)
21. Tiwari, S., Singh, A., Singh S.K.: Newborn's ear recognition: can it be done?. In: International Conference on Image Information Processing (ICIIP), JUIT India, 3–5 November 2011
22. Keogh, E., Wei, L.: Semi-supervised time series classification. In: SIGKKD (2006)
23. Chapelle, O., Scholkopf, B., Zien, A.: Semi-Supervised Learning. The MIT Press, Cambridge (2006)
24. Dillon, J.V., Balasubramanian, K., Lebanon, G.: Asymptotic analysis of generative semi-supervised learning. In: Appearing in Proceedings of the 27th International Conference on Machine Learning, Haifa, Israel (2010)
25. Bay, H., Ess, A., Tuytelaars, T., Gool, L.V.: Surf: speeded up robust features. Comput. Vis. Image Underst. **110**(3), 346–359 (2008)
26. Giraldo, J., Vivas, N.M., Vila, E., Badia, A.: Assessing the (a) symmetry of concentration-effect curves: empirical versus mechanistic models. Pharmacol. Ther. **95**(1), 21–45 (2002)
27. Richards, F.J.: A flexible growth function for empirical use. J. Exp. Bot. **10**, 290–300 (1959)

# Implementation of Data Analytics for MongoDB Using Trigger Utility

Kalpana Dwivedi and Sanjay Kumar Dubey

**Abstract** SQL based traditional databases like Oracle; SQL server offers the capability to develop programs like Trigger. Trigger is a very important feature provided by many databases, especially useful in monitoring, rule enforcement, data validation and data analytics etc. MongoDB is non SQL document oriented database. MongoDB is the fastest growing and most demanding non SQL database. Since MongoDB primarily is operated using its out of box tools like mongo, mongos, bsondump, mongod, mongoexport and Java script function. MongoDB does not provide in-built feature for triggers which is very efficient in data analytics, monitoring and reporting purpose. Paper presents two utility in which one utility is a listener or poller utility which is developed to give similar feature like trigger and after that second utility is developed which gives historical data analytic capability on Mongo database by using the trigger utility. It pulls the data from analytic collection and generates the graph. Data analytic tools plays vital role in decision making in today's complex business environment where data size is very huge and unstructured by nature.

**Keywords** Trigger · MongoDB · Data analytics · JSON · HDFS

## 1 Introduction

Since last about a decade or so, trillions of terabyte data is generating from various social media, discussion forums, and also from traditional business. Hadoop and Cassandra are addressing the problems arrived due to heavy volume of data. Hadoop uses HDFS and Cassandra uses CFS for storing the data on storage media.

K. Dwivedi (✉) · S.K. Dubey
Amity University, Sec-125, Noida, U.P., India
e-mail: dwivedi.dolly@gmail.com

S.K. Dubey
e-mail: skdubey1@amity.edu

© Springer India 2016                                                                                      39
H.S. Behera and D.P. Mohapatra (eds.), *Computational Intelligence in Data Mining—Volume 1*, Advances in Intelligent Systems and Computing 410, DOI 10.1007/978-81-322-2734-2_5

Both Hadoop and Cassandra play vital role in big data area. Since HDFS and CFS both are filed system and not database, therefore these two does offer the advantages offered by a database. Many business houses have huge amount of data typically unstructured by nature, stored in Hadoop or Cassandra file systems, but for many data manipulation purpose, data has to be imported in a database which supports both some database feature as well capable to handle large amount of data and the best fit solution is MongoDB.

MongoDB is also fastest growing NON SQL, document oriented database. MongoDB offers very flexible and dynamic schema to store documents in JSON format and is selected for performing data analytics. MongoDB is the most popular NoSQL database Since live data analytic requires consistent monitoring of new data added in database and trigger like feature are greatly helpful for this purpose.

In this paper the following steps are taken to perform the analytics. First install MongoDB. Then prepare mongo cluster that is having master/slave architecture. Data is added or imported into master node and replicated to all slave nodes. Trigger is developed and enabled on a collection on master node. Trigger fetch the analytic data and store in a separate collection. An analytic utility keeps on posting huge amount of data into the database and trigger utility which is consistently listening on an analytic database extracts necessary information required for the analytics from the analytic database and stores in a separate database. Further a java tool is developed which pulls the data from analytic collection and generate the graph. The reporting tool generates a real time graph on the analytic data into the database. Since currently at present there is no access to corporate database to perform the analytics that is why analytic application is developed to post requirement data into the database and once the access to any production system analytic application can be replaced by some smart import tool that migrates the legacy business data from variety of system and coming in variety of format into MongoDB as JSON document and after the reporting tool generates the graph as per the requirement.

## 2  MongoDB

In 1970 flat file system was created but the problem is that there is no standard implementation. Relational database was an answer and were created [1]. But the problem is that they could not handle big data. NoSQL database was an answer [2]. There are three categories of storing NoSQL database mechanism: key value store, tabular, and document oriented. Document oriented examples are MongoDB, C Lightweight couchDB, Cloudant and key value [3]. NoSQL is mainly used for scalability, high availability and performance and not used for complex transactions, Joins and Constraints [4, 5]. In RDBMS records are converted into objects in case of oops language but in MongoDB there is no conversion required [6]. Replication and Sharding are two most important features of MongoDB. MongoDB is a Document Oriented which means not .pdf or .doc files and work with JSON like Objects called BSON [7]. MongoDB is a Full-Featured database because it supports rich querying, real time aggregation, traditionally consistent. Others features are fully-consistent reads, secondary indexes, atomic write and query language [8]. MongoDB offers very flexible and dynamic schema to store documents in JSON format [9]. Arrays within other documents and embedding of documents are supported by MongoDB BSON [10–12]. For MongoDB the primary data representation is traversal. C data types are used mostly in languages for performing decoding from BSON and encoding data to BSON [13]. MongoDB is good at Documents/Content Management, Archiving, Event Logging; Real Time Stats/Analysis [14]. MongoDB uses dynamic schemas. Without defining the structure collections can be created. By deleting existing ones or by adding new fields the documents structure can simply change [15]. Documents are all records, or data, in MongoDB because MongoDB is a document-based database system [16–18]. Other features of MongoDB are Grid FS API, Capped Collection that works like a circular Queue, aggregation which Offer similar functionality as SQL Group by clause, Batch processing of data [19–21]. It understands Latitude and Longitude that means give special support for locations [22]. Automatic load balancing configuration is easy to deploy in MongoDB [23]. Applications of MongoDB are configuration, forms data user accounts, geo data, content management system, log data [24]. The three most important features of MongoDB are profiling, replication and sharding [25].

## 3 Related Work

MongoDB is a Document Oriented, High Performance, Horizontally-Scalable, Full-Featured database [26]. MongoDB is an Open Source NOSQL database developed and supported by 10gen and Commercial Licenses are also available [27]. Chitra [28] compared and analyzed the nonfunctional requirements of NOSQL databases. In] paper a new methodology produced where MapReduce paradigm along with online aggregation is used with MongoDB [29]. In paper by Dede [30], MongoDB with Hadoop has been evaluated for the performance, scalability and fault-tolerance features. Khan [31] attempted to replace the relational database with NoSQL database. In paper by Liu [32] cluster's concurrent reading and writing performance is improved and as well as it balances the data among the shards. Boicea [33] presents differences between Oracle Database and MongoDB. Zhao [34] compared MongoDB, with relational algebra. Author investigates some issues related to the feasibility of the migration of traditional relational databases to NoSQL databases. In paper by Kanade [35], a methodology in MongoDB is used to find the extent of embedding and normalization. Murugesan [36] presented a simple model that aggregates the two techniques for management of logs. Xinwei [37] presented a structure of RFID that is used in order to manage data in MongoDB. The detail about Hadoop is also presented for analytical approach [38]. In paper by Parker [39], the performance is compared in terms of runtime, of the two databases NoSQL MongoDB and SQL DB. Abramova [40] compared and evaluated two most popular NoSQL databases: MongoDB and Cassandra.

## 4 Implementing the Trigger Utility in MongoDB

MongoDB does not support triggers. To achieve the functionality of trigger a utility is developed that keeps on listening any event on Mongo collection and perform the designated task. Basically any JS function can be executed in MongoDB. MongoDB provides us the provision to execute any task as JS function and JS function is executed by passing it as a parameter to mongo command. So passing JS function to mongo client while starting it. Mongo client executes the JS function on Mongo server. There are three steps and commands for implementing the trigger as follows.

Creating Capped Collection:
```
db.createCollection ( "MyCappedCollection", {capped: true, size: 10000 } )
```
Creating Trigger:

```
var coll = db.MyCappedCollection;
var backupColl = db.AnalyticData;
var lastTimeStamp = coll.find().sort({ '$natural' : -1 })[0].OperationTime;
while(1){
cursor = coll.find({ OperationTime: { $gt: lastTimeStamp } });
cursor.addOption( 32 );
while( cursor.hasNext() ){
var doc = cursor.next();
lastTimeStamp = doc.OperationTime;
backupColl.insert(doc);}
}
```

Executing the Trigger:

```
Mongo localhost: 27017/CDS UserScripts/TriggerScript.js
```

## 5 Experimental Work

For implementation of data analytics, one company is chosen let the company be the RTCCM (Round The Clock Coverage Media) Pvt. Ltd that maintains and publish various articles. Company is very much interested in collecting the important information regarding its articles so that in future it can improve the way and category of articles that it publishes. RTCCM every day publishes new articles on its web site and interested to collect the information like which articles have most HIT count, what is the total hit count of all articles for any day, what is total hit count of any specific article for given period of time, what is total hit count of all articles on any minutes or hour of the given date and so on. This kind of information, RTCCM Pvt. Ltd. is interested in because based on these data point it will try to improve the category of articles and content, it publishes. While performing data analytics for RTCCM Pvt. Ltd. for calculating HIT count for various articles, there only one document is created for each date i.e. there are only 365 (Three hundred sixty five) documents is created for entire 1 year. It means if there are 1 lac hit count for all articles every day and category of article also varying every day, then it is very expensive and lengthy to perform such analytics in SQL DBs because every article is having its own entry in transaction table. For every day a new document with above structure is inserted into the Mongo database. This document maintains daily hit count, every minute hit count, every hour hit count and article wise hit count. Following Figs. 1, 2 and 3 shows the graphical representation of data analytics for article wise hit count, hourly hit count and minutely hit count for the company.

Whenever any article is hit by the user relevant hit count is either inserted or updated into the document. Next day new such document will be created to store the analytic information for that day.

Similarly for performing data analytics for RTCCM Pvt. Ltd the document used here is for maintaining every hour hit count for the various categories of articles and improving the content of it.

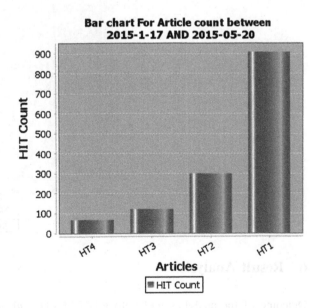

**Fig. 1** Mongo analytics chart of article wise hit count between 17-1-2015 and 20-5-2015 for RTCCM Pvt. Ltd.

**Fig. 2** Mongo analytics chart of hourly hit count for today for RTCCM Pvt. Ltd.

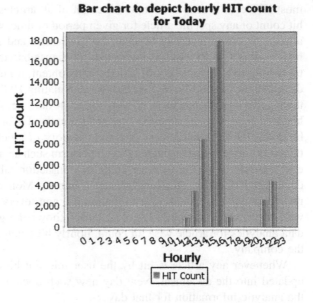

**Fig. 3** Mongo analytics chart of minutely hit count for given minutes for RTCCM Pvt. Ltd.

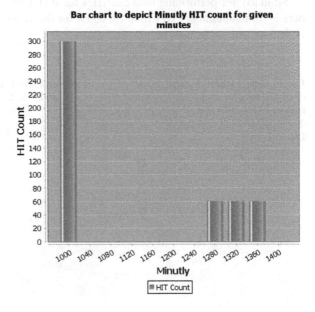

## 6　Result Analysis

Outcome of the analytics will help RTCCM Pvt. Ltd. to analyze which type of articles is most and least preferred among users. What time, the articles are more frequently accessed so that more numbers of nodes in the cluster can be made running during that time. Also, such analytics will help RTCCM Pvt. Ltd. to identify the taste and mood of the reader which in turn will help its business to take appropriate decision. Analytics is possible in MongoDB database because MongoDB support dynamic structure and upsert operation of the document in its database and will also store huge number of records into analytic table because of varying category of the articles but Mongo's dynamic schema support helps to store all analytics information for all articles irrespective of their number, into same document.

## 7 Conclusion and Future Work

Outcome of the research can be divided into two parts. First outcome is a poller or listener utility having same capability as trigger in traditional databases. This trigger consistently monitors the database (collection) for any operation on it Second outcome is an analytic tool capable of conducting data analytic. Initially entire focus is for MongoDB but in future, tool's analytic capability extended to cover other major big data tools like Hadoop, Cassandra etc. Data analytics can be applied on data imported either from HDFS or CFS. MongoDB offers powerful MapReduce tool similar to Hadoop. It also offers MongoDB-Hadoop connector to connect with Hadoop framework.

## References

1. Chang, F., Dean, J., Ghemawat, S., Hsieh, W., Burrows, M., Chandra, T., Andrew Fikes, A., Gruber, R.: Bigtable: a distributed storage system for structured data OSDI'06. In: Seventh Symposium on Operating System Design and Implementation, Seattle, WA (2006)
2. Han, J., Le, G.: Survey on NoSQL database in IEEE (2011). 978-1-4577-0208-2
3. Abouzeid, A.: Hadoop DB: an architectural hybrid of mapreduce and DBMS technologies for analytical workloads. In: Proceedings of the VLDB End
4. Levitt, N.: Will NoSQL databases live up to their promise. IEEE Comput. Soc. **43**(2), 922–933 (2009)
5. Han, I., Haihong, E., Le, G., Du, J.: Survey on NoSQL database. In: 6th International Conference on Pervasive Computing and Applications, pp. 363–366 (2011)
6. Padhy, R., Patra, M., Satapathy, S.: RDBMS to NoSQL: Reviewing some next-generation non-relational database's. Int. J. Adv. Eng. Sci. Technol. **11**(1), 015–030 (2011)
7. Rutishauser, N.: TPC-H applied to MongoDB: how a NoSQL database performs supervised by Prof. Dr. Michael B"ohlen. Amr Noureldin 349 (2012)
8. DeCandia, G., Hastorun, D., Jampani, M.: Dynamo: amazon's highly available key-value store. SOSP'07. Stevenson, Washington, USA (2007)
9. Tudorica, B., Bucur, C.: A comparison between several NoSQL databases with comments and notes. In: Roedunet International Conference (2011)
10. Kanade, A., Gopal, A.: An experimental study on open source RDBMS FDI issues and prospects. In: National Conference, pp. 203–206 (2013)

11. Hecht, R., Jablinski, S.: NoSQL evaluation a use case oriented survey. In: Proceedings International Conference on Cloud and Service Computing, pp. 12–14 (2011)
12. Adam, L., Jakob Mattson, J.: Investigating storage solutions for large data: a comparison of well performing and scalable data storage solutions for real time extraction and batch insertion of data (2011)
13. Yang, H., Dasdan, A., Hsiao, R., Parker, D.: Map-reduce-merge: simplified relational data processing on large clusters. In: Proceedings of the ACM SIGMOD International Conference on Management of Data (SIGMOD '07), pp. 1029–1040. ACM, New York, NY, USA (2007)
14. Veen, J., Waaij, B., Meijer, R.: Sensor data storage performance: SQL or NoSQL. In: Physical or Virtual Fifth International Conference on Cloud Computing, pp. 431–438. IEEE (2012)
15. Sanders, G., Shin, S.: Denormalization effects on performance of RDBMS. In: Proceedings of the 34th Hawaii International Conference on System Sciences, pp. 1–9. IEEE (2001)
16. Cruz, F., J. Pereira, J., Oliveira, R.: An effective scalable SQL engine for NoSQL databases in distributed applications and interoperable systems, pp. 155–168. Springer (2013)
17. Calil, A., Mello, R.: SimpleSQL: a relational layer for SimpleDB in ADBIS, pp. 99–110 (2012)
18. Stonebraker, M.: SQL databases v. NoSQL databases. Commun. ACM 53(4), 10–11 (2010)
19. Roijackers, J., Fletcher, L.H.G.: On bridging relational and document-centric data stores in BNCOD, pp. 135–148 (2013)
20. Hacigumus, H., Tatemura, J., PHsiung, W., Moon, H., Chi, Y.: CloudDB: one size fits all revived in services, pp. 148–149 (2010)
21. Ameri, P., Grabowski, U., Meyer, J., Streit, A.: On the application and performance of MongoDB for climate satellite data. In: Proceedings of 13th International Conference on Trust, Security and Privacy in Computing and Communications in IEEE (2014)
22. Wu, S., Jiang, S., Ooi, B., Tan, K.: Distributed online aggregations. In: Proceedings VLDB, pp. 443–454 (2009)
23. Pavlo, A.: A comparison of approaches to large-scale data analysis. In: Proceedings of the ACM SIGMOD, pp. 165–178 (2009)
24. Padhy, R., Patra, M., Satapathy, S.: RDBMS to NoSQL: reviewing some next-generation non-relational database's. Int. J. Adv. Eng. Sci. Technol. 11(1), 015–030 (2011)
25. Banker, K.: MongoDB in action (2011)
26. Huang, S., Cai, L., Liu, Z., Hu, Y.: Non-structure data storage technology: a discussion computer and information science (ICIS), pp. 482–487 (2012)
27. Tauro, C.S.A.: Comparative study of the new generation, agile, scalable, high performance NOSQL database. Int. J. Comput. Appl. 48(20), Commun. ACM, 25(4), 0975–888 (2012)
28. Chitra, K., Jeevarani, B.: Study on basically available, scalable and eventually consistent NOSQL databases. Int. J. Adv. Res. Comput. Sci. Softw. Eng. 3(7) (2013)
29. Mohan, B., Govardhan, A.: Online aggregation using MapReduce in MongoDB. Int. J. Adv. Res. Comput. Sci. Softw. Eng. 3(9) (2013)
30. Dede, E., Govindraju, M.: Performance evaluation of a MongoDB and Hadoop platform for scientific data analysis. In: Proceedings of the 4th ACM Workshop on Scientific Cloud Computing (2013)
31. Khan, S., Mane, V.: SQL support over MongoDB using metadata. Int. J. Sci. Res. Public. 3(10) (2013)
32. Liu, Y., Wang, Y., Jin, Y.: Research on the improvement of MongoDB auto-sharding in cloud environment in computer science and education (ICCSE), pp. 851–854 (2012)
33. Boicea, A., Radulescu, F., Agapin, L.: MongoDB vs Oracle: database comparison. In: Third International Conference on Emerging Intelligent Data and Web Technologies (2012)
34. Zhao, G., Huang, W., Liang, S., Tang1, Y.: Modeling MongoDB with relational model. In: Fourth International Conference on Emerging Intelligent Data and Web Technologies (2013)
35. Kanade, A., Gopal, A., Kanade, S.: A study of normalization and embedding in MongoDB in IEEE (2014)
36. Murugesan, P., Ray, I.: Audit log management in MongoDB. In: IEEE 10th World Congress on Services (2014)

37. Xinwei, J., Guicheng, S.: Managing RFID data in MongoDB Workshop on Advanced Research and Technology in Industry Applications (WARTIA) in IEEE (2014)
38. Dwivedi, K., Dubey S.K.: Analytical review on Hadoop distributed files system. In: Proceeding of 5th International Conference on the Next Generation Information Technology Summit (Confluence), pp. 174–181. IEEE (2014)
39. Parker, Z., Poe, S., Vrbsky, S.: Comparing NoSQL MongoDB to an SQL DB. In: Proceedings of the 51st ACM Southeast Conference (2013)
40. Abramova, V., Bernardino, J.: NoSQL Databases: MongoDB vs. Cassandra. In: Proceedings of the International C* Conference on Computer Science and Software Engineering (2013)

# Heart Disease Prediction Using k-Nearest Neighbor Classifier Based on Handwritten Text

Seema Kedar, D.S. Bormane and Vaishnavi Nair

**Abstract** Heart diseases are the major cause of mortality in developed as well as developing countries. Early prediction of heart disease is required to reduce the number of deaths occurring due to it and by using handwriting analysis we can achieve this. Handwriting of an individual shows the presence of a heart disease even before physical symptoms appear. The proposed system predicts presence of heart disease based on handwriting analysis using k-Nearest Neighbor classifier. It extracts ten writing features namely right slant, left slant, vertical lines, horizontal lines, total length of vertical baselines, total number of left slant lines, total length of horizontal baselines, total number of right slant lines, pen pressure and size from a writing sample and using this information it predicts heart disease and risk factors for heart disease like low blood pressure, and diabetes. The proposed system provides 75 % accuracy.

**Keywords** Heart disease · Handwriting analysis · Writing features · k-NN

## 1 Introduction

Heart disease is number one killer across the world and also in India [1–3]. Heart disease is slated to be the number one global killer among the different communicable and non-communicable diseases [4]. Every year 17.3 million people die from heart disease, which represents 30 % of all global deaths. This number is

S. Kedar (✉) · D.S. Bormane · V. Nair
JSPM's Rajarshi Shahu College of Engineering, Savitribai Phule
Pune University, Pune 411033, India
e-mail: seemaahkeddar@gmail.com

D.S. Bormane
e-mail: bdattatraya@yahoo.com

V. Nair
e-mail: vaishnavia26@yahoo.co.in

© Springer India 2016
H.S. Behera and D.P. Mohapatra (eds.), *Computational Intelligence in Data Mining—Volume 1*, Advances in Intelligent Systems and Computing 410, DOI 10.1007/978-81-322-2734-2_6

expected to grow to >23.6 million by 2030 [5]. In India every year over 3 million deaths occur due to heart diseases which is more than any other country. Hence, India is called as 'heart disease capital of the world' [6–9]. The major reason for this is a change in the lifestyle patterns of the people such as a sedentary lifestyle and a general lack of awareness about the disease. Sedentary lifestyles double the risk of cardiovascular diseases, and the factors causing the heart disease such as diabetes, obesity, high blood pressure, depression and anxiety [5].

Heart disease is a silent disease. It slowly takes root in the body and comes out all of a sudden catching the person off-guard. Due to its sudden occurrence, it proves to be fatal to the individual. Hence, it becomes even more essential to detect heart disease in its early stage. Various methods used for heart disease detection are ECG scanning, blood sample analysis, stress test, heart MRI, chest X-Ray and CT Heart scan [10]. But these approaches help in detection only after physical symptoms of heart disease occurs [11]. Handwriting analysis is a technique through which we can detect heart disease at early stage and save more number of lives.

In 1999 the University of Plymouth, England carried out research study on graphology for health purposes and found that it is a useful tool for detecting health conditions before they become too critical. It is said that handwriting of a person gives a personal message about his health [12]. The PRE-ILLNESS warnings can be obtained from handwriting much before the actual symptoms of disease can be detected by any tests [13].

The handwriting is done by the brain and not by the hand or by the feet. Hence handwriting is also known as 'brain writing' [14]. Scientists in the neuromuscular research field state that personality is associated with some tiny neuromuscular movements. Each trait of personality is shown by a neurological brain pattern. Each neurological brain pattern produces an exclusive neuromuscular movement which is same for all persons with that personality trait [15, 16]. These minute movements arise unconsciously while writing. Using handwriting analysis these movements are identified and the related disease is predicted.

Research has shown that handwriting shows features of the occurrence of a disease way before it shows visible physical symptoms. It has been observed that all people afflicted with heart disease have similar handwriting features [17]. These features can be analyzed and studied to detect heart disease [18]. Compared to medical methods, it is more cost effective and also efficient. It provides instant and early prediction unlike other medical approaches [19].

A system for predicting the presence of heart disease in people by analyzing their handwriting features is proposed here. The proposed system analyzes ten writing features namely right slant, left slant, vertical lines, horizontal lines, total length of vertical baselines, total number of left slant lines, total length of horizontal baselines, total number of right slant lines, pen pressure and size from writing sample for predicting heart disease. The system uses k-Nearest Neighbor (k-NN) classifier to implement the same. We claim that this is first computerized system to predict heart disease based on handwriting analysis.

The organization of the remainder of the paper is as follows: literature review related to heart disease prediction is given in Sect. 2, proposed system is described in Sect. 3, the experimental results are presented in Sect. 4, and conclusion is given in Sect. 5.

## 2 Literature Review

As heart disease is number one killer worldwide, various techniques have been proposed in literature to predict it. Kavitha et al. [20] have proposed a system for heart disease identification using feed forward architecture and genetic algorithm. They have prepared data set of 270 patient's record. Thirteen parameters are used to define each patient condition. The training data set contains 150 patient's record and remaining 120 are taken for testing. Hybridization is used for training the neural network using Genetic algorithm which provides better results than usual back propagation.

Elbedwehy et al. [21] have used binary particle swarm optimization method, leave-one-out-cross-validation (LOOCV) algorithm and combination of support vector machine and KNN. They have used 198 sound signals of heart to form the dataset. They have collected these sound signals from control group patients and from the patients suffering from heart valve diseases such as mitral regurgitation, aortic stenosis, mitral stenosis and aortic regurgitation. Support Vector Machine is used to classify these sound signals into healthy cases and cases with above mentioned heart valve disease.

Kumar et al. [22] have used Fuzzy expert system to detect heart disease. Six input parameters namely blood pressure, old peak, cholesterol, blood sugar, chest pain type, and maximum heart rate are considered to detect heart disease. The approach provides 92 % accuracy.

Dangare et al. [10] have developed a system for predicting heart disease using neural network. It provides 99.25 % accuracy by using 13 parameters such as sex, cholesterol, resting electrographic results, fasting blood sugar, blood pressure, exercise induced angina, age, chest pain, maximum heart rate achieved, number of major vessels colored by fluoroscopy, ST depression induced by exercise relative to rest, defect type and slope of the peak exercise ST segment. The authors obtained 100 % accuracy by adding two more parameters i.e. smoking and obesity.

A decision support system is developed by Vanisree et al. for detecting congenital heart disease [23]. They have used back propagation neural network model which contains thirty six input nodes, ten hidden nodes and one output node. The network is educated using a supervised training and a delta learning rule. To train and test the network 200 samples are used. Using proposed system 90 % accuracy is obtained.

The paper by Strang [17] is one of the pioneering works establishing a relationship between heart disease and handwriting analysis. The author herself is a renowned graphologist and has analyzed various handwriting samples for features

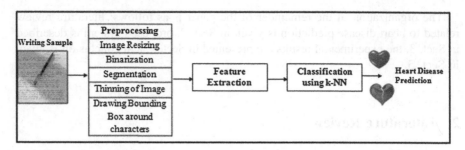

**Fig. 1** Block diagram of the proposed system

that suggest a possibility of a heart disease in the writer. This paper states certain writing features like half formed letters, heart shaped letter 'o', jerky movements, resting dots etc. predicts the possibility of heart disease. The study proves that handwriting can be used for early detection of heart disease.

# 3 Proposed System

The block diagram of proposed system is shown in following Fig. 1. It has three main steps such as pre-processing, feature extraction and classification. The handwriting sample is collected on a plain A4 size white paper. It is then scanned using a scanner. Thus, a JPEG image of the handwriting sample is generated and it is given as input to the system. The output is prediction about the heart disease.

## 3.1 Preprocessing

The preprocessing of an input image is done in order to make the feature extraction stage easy and efficient to implement. It enhances the image so that accurate values for features will be obtained. Initially the scanned input image is converted into grayscale image which is then resized to standard [512, 512] image. The resized image is converted into binary image. Here, all background pixels are set to 0 and foreground pixels to 1. The binary image is then segmented. During segmentation the image is partitioned into multiple segments. This is done in order to get only the relevant information i.e. text and cancel out all unwanted information like the background color of the image. The thinning operation is performed on segmented image. This process removes the selected foreground pixels from the binary image. It is a morphological operation. It reduces all lines of characters into single pixel thickness. It is used for skeletonization. At last bounding boxes are drawn around each and every character to recognize them and to perform feature extraction.

## 3.2 Feature Extraction

In this step, initially the universe of discourse of the preprocessed image is selected. Then the image is segmented into nine zones of equal size and writing features such as right slant, left slant, vertical lines, horizontal lines, total length of vertical baselines, total number of left slant lines, total length of horizontal baselines, total number of right slant lines, pen pressure and size are extracted from each zone. This is done to get fine details from the image [24]. These extracted features are then summed together and stored in a feature array which is further used for classification purpose.

## 3.3 Classification

The proposed system uses k-NN algorithm for classification as it reduces the misclassification error, when training dataset contains number of samples. On similarity measure, it classifies new unclassified data points [25]. It is a lazy learning algorithm as no actual learning is performed during the training phase and all computations are deferred until classification. The distance metrics selected to calculate nearest distance and the value of K affects the accuracy of classification. The different distance metrics are city block, Euclidean distance, hamming, correlation, and cosine. In this work the distance between two neighbors is found using the Euclidean distance:

$$Euclidean\ dist. = \sqrt{\sum_{i=1}^{k} (x_i - y_i)^2} \tag{1}$$

When a new unknown sample is presented for evaluation, the algorithm computes its 'K' closest neighbors, and assigns appropriate class by voting among those neighbors.

## 3.4 Algorithm

1. Prepare training feature vector by applying steps 1.1 to step 1.3 on all the images selected for training purpose.

   1.1 Select image.
   1.2 Pre-process the image by applying following steps:

      (a) Convert the input image into grayscale image.
      (b) Resize the image to standard [512, 512] size.
      (c) Convert grayscale image into binary image.

**Table 1** Multilevel confusion matrix

| | | Predicted class | | | | Marginal sum of actual values |
|---|---|---|---|---|---|---|
| | | Heart disease | Low blood pressure | Diabetes | Control group | |
| Actual class values | Heart disease | 7 | 1 | 0 | 0 | 8 |
| | Low blood pressure | 0 | 0 | 0 | 2 | 2 |
| | Diabetes | 0 | 0 | 2 | 1 | 3 |
| | Control group | 1 | 0 | 0 | 6 | 7 |
| Marginal sum of predicated values | | 8 | 1 | 2 | 9 | T = 20 |

    (d)  Segment the binary image.

    (e)  Thin the segmented image.

    (f)  Draw bounding box around each and every character in thinned image.

  1.3  Extract features from the image by applying following steps:

    (a)  Skeletonize the image.

    (b)  Select the universe of discourse.

    (c)  Partition the image into nine zones.

    (d)  Compute right slant, left slant, vertical lines, horizontal lines, total length of vertical baselines, total number of left slant lines, total length of horizontal baselines, total number of right slant lines, pen pressure and size for each zone.

    (e)  Sum up together the extracted features of each zone and store them in feature vector.

2. Select image for testing.
3. Prepare test image feature vector by applying steps 1.2 and step 1.3 on it.
4. Find appropriate disease category i.e. training class for test image feature vector using k-NN classifier.
5. Display selected disease category.

# 4  Experimental Results

For performing experimentation the data set containing 60 handwriting samples of both control group and heart disease patients is used. The patient's age ranging from 38 to 81 comprising of both male and female are taken for this work. The

**Fig. 2** Recognition rate for different types of handwriting samples

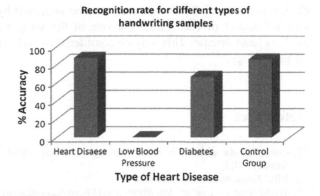

training dataset is formed by using 40 samples which are divided into four classes such as heart disease, diabetes, low blood pressure, and control group. The testing dataset contains 20 samples. The proposed system provides 75 % accuracy. It is calculated using following Eq. (1):

$$Accuracy = \frac{Correctly\ predicted\ samples}{Total\ number\ of\ samples} * 100$$
$$= \frac{15}{20} * 100$$
$$= 75\ \%$$

The multilevel confusion matrix with actual class values and predicted class values is shown in following table Table 1.

The graph in Fig. 2 shows disease wise recognition rate. The training data set of heart disease class contains 13 samples and 13 samples of control group class, hence we obtained 87.5 and 85.71 % recognition rate for these classes respectively. As training data set for diabetes class contains 10 samples and 4 samples of low blood pressure class, we obtained 66.66 and 0 % recognition rate for these classes respectively.

## 5 Conclusion

The approach for prediction of heart disease in people based on handwriting analysis is proposed here. This is first paper for predicting heart disease based on handwritten text as per our knowledge. A training dataset contains 40 samples and testing dataset contains 20 samples. The proposed system gave an overall accuracy of 75 % by using the k-NN algorithm for classification. We observed that if the training samples of particular class are more, the recognition rate of that class is also more. As the training samples of heart disease and control group class are more as compared to low blood pressure and diabetes class, the recognition rate of these classes is above 85 %. The accuracy can be increased by adding more samples in training dataset of each class. For one of the sample of control group, system predicts heart disease. This may be considered as early sign of heart disease based on graphology.

# References

1. Cardiovascular diseases in India Challenges and way ahead. International Heart Protection Summit (2011)
2. BBC News: http://news.bbc.co.uk
3. World Heart Federation: http://www.world-heart-federation.org
4. Gupta, R., Guptha, S., Sharma, K.K., Gupta, A., Deedwania, P.: Regional variations in cardiovascular risk factors in India: India heart watch. World J. Cardiol. 112–120. ISSN:1949-8462. © 2012 Baishideng (2012)
5. World Health Organization: http://www.who.int
6. Chauhan, S., Aeri, B.T.: Prevalence of cardiovascular disease in India and it is economic impact-a review. Int. J. Sci. Res. Publ. 3(10) (2013). ISSN:2250–3153
7. The Indian Express Archive: http://archive.indianexpress.com
8. India Today: http://indiatoday.intoday.in
9. Times of India: http://timesofindia.indiatimes.com
10. Dangare, C.S., Apte, S.S.: A data mining approach for prediction of heart disease using neural networks. Int. J. Comput. Eng. Technol. (IJCET) 3(3), 30–40 (2012). ISSN:0976 – 6367 (Print). ISSN:0976 – 6375
11. British Heart Foundation: http://www.bhf.org.uk
12. Zoom Health: http://www.zoomhealth.net/WhatYourHandwritingSaysAboutYourHealth.html
13. GROW: http://www.handwritingexplore.com/grow-paid-services.html
14. Johan Antony, D., Cap, O.F.M.: A textbook of handwriting analysis, personality profile through handwriting analysis. Anugraha Publications, India (2008)
15. Abdul Rahiman, M., Diana, V., Manoj Kumar, G.: HABIT-handwriting analysis based individualistic traits prediction. Int. J. Image Process. (IJIP) 7(2) (2013)
16. Kedar, S.,. Nair, V., Kulkarni, S.: Personality identification through handwriting analysis: a review. Int. J. Adv. Res. Comput. Sci. Softw. Eng. 5(1), (2015)
17. Strang, C.: Handwriting in the early detection of disease—a research study. Graphol.: J. British Inst. Graphol. 31(1), 118 (2013)
18. Natural News: http://www.naturalnews.com
19. Dr. Manoj Thakur Blog: http://drmanojtanurkar.hubpages.com
20. Kavitha, K.S., Ramakrishnan, K.V., Singh, M.K.: Modeling and design of evolutionary neural network for heart disease detection. IJCSI Int. J. Comput. Sci. Issues 7(5), 272–283 (2010)
21. Elbedwehy1, M.N., Zawbaa†, H.M., Ghali‡, N., Hassanien†, A.E.: Detection of heart disease using binary particle swarm optimization. Egypt Scientific Research Group in Egypt
22. Kumar, S., Kaur, G.: Detection of heart diseases using fuzzy logic. Int. J. Eng. Trends Technol. (IJETT) 4(6) (2013). ISSN:2231-5381
23. Vanisree, K., Singaraju, J.: Decision support system for congenital heart disease diagnosis based on signs and symptoms using neural networks. Int. J. Comput. Appl. 19(6), 0975–8887 (2011)
24. Dileep, D.: A feature extraction technique based on character geometry for character recognition. Department of Electronics and Communication Engineering, Amrita School of Engineering, Kollam, India (2012)
25. Jiawei, H., Micheline, K.: Data mining concepts and techniques, 2nd edn. Elsevier Inc (2006). ISBN:978-1-55860-901-3

# Optimizing a Higher Order Neural Network Through Teaching Learning Based Optimization Algorithm

Janmenjoy Nayak, Bighnaraj Naik and H.S. Behera

**Abstract** Higher order neural networks pay more attention due to greater computational capabilities with good learning and storage capacity than the existing traditional neural networks. In this work, a novel attempt has been made for effective optimization of the performance of a higher order neural network (in particular Pi-Sigma neural network) for classification purpose. A newly developed population based teaching learning based optimization algorithm has been used for efficient training of the neural network. The performance of the model has been benchmarked against some well recognized optimized models and they have tested by five well recognized real world bench mark datasets. The simulating results demonstrated favorable classification accuracies towards the proposed model as compared to others. Also from the statistical test, the results of the proposed model are quite interesting than others, which analyzes for fast training with stable and reliable results.

**Keywords** Higher order neural network · Pi-Sigma neural network · Teaching learning based algorithm (TLBO)

## 1 Introduction

Think a computer can do almost everything, what we think! Is it possible? The development of a neural network is based on this critical assumption. Neural networks have been replaced with the conventional computers only due to their strong

J. Nayak (✉) · B. Naik · H.S. Behera
Department of Computer Science Engineering and Information Technology,
Veer Surendra Sai University of Technology, Sambalpur, Burla 768018
Odisha, India
e-mail: mailforjnayak@gmail.com

B. Naik
e-mail: mailtobnaik@gmail.com

H.S. Behera
e-mail: mailtohsbehera@gmail.com

© Springer India 2016
H.S. Behera and D.P. Mohapatra (eds.), *Computational Intelligence in Data Mining—Volume 1*, Advances in Intelligent Systems and Computing 410, DOI 10.1007/978-81-322-2734-2_7

ability to understand the imprecise data and extract patterns and do not use the algorithmic approach with a predefined set of instructions which are to be executed. In conventional systems, until the steps for solving the problem is not fed to the system, it cannot solve the problem efficiently and restricts the capacity. But, a neural network is far away from such problems and a trained neural network acts as an expert for solving a particular problem which helps later to give some new projections to the problem with the analysis of if-else situations. In 1943, McCulloch and Walter Pits first coined the concept of artificial neuron and neural network. However, in 1969, Minsky and Papert have raised their frustration on neural network with some negative feedbacks. But after a rigorous analysis, in the late 1980's Multilayer Perceptron overcomes those limitations. Since then, it has never seen back and the developments from a single perceptron or adaptive resonance theory (ART) to Higher order neural network or Pulse Neural network or Fuzzy based Neural network are being a frequent interest of research of all age researchers. Various types of earlier developed neural networks such as HEBB net, ADALINE, MADALINE, Perceptron Layer Network (PLN), Associative memory, Self-organizing map (SOM), Back Propagation Network, Adaptive Resonance Theory (ART), Probabilistic Networks, Optical Neural Network, Holographic Neural Network etc. are based on some specific applications and these networks performed well in diversified applications. But due to some limitations (Table 1) like complex architecture, complicated mathematical functional calculations for the neuron units etc., researchers move towards a new type of neural network such as higher order neural network which is efficient and able to learn more quickly than the above described networks.

Higher order neural networks (HONN) came into picture in late 80's period with addition of some higher order logical processing units with the earlier feed forward networks. They are quite effective in solving various non linear problems. The first ever higher order neural network with the capacity to approximate the polynomial functions had been proposed by Ivakhnenko [1]. The first research paper was introduced by Giles and Maxwell [2] and a book on higher order neural network was written by Bengtsson [3]. He suggested that a higher order network can be easily implemented with more accurate information as the hidden layer is absent as like Hebbian and perceptron learning rules. Zhang et al. [4] described that HONNs can lead to fast convergence due to the exclusion of hidden layer and can be more accurate in curve fitting with a abridged network architecture. The main purpose behind the development of HONN is to increase in the generalization capability with less complex network architecture. However, the reason behind the development was successful and as a result, today HONNs can be found to be extensively used in various real life applications like data mining [5, 6], signal processing, speech recognition, medicine, intelligent control, function approximation, financial forecasting, condition monitoring, process monitoring and control, pattern analysis etc.

This work is mainly focused on a special type of HONN called Pi-Sigma neural network (PSNN) developed by Shin and Ghosh [7]. Since its inception, researchers have frequently used in image segmentation [8], solving memory requirements [9], transformation problem [10], financial time series prediction [11], function

**Table 1** Various types of neural networks with their advantages and limitations

| Name of the NN | Year of development | Advantages | Limitations |
|---|---|---|---|
| McCulloch-Pitt | 1943 | • Simple<br>• Substantial computing power | • Fixed value of weights and threshold units |
| HEBB net | 1988 | • Strong correlational structure to the inputs<br>• Good generalization capability | • Unable to learn if the input pattern is in binary form |
| ADALINE and MADALINE | 1960 | • Better convergence<br>• Noise correction<br>• Good for non-separable problems | • Working speed is less as compared to other networks |
| Perceptron layer network | 1957 | • Suitable for binary and bipolar input vectors | • Nonlinearly separable |
| Associative memory | 1982 | • Recurrent neural processing<br>• Noise control | • Capacity<br>• Memory convergence |
| Back propagation network | 1986 | • Simple Implementation<br>• Doesn't require any special kind of function evaluations | • Slow and not efficient<br>• Suck at local minima<br>• Network complexity is more |
|  |  | • Batch update of weights | • Outputs may be non-numeric or fuzzy type |
| ART | 1987 | • Controls degree of similarity between patterns<br>• Good for clustering | • Result is depending upon the processed training data<br>• Do not possess the statistical property of consistency |
| Feed forward network | 1985–1990 | • Fixed computation time<br>• Speed of computation is high | • Small error is associated with each output<br>• Errors may vary, based on situations<br>• Training time is more |

optimization [12], cryptography [13], classification [14–17] etc. In this paper, PSNN has been used with a recently developed nature inspired optimization algorithm called teaching learning based algorithm (TLBO) for non linear data classification. Teaching Learning based Optimization is a recently developed metaheuristic proposed by Rao et al. [18, 19] inspired by the teaching and learning process of teacher and students or learners. Like other optimization techniques, it does not require any controlling parameters settings which make the algorithm simpler and effective to use in the real life domains.

The performance of the proposed model has been tested with various benchmark datasets from UCI machine learning repository and compared with the resulting performance of other approaches. The model has been implemented in MATLAB and the accuracy measures have been tested through a statistical technique to show the technique is statistically valid. Experimental results reveal that the proposed approach is steady as well as reliable and provides better classification accuracy than others.

The rest of the paper is organized as follows: Section 2 elaborates some basic preliminary concepts. Section 3 describes the proposed method and Sect. 4 outlines the details about the experimental set up consisting of the environment set up, parameter settings and the data set information. Section 5 analyses the simulation results obtained from the experiment. Section 6 gives the details about the statistical significance of the proposed method compared to other methods and finally Sect. 7 outlines the conclusion with some future directions.

## 2  Preliminaries

### 2.1  Pi-Sigma Neural Network (PSNN)

PSNN is a special kind of multilayered feed forward network with less number of synaptic weights than other HONNs. However, they are quite efficient to manage the capabilities of HONNs and have effectively addressed some complex problems. The main feature of PSNN is the use of product cells for processing units at output layer, instead of sum of product of inputs as other networks. It has an input layer, a single middle layer and an output layer consisting of product units of the inputs. During the learning of the network, the weight values from the input to the middle layer are adapted, while the weight values from the middle to output are fixed to 1. Due to this reason the complexity of the hidden layer can be dramatically reduced by the number of tunable weights, for which the model can be easily implementable and accelerated. By using fewer weights vectors and processing units these are capable of quick learning which makes them more accurate and tractable than the other neural networks. Figure 1 depicts the 3-layered architecture of PSNN, where $x_1, x_2 \ldots x_n$ denotes the input vectors. The output at the hidden layer $h_j$ and output layer O can be computed by Eqs. (1) and (2) respectively.

$$h_j = B_j + \sum w_{ji} x_i \tag{1}$$

where B is the bias, $w_{ij0}, w_{ij1}, w_{ij2} \ldots w_{ijn}$ are the weight units, where $i = 1 \ldots k$ and k is the order of the network.

$$O = f(\prod_{j=1}^{k} h_j) \tag{2}$$

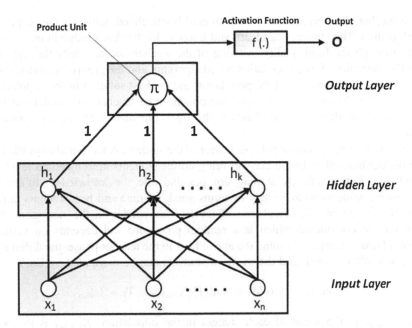

**Fig. 1** Architecture of pi-sigma network

where f(.) is an suitable activation function. Here it should be noted that the weight from the hidden layer to output layer is fixed to 1.

The network is trained with the production of errors and the overall estimated error function E can be calculated as in Eq. (3).

$$E_j(t) = d_j(t) - O_j(t) \tag{3}$$

where $dj(t)$ indicates the final desired output at time $(t-1)$. At each time $(t-1)$, the output of each $Oj(t)$ is calculated.

## 2.2 Teaching Learning Based Optimization Algorithm (TLBO)

Inspired from the teaching and learning process in a classroom environment, TLBO was developed to solve multimodal problems having less algorithm specific parameters as compared to the other evolutionary algorithms. Basically, the working principle of TLBO is based on the influence or impact of the teaching process of a teacher on the students in the classroom. Based on the quality teaching of the teacher, the performance of the student is calculated, which is the output of the algorithm. The students can learn and improve his/her knowledge in two ways: firstly through the high quality teaching of the teacher and secondly through the friend's interaction. It is a population based algorithm and the groups of students are considered as the members of the population and the best solution in the population is the teacher. There are two major components like teacher and student in the algorithm. Also, there are two phases in the algorithm such as teacher and student phase.

The teacher phase simulates the behavior of the student. A teacher always tries to give his/her best effort in the class to bring all the students upto a certain level of knowledge. But practically it may not possible due the knowledge difference between the students as there may be terms good, average and best students in the class. So, for an overall calculation of level of knowledge in the classroom, the mean can be considered which is a random procedure and depends on various external factors. Keeping in mind the above facts in the teacher phase, the difference between teacher's result and the mean of students can be computed by Eq. (4).

$$X_i(new) = X_i(old) + rand(1)(X_{teacher} - T_F * X_{mean}) \qquad (4)$$

where $X_{mean}$ is the mean of each student in the population, $X_{teacher}$ is the best solution in the population, $T_F$ is a teaching factor and rand is a random number. $T_F$ is the decisive factor for the changing value of mean and its value is generated randomly with an equal probability of either 0 or 1. The value for rand (1) is selected in between the range [0, 1]. Note that, $T_F$ is not used as an input parameter to the algorithm, rather its value is selected randomly.

On the other hand, the learning phase simulates the behavior of the students through the interaction or discussion of their knowledge with other students or friends in the class. Students may acquire some knowledge on a concerned subject from their friends by the method of discussion or interaction. Students can also acquire some new knowledge from their friends if their friends have more expertise on the concerned subject. From a random selection of the solutions in the population, the best solution in the learning phase can be calculated as in Eqs. (5) and (6). The steps of the TLBO algorithm can be depicted in Fig. 2.

$$X_i(new) = X_i(old) + rand(1)(X_j - X_i). \qquad (5)$$

$$X_j(new) = X_j(old) + rand(1)(X_i - X_j) \qquad (6)$$

where $X_i$ and $X_j$ are the two weight sets, those are randomly selected from the population. If the fitness of the $X_i$ will be better than $X_j$, then Eq. (6) is performed else Eq. (5) is performed.

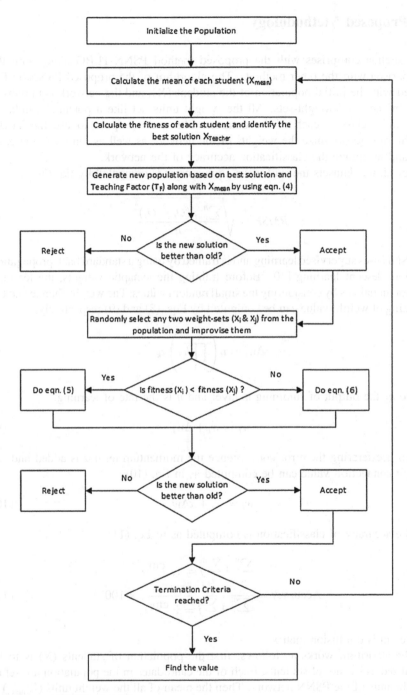

**Fig. 2** Steps of TLBO algorithm

## 3  Proposed Methodology

This section comprises with the proposed method PSNN-TLBO along with the comparison with the other methods. The algorithm of the proposed PSNN-TLBO started with the initial population of the students(X). and the network is initialized with 'n' no. s of weight-sets. All the weight units act like a potential candidate solution to classify each of the datasets. The main purpose to use the TLBO algorithm is to optimize the weights of the network as well as minimize the error rate and maximize the classification accuracy of the network.

For all the datasets the root mean square error is computed by Eq. (7).

$$RMSE = \sqrt{\frac{\sum_{i=1}^{n}(O_i - \hat{O}_i)^2}{n}} \tag{7}$$

PSNN uses supervised learning and is trained by using a standard back propagation gradient descent learning [20]. Before training the synaptic weights, the learning method initializes by considering the small random values. The weight change and the updating of weight value can be computed by Eqs. (8) and (9) respectively.

$$\Delta w_j = \eta \left( \prod_{j\neq 1}^{m} h_{ji} \right) x_k \tag{8}$$

where $h_{ji}$ the output of summing is layer and $\eta$ is the rate of learning.

$$w_i = w_i + \Delta w_i \tag{9}$$

For accelerating the error convergence the momentum term $\alpha$ is added and the weight connection value can be computed as in Eq. (10).

$$w_i = w_i + \alpha \Delta w_i \tag{10}$$

The accuracy of classification is computed as in Eq. (11).

$$Accuracy = \frac{\sum_{i=1}^{n} \sum_{j=1, \atop i==j}^{m} cm_{i,j}}{\sum_{i=1}^{n} \sum_{j=1}^{m} cm_{i,j}} \times 100. \tag{11}$$

where cm is confusion matrix.

The algorithm works as follows: first the population of students (X) is to be initialized as 'n' no. of students. Each of the candidates in the population are set as weight units of the PSNN network. Then the mean of all the weight units ($X_{mean}$) is computed. The fitness from X is to be calculated by the complete network functions (Eqs. 1–3 and 11) to find out the best teacher having the best fitness value. Then the next population is generated and the old will be replaced with the new one with better solutions. This process will be continued till the best value is obtained or upto the maximum no of iterations. The detailed pseudo code of the proposed algorithm is given below.

---

**Algorithm – 1 PSNN-TLBO**

INPUT: Dataset with target vector 't' , initial population of weight-units 'X', Bias B.
OUTPUT: PSNN with optimized weight-set 'w'.
1. Initialize the population of students as : $X = \{x_1, x_2, ...,x_n\}$, where each $x_i$ is a randomly initialized potential weight-unit of PSNN network as $x_i = \{w_{i,1}, w_{i,2}, ...,w_{i,n}\}$.
2. Calculate $(x_{mean})$ .
3. Evaluate the fitness of all the weight-sets by using Algorithm-2.
4. Select $(x_{teacher})$ as best weight-set based on the maximum fitness value.
5. Generate the next population $X_{next}$ by using weight-sets in the old population X, $X_{mean}$ , $x_{teacher}$ and $T_F$.

    **for** i=1:1: nos of weight-set in  X
        Compute $T_F$.
        $X_i$(new) is computed by using eq. (4).
        $X_{next}(i) = x_i$
    **endfor**

6. Compare the fitness of weight-sets in X and X$_{next}$. Accordingly, update the weight-sets in the population.

    **for** i=1:1: nos of weight-set in X
        **if** (X(i) < X$_{next}$(i) )
            X(i)= X$_{next}$(i)
        **endif**
    **endfor**

7. Randomly choose any  two weight-set ($x_i$ and $x_j$) in $X$ and improvise them.

    **for** k =1:1: nos of weight-set in $X$
        Calculate the fitness F$_i$ & F$_j$ of $x_i$ and $x_j$ respectively by using
                algorithm-2 .
        **If** (F$_i$ > F$_j$ )
        Then $X_j$(new) is computed by eq.(6).
        **Else**
        $X_i$(new) is computed by eq.(5).
        **Ifend**
    **Endfor**

8. The best solutions are replaced by previous worst solutions in $X$.
9. **if** (maximum no. of generation reached **OR** 95% of weight-set in $X$ are equivalent )

        then goto step-10.
    **else**
        goto step-2
    **endif**
10. **Exit**

***

***Algorithm – 2:  Fitness From Training Algorithm***

1. **FUNCTION** F= **Fitness-From-Training** (x, w, t, B)
2. *for*  i = 1 to n, n is the length of the dataset.
3.     PSNN's hidden layer output is calculated by using eq.  (1).
4.     PSNN's overall output is calculated by using (2).
5.     Calculate the error term by using eq. (3) and compute the fitness F(i)=1/RMSE .
6. *end for*
7.   The RMSE is calculated by using eq. (7) from the target value and output.
8.   The modifications in the weight-set is computed by using (8).
9.   Update the weight by using eq. (9).
10. After adding the momentum term, the weight value changes as per  eq. (10).
11. **IF**
         the stopping criteria reached,
                 Then Stop.
     **Else** repeat the step from 2 to 11.
12. **END**

**(Note: The stopping criteria is training error or maximum no. of epochs)**

***

# 4  Experimental Set up

## 4.1  Environment Set up

The environment for this experiment is MATLAB 9.0 on a system with an Intel Core 2 Duo CPU T5800, 2GHz processor, 2 GB RAM and Microsoft Windows-2007 OS.

## 4.2  Parameter Settings

A few of the parameters have been set for PSNN network as well as TLBO algorithm. The weight sets have been initialized between −1 to 1 and the values for the hidden layer to the output layer is fixed to unity. The number of epochs for the training of network is set to 1000. In case of the TLBO algorithm as there are no specific algorithm specific parameters, only the value of $T_F$ is set as 1 or 2 with a chance of equal probability. Initially the students population size has been set to 40 and 100 no. of generations are performed. As earlier mentioned the stopping criterion is the maximum no. of iteration.

## 4.3  Data Sets

For this experiment, the data sets have been considered from the UCI repository and the used datasets are processed by KEEL software [21]. The dataset considered in this work are prepared by using 5-fold cross validation technique. i.e. a dataset has

**Table 2**  Details about the data sets

| Dataset | Number of pattern | Number of features/attributes | Number of classes |
|---------|-------------------|-------------------------------|-------------------|
| PIMA | 768 | 09 | 02 |
| ECOLI | 336 | 07 | 08 |
| VEHICLE | 846 | 18 | 04 |
| BALANCE | 625 | 04 | 03 |
| HAYESROTH | 160 | 05 | 03 |

been prepared by splitting into 5-folds, out of which 4 folds are used for training and 1 fold is used for testing. More details about the dataset information is given in Table 2.

# 5  Result Analysis

The main aim of this work is to show the performance of the proposed PSNN-TLBO and also to compare its performance with some other methods like PSNN, PSNN-GA, and PSNN-PSO. For all the methods, the datasets described in Table 2 has been run and the results of the networks in terms of classification accuracy is shown in Table 3. During the simulation, a frequent observation on the changes of weight set have been made and each time the value have been stored for checking closer value to the target output. Simulation result reveals that in all the cases (dataset) the proposed method outperforms well over the other described method. For PSNN, GA-PSNN and PSO-PSNN, the experimental set up procedure of the author's work [22] have been followed. The classification accuracies of the proposed TLBO-PSNN in training of five datasets are 96.365, 94.834, 96.368, 98.935 and 96.663 respectively. Similarly in case of testing, the resulting values are 96.338, 94.768, 96.277, 98.942 and 95.703. In both training and testing case, all the accuracies are quite better (Table 3) than the results of the other methods. The most less accuracy is found only with the PSNN's performance and when no any optimization algorithm has been used for the weights.

**Table 3**  Average classification accuracy results of the classifiers

| Dataset | Average classification accuracy (in %) | | | | | | | |
|---------|------------|--------|----------|--------|---------|--------|--------|--------|
| | TLBO-PSNN | | PSO-PSNN | | GA-PSNN | | PSNN | |
| | Train | Test | Train | Test | Train | Test | Train | Test |
| PIMA | **96.365** | **96.338** | 91.273 | 91.315 | 90.244 | 89.382 | 88.214 | 87.508 |
| ECOLI | **94.834** | **94.768** | 91.003 | 91.018 | 90.365 | 90.333 | 88.001 | 88.101 |
| VEHICLE | **96.368** | **96.277** | 91.607 | 91.552 | 90.333 | 90.398 | 88.006 | 87.100 |
| BALANCE | **98.935** | **98.942** | 95.206 | 95.094 | 94.129 | 93.795 | 89.131 | 88.905 |
| HAYESROTH | **96.663** | **95.703** | 91.292 | 90.278 | 90.200 | 90.234 | 89.151 | 89.312 |

# 6   Statistical Analysis

To prove the performance of the proposed method, in this section statistical tool called ANOVA has been used to validate the model statistically. By using one way ANOVA test the null hypothesis can be tested and the variability in the performance of different models is estimated. The null hypothesis can be rejected to get some difference among the classifiers, based on some marginal better variability of the between classifier in compared to error variability. The result of one way ANOVA in Duncan multiple test range with 95 % confidence interval, 0.05 significant level and linear polynomial contrast has been shown in Fig. 3, the result of posthoc test is shown in Fig. 4 and the result of Tukey and Dunnett Tests have been shown in Fig. 5.

**Descriptives**

Result

| | N | Mean | Std. Deviation | Std. Error | 95% Confidence Interval for Mean | | Minimum | Maximum |
|---|---|---|---|---|---|---|---|---|
| | | | | | Lower Bound | Upper Bound | | |
| PSNN | 10 | 88.3429 | .75105 | .23750 | 87.8056 | 88.8802 | 87.10 | 89.31 |
| GA-PSNN | 10 | 90.9413 | 1.62069 | .51251 | 89.7819 | 92.1007 | 89.38 | 94.13 |
| PSO-PSNN | 10 | 91.9638 | 1.71969 | .54381 | 90.7336 | 93.1940 | 90.28 | 95.21 |
| TLBO-PSNN | 10 | 96.5193 | 1.43130 | .45262 | 95.4954 | 97.5432 | 94.77 | 98.94 |
| Total | 40 | 91.9418 | 3.29273 | .52063 | 90.8888 | 92.9949 | 87.10 | 98.94 |

**ANOVA**

Result

| | | | Sum of Squares | df | Mean Square | F | Sig. |
|---|---|---|---|---|---|---|---|
| Between Groups | (Combined) | | 349.071 | 3 | 116.357 | 56.783 | .000 |
| | Linear Term | Contrast | 326.445 | 1 | 326.445 | 159.306 | .000 |
| | | Deviation | 22.626 | 2 | 11.313 | 5.521 | .008 |
| Within Groups | | | 73.770 | 36 | 2.049 | | |
| Total | | | 422.841 | 39 | | | |

**Fig. 3** Results of ANOVA test

**Result**

| | | | Subset for alpha = 0.05 | | |
|---|---|---|---|---|---|
| | Algo | N | 1 | 2 | 3 |
| Tukey HSD[a] | PSNN | 10 | 88.3429 | | |
| | GA-PSNN | 10 | | 90.9413 | |
| | PSO-PSNN | 10 | | 91.9638 | |
| | TLBO-PSNN | 10 | | | 96.5193 |
| | Sig. | | 1.000 | .393 | 1.000 |
| Duncan[a] | PSNN | 10 | 88.3429 | | |
| | GA-PSNN | 10 | | 90.9413 | |
| | PSO-PSNN | 10 | | 91.9638 | |
| | TLBO-PSNN | 10 | | | 96.5193 |
| | Sig. | | 1.000 | .119 | 1.000 |

Means for groups in homogeneous subsets are displayed.

a. Uses Harmonic Mean Sample Size = 10.000.

**Fig. 4** Results of post hoc Tests

**Multiple Comparisons**

Dependent Variable:Result

| (I) Algo | | (J) Algo | Mean Difference (I-J) | Std. Error | Sig. | 95% Confidence Interval | |
|---|---|---|---|---|---|---|---|
| | | | | | | Lower Bound | Upper Bound |
| Tukey HSD | PSNN | GA-PSNN | -2.59840* | .64018 | .001 | -4.3226 | -.8742 |
| | | PSO-PSNN | -3.62090* | .64018 | .000 | -5.3451 | -1.8967 |
| | | TLBO-PSNN | -8.17640* | .64018 | .000 | -9.9006 | -6.4522 |
| | GA-PSNN | PSNN | 2.59840* | .64018 | .001 | .8742 | 4.3226 |
| | | PSO-PSNN | -1.02250 | .64018 | .393 | -2.7467 | .7017 |
| | | TLBO-PSNN | -5.57800* | .64018 | .000 | -7.3022 | -3.8538 |
| | PSO-PSNN | PSNN | 3.62090* | .64018 | .000 | 1.8967 | 5.3451 |
| | | GA-PSNN | 1.02250 | .64018 | .393 | -.7017 | 2.7467 |
| | | TLBO-PSNN | -4.55550* | .64018 | .000 | -6.2797 | -2.8313 |
| | TLBO-PSNN | PSNN | 8.17640* | .64018 | .000 | 6.4522 | 9.9006 |
| | | GA-PSNN | 5.57800* | .64018 | .000 | 3.8538 | 7.3022 |
| | | PSO-PSNN | 4.55550* | .64018 | .000 | 2.8313 | 6.2797 |
| Dunnett t (2-sided) a | PSNN | TLBO-PSNN | -8.17640* | .64018 | .000 | -9.7462 | -6.6066 |
| | GA-PSNN | TLBO-PSNN | -5.57800* | .64018 | .000 | -7.1478 | -4.0082 |
| | PSO-PSNN | TLBO-PSNN | -4.55550* | .64018 | .000 | -6.1253 | -2.9857 |

*. The mean difference is significant at the 0.05 level.

a. Dunnett t-tests treat one group as a control, and compare all other groups against it.

**Fig. 5** Results of Tukey and Dunnett tests

# 7 Conclusion and Future Work

In this work, to optimize the performance of a higher order neural network teaching learning based optimization algorithm is used. The weights of the HONN Pi-Sigma neural network are optimized and the network is able to classify the real world data. The performance of the proposed algorithm is tested with five bench mark datasets and the results are compared with some other standard methods including GA and PSO. Simulation result show that the performance of the PSNN-TLBO outperforms the other in terms of classification accuracy. In some standard evolutionary algorithms like GA and PSO, a no. of algorithmic input specific parameters (mutation and crossover in GA, inertia weight, position and velocity in PSO) are supplied. But in TLBO there are no use of such dependent parameters, still it performs well in with better convergence than others. Inspiring by the performance and optimizing capability of TLBO algorithm, our future work may comprise on application of this algorithm in some other popular HONN and to test the performance with PSNN. Also, some of the modifications in greedy selections in both the teacher and student phase will be the future work.

**Acknowledgments** This work is supported by Department of Science and Technology (DST), Ministry of Science and Technology, New Delhi, Govt. of India, under grants No. DST/INSPIRE Fellowship/2013/585.

# References

1. Ivakhnenko, A.G.: Polynomial theory of complex systems polynomial theory of complex systems. IEEE Trans. Syst. Man Cybern. **1**(4), 364–378 (1971)
2. Giles, C.L., Maxwell, T.: Learning, invariance, and generalization in high-order neural networks. Appl. Optics, **26**(23), 4972–4978 (1987). ISI:A1987L307700009
3. Bengtsson, M.: Higher order artificial neural networks. Diane Publishing Company, Darby PA, USA (1990). ISBN 0941375927
4. Zhang, M., Xu, S.X., Fulcher, J.: Neuron-adaptive higher order neural-network models for automated financial data modeling. IEEE Trans. Neural Netw. **13**(1), 188–204 (2002). WOS: 000 173440100016
5. Naik, B., Nayak, J., Behera, H.S.: A honey bee mating optimization based gradient descent learning–FLANN (HBMO-GDL-FLANN) for Classification. In: Emerging ICT for Bridging the Future-Proceedings of the 49th Annual Convention of the Computer Society of India CSI, vol. 2, pp. 211–220. Springer International Publishing (2015)
6. Naik, B., Nayak, J., Behera, H.S., Abraham, A.: A harmony search based gradient descent learning-FLANN (HS-GDL-FLANN) for Classification. In: Computational Intelligence in Data Mining, vol. 2, pp. 525–539. Springer India (2015)
7. Shin, Y., Ghosh, J.: The pi-sigma networks: an efficient higher order neural network for pattern classification and function approximation. In: Proceedings of International Joint Conference on Neural Networks, vol. 1, pp. 13–18. Seattle, Washington, July 1991
8. Hussain, A.J., Liatsis, P.: Recurrent pi-sigma networks for DPCM image coding. Neurocomputing **55**, 363–382 (2002)
9. Li, C.-K.: Memory-based sigma-pi-sigma neural network. IEEE SMC, TP1F5, pp. 112–118 (2002)
10. Weber, C., Wermter, S.: A self-organizing map of sigma–pi units. Neurocomputing **70**, 2552–2560 (2007)
11. Ghazali, R., Hussain, A., El-Deredy, W.: Application of ridge polynomial neural networks to financial time series prediction. In: 2006 International Joint Conference on Neural Networks, pp. 913–920, July 16–21, 2006
12. Nie, Y., Deng, W.: A hybrid genetic learning algorithm for Pi-sigma neural network and the analysis of its convergence. In: IEEE Fourth International Conference on Natural Computation, 2008, pp. 19–23
13. Song, G., Peng, C., Miao, X.: Visual cryptography scheme using pi-sigma neural networks. In: 2008 International Symposium on Information Science and Engineering, pp. 679–682
14. Nayak, J., Naik, B., Behera, H.S.: A novel chemical reaction optimization based higher order neural network (CRO-HONN) for nonlinear classification. Ain Shams Eng. J. (2015)
15. Nayak, J., Naik, B., Behera, H.S.: A novel nature inspired firefly algorithm with higher order neural network: performance analysis. Eng. Sci. Technol. Int. J. (2015)
16. Nayak, J., Naik, B., Behera, H.S.: A hybrid PSO-GA based Pi sigma neural network (PSNN) with standard back propagation gradient descent learning for classification. In: 2014 International Conference on Control, Instrumentation, Communication and Computational Technologies (ICCICCT). IEEE (2014)
17. Nayak, J., Kanungo, D.P., Naik, B., Behera, H.S.: A higher order evolutionary Jordan Pi-Sigma neural network with gradient descent learning for classification. In: 2014 International Conference on High Performance Computing and Applications (ICHPCA), pp. 1–6. IEEE (2014)
18. Rao, R.V., Savsani, V.J., Vakharia, D.P.: Teaching–learning-based optimization: an optimization method for continuous non-linear large scale problems. Inf. Sci. **183**(1), 1–15 (2012)
19. Rao, R.V., Kalyankar, V.D.: Parameter optimization of modern machining processes using teaching–learning based optimization algorithm. Eng. Appl. Artif. Intel. **26**(1), 524–531 (2013)

20. Rumelhart, D.E., Hinton, G.E., Williams, R.J.: Learning representations by back-propagating errors. Nature **323**(9), 533–536 (1986)
21. Alcalá-Fdez, J., Fernandez, A., Luengo, J., Derrac, J., García, S., Sánchez, L., Herrera, F.: KEEL data-mining software tool: data set repository, integration of algorithms and experimental analysis framework. J. Multiple-Valued Logic Soft Comput. **17**(2–3), 255–287 (2011)
22. Nayak, J., et al.: Particle swarm optimization based higher order neural network for classification. Comput. Intell. Data Mining. Springer India. **1**, 401–414 (2015)

# Deploying a Social Web Graph Over a Semantic Web Framework

Shubhnkar Upadhyay, Avadhesh Singh, Kumar Abhishek
and M.P. Singh

**Abstract** In this paper, the authors have deployed a Social Web Graph over a
Semantic Web Framework. A social web graph is a collection of nodes, connected
via directed edges. A graph simplifies the search by reducing the sample space by
half after every node. But this is a keyword based search and is not rather intelligent. On the other hand a Semantic web framework organizes information in triples
in addition to having the benefits of a graph. This makes the information more
understandable and easily accessible.

**Keywords** Semantic web · Ontology · OWL (Web ontology language) · Social
web graph · DL (Description logic)

## 1 Introduction

The World Wide Web as seen today has changed the way people access information. Since its inception the Web has grown uncontrollably and in a very chaotic
way. Due to this unorganized form of data it is very difficult to retrieve relevant
information. This is seen clearly in the case of social media, where individuals are

S. Upadhyay (✉) · A. Singh · K. Abhishek · M.P. Singh
Department of Computer Science & Engineering, National Institute
of Technology Patna, Patna, India
e-mail: shubhnkar123988@nitp.ac.in

A. Singh
e-mail: avadhesh123979@nitp.ac.in

K. Abhishek
e-mail: kumar.abhishek@nitp.ac.in

M.P. Singh
e-mail: mps@nitp.ac.in

© Springer India 2016
H.S. Behera and D.P. Mohapatra (eds.), *Computational Intelligence
in Data Mining—Volume 1*, Advances in Intelligent Systems
and Computing 410, DOI 10.1007/978-81-322-2734-2_8

linked randomly and without any semantic meaning. This paper proposes a solution of linking the individuals with each other through proper semantics. This is done by creating a small graph based structure of a social network. This graph is then implemented on a Semantic Web framework using OWL [1]. All the nodes are represented as classes and relations between these nodes are represented as object properties.

Various queries are implemented on the ontology thus created, and its results are shown. The queries are executed on both the graph as well as class structure of the ontology. Both, the quantitative result and the graphical result show the same output for a query.

## 1.1 Motivation

The Social Network at the present time does not employ a proper graphical structure. It is rather quite unorganized and chaotic. This is taken care of when using Semantic technologies. The classes that are created automatically form a well-defined graphical structure. The semantics of the structure are not clear. This means that the inherit meaning of the relationships between the nodes are unclear. This is rectified using object properties in OWL. The search is based on keywords. The query searches the similar keywords in the structure and returns the required results, while in the Semantic web the search is predicate based. In predicate based search the search is refined by using the properties between the resources in the search query. The reachability in the current Social network is not defined properly. There is no way of relating the ancestors with their children. Anyone can reach into anybody's profile because the relationships are not defined properly. This is ensured not to happen in Semantic web by using inverse and transitive properties.

## 2  Graphical Representation

For this paper a sample graph is considered which represents a social network. The graph is based on paper [2] and depicts a Social Network. It consists of 8 nodes which are connected with each other through directed edges. The whole graph is subdivided into two smaller graphs which are connected through only one directed edge (Fig. 1).

The graph consists of strongly connected components (SCC) with same edge types and mixed edge types. The component (1-2-3) is an SCC with same edge type, whereas the component (1-2-3-4) is an SCC with mixed edge type [2].

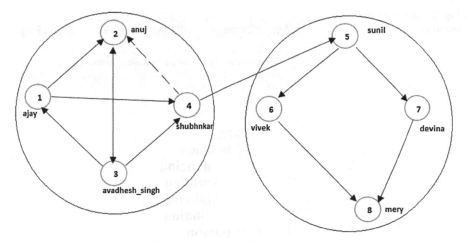

**Fig. 1** Sample graph structure

The graph also consists of cycles which are represented irrespective of the edge type present in the cycle. The component (1-2-3-1) is a cycle and so is (2-3-4-2). Nested cycles are also possible in the graph structure. The cycle (2-3-2) is also a reciprocal edge of the graph structure.

The graph structure shown also allows to define the reachability of various nodes. For example, in the graph "vivek" and "devina" can reach "mery" but the vice versa is not true. This means that "mery" cannot reach the above two nodes.

## 3 Ontology Development

The graphical structure mentioned in the previous section is now converted to an ontology. For this purpose classes, object properties and data properties are created and the individuals are set as instances of these classes.

Object properties are analogous to directed edges in graphs, as they define a particular relationship between two classes. The data properties are used to set the data values of the instances. The following figures clearly depict the classes and object properties for the ontology and a graph structure is also shown for all the instances of the ontology (Figs. 2 and 3).

In the graph of Fig. 4 there are two types of edges used, one is yellow colored and depicts isFriendOf object property and the other is green colored which depicts hasFriend object property. It is because of these properties only that the reachability is properly defined for the graph structure.

**Fig. 2** Classes in the
ontology

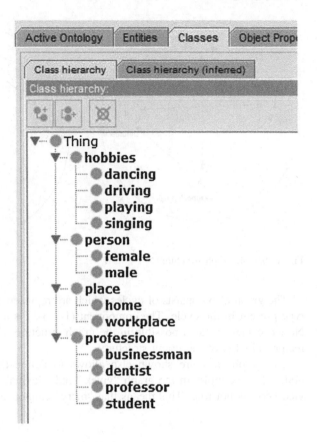

## 3.1   Searching the Graph in Ontology

The searching performed in OWL is predicate based searching, this means that the query contains the objects to be searched as well as their properties. This refines the search by a great amount and also helps in filtering unwanted results.

Following are some queries which were executed on the considered ontology. There are 2 figures associated with each query, the first one is the Description Logic (DL) query output in Protégé 5.0 and the second is the graph of that query result.

(1) DL Query 1—Fetching all the individuals that like singing (Figs. 5 and 6).
(2) DL Query 2—Showing all the properties of Ajay (Figs. 7 and 8).
(3) DL Query 3—Finding Ancestors of individuals using hasFriend property (Figs. 9 and 10).
(4) DL Query 4—Finding child of individuals using isFriendOf object property (Figs. 11 and 12).
(5) DL Query 5—Reachability is defined (Fig. 13).

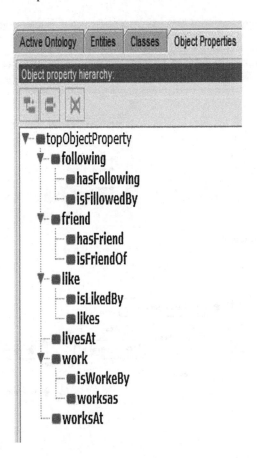

**Fig. 3** Object properties in the ontology

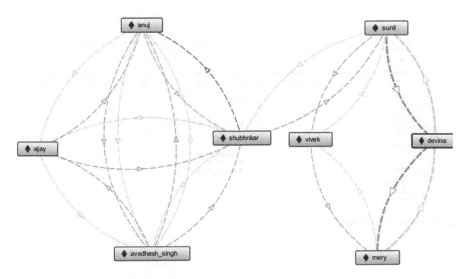

**Fig. 4** Sample graph of the ontology

**Fig. 5** Query 1 in Protégé

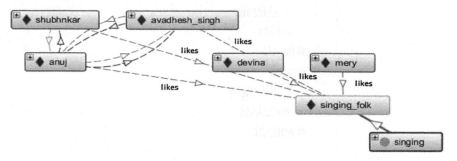

**Fig. 6** Graph of the query

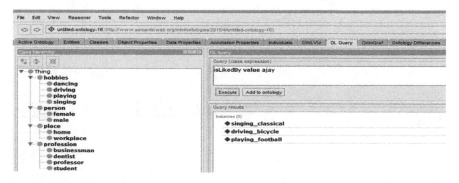

**Fig. 7** Query 2 in Protégé

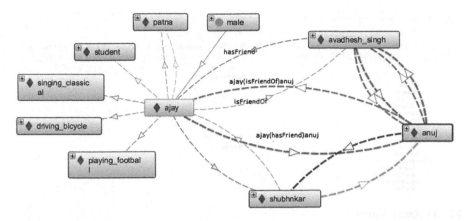

**Fig. 8** Graph of the query

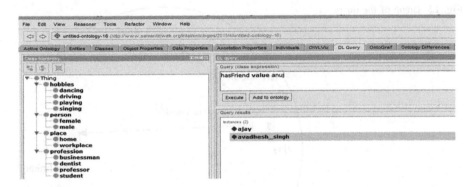

**Fig. 9** Query 3 in Protégé

**Fig. 10** Graph of the query

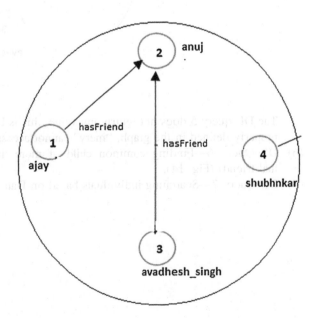

**Fig. 11**  Query 4 in Protégé

**Fig. 12**  Graph of the query

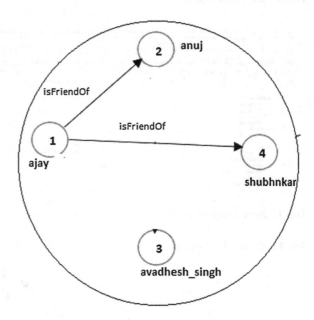

The DL query 5 does not return any value, this is because the reachability is properly defined in the graph, "mery" cannot access her ancestor.

(6)  DL Query 6—Finding common child of more than two individuals'(mutual-friend) (Fig. 14).

(7)  DL Query 7—Searching individuals based on Data properties (Fig. 15).

**Fig. 13** Query 5 in Protégé

**Fig. 14** Query 6 in Protégé

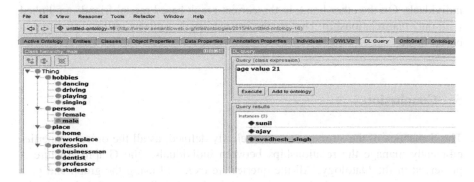

**Fig. 15** Query 7 in Protégé

# 4 Performance Analysis

The sample graph considered here consisted of nodes and well-defined directed edges. Through constructing efficient algorithms the query on this graph would have been possible. But since this graph was deployed over an ontology, hence no algorithms were needed for querying purposes. All the querying was done with the help of Description Logic (DL) query. A total of seven different queries were executed using DL query and the results were noted. These results were then checked with the graph structure. All the query results were in accordance with their graph results. Hence, it can be concluded that the deploying of the graph over the ontology was successful.

Moreover some of the queries showed here output absurd results when implemented on the existing graph model. This is because the current social graph relies on keyword based search while the ontology model uses the predicate based search. This ensures that the results are calculated on the basis of their relationships and not just the keyword.

For example, in the previous section, query number 7 searches the individuals who are 21 years of age. Now, when this query is implemented on the existing node based graph model, it only searches the keywords like "age" and "21" from nodes and returns it. But in semantic search, these are checked with data properties and only then the results are returned. Although, these type of search queries do return quite accurate results in the modern keyword based graph searches, but still these are keyword based searches and do not take in account the semantic of the query. Complex queries which have nested conditions do not return accurate results in keyword based search.

The reachability in case of ontology is well defined, as contrary to current graph based model. For example, in the ontology used in this paper two object properties "hasFriend" and "isFriendof" are used, that define a relationship between two individuals. On the other hand, in the current graph based model friends are linked with only one relationship. Due to this reachability is an issue in the current graph based model.

# 5 Conclusion

The Semantics of the structure are now clearly defined, as all the object properties efficiently manage the relationships between individuals. The Graph structure is preserved in the Ontology. All the queries are executed using the graph and give results in accordance with the graph structure. Reachability is well-defined in the ontology. This is done with the help of object properties which restrict certain individuals from accessing others.

While searching, the data properties of the individuals can also be used. The query can be constructed in such a way that the data properties of individuals are used, for example "person with age 21" or "person with certain birthdate".

# References

1. Antoniou, G., van Harmelen, F.: A Semantic Web Primer. MIT Press, Cambridge (2008)
2. Mitra, S., Bagchi, A., Bandyopadhyay, A.K.: Complex query processing on web graph: a social network perspective. In Journal of Digital Information Management (2006)
3. Semantic Web. https://en.wikipedia.org/wiki/Semantic_Web
4. Graph Theory. https://en.wikipedia.org/wiki/Graph_theory
5. Tserpes, K., Papadakis, G., Kardara, M., Papaoikonomou, A., Aisopos, F., Sardis, E., Varvarigou, T.: An ontology for social networking sites interoperability. In: International Conference on Knowledge Engineering and Ontology Development, pp 245–250
6. Erétéo, G., Buffa, M., Gandon, F., Corby, O.: Analysis of a real online social network using semantic web frameworks. In: ISWC 2009, LNCS 5823, pp. 180–195 (2009)
7. Wennerberg, P.O.: Ontology based knowledge discovery in social networks. In: JRC Joint Research Center European Commission, Institute for the Protection and Security of the Citizen (IPSC)
8. Mika, P.: Ontologies are us: a unified model of social networks and semantics. Web Semant. Sci. Serv. Agents World Wide Web 5, 5–15 (2007)

# Detecting and Analyzing Invariant Groups in Complex Networks

Dulal Mahata and Chanchal Patra

**Abstract** Real-world complex networks usually exhibit *inhomogeneity in* functional properties, resulting in densely interconnected nodes, *communities*. Analyzing such communities in large networks has rapidly become a major area in network science. A major limitation of most of the community finding algorithms is the dependence on the ordering in which vertices are processed. However, less study has been conducted on the effect of vertex ordering in community detection. In this paper, we propose a novel algorithm, *DIGMaP* to identify the *invariant groups* of vertices which are not affected by vertex ordering. We validate our algorithm with the actual community structure and show that these detected groups are the core of the community.

**Keywords** Permamence · DIGMaP · Invariant community structure · Complex network · Community detection

## 1 Introduction

Complex networks are the abstract representation of the real-world complex phenomena. These networks can generally be represented by graphs, where nodes represent entities and edges indicate relations between entities. For example, a social network can be represented by a graph where nodes are users and edges represent social relationships such as friendship. The distribution of links also shows *inhomogeneity*, both globally and locally, describing the phenomenon that nodes naturally cluster into groups and links are more likely to connect nodes within the same

D. Mahata (✉) · C. Patra
Department of Computer Science, Vidyasagar University, Midnapur
West Bengal, India
e-mail: dulalmahata1@gmail.com

C. Patra
e-mail: chanchalpatra89@gmail.com

© Springer India 2016
H.S. Behera and D.P. Mohapatra (eds.), *Computational Intelligence in Data Mining—Volume 1*, Advances in Intelligent Systems and Computing 410, DOI 10.1007/978-81-322-2734-2_9

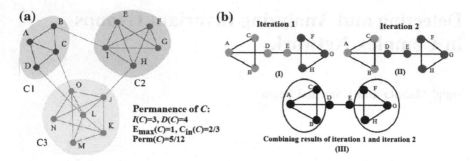

**Fig. 1** **a** (Color online) A schematic diagram of a network with three communities C1, C2 and C3. Permanence of vertex $C$ (see Sect. 5.2) is also shown here. **b** Toy example of the formation of invariant groups (communities are indicated by *red* and *green* colors)

group. Network scientists call this organization as the *community structure* of networks. A figurative sketch of community structure is depicted in Fig. 1a.

Detecting communities is of prime importance in interdisciplinary research where systems are often represented as graphs. This problem is a difficult one and has not been solved satisfactorily over the past one and half decades (see [1] for the reviews). The existing community discovery algorithms are optimizing a certain metric (such as modularity [2]). Therefore, a minor change in the vertex order can change the assignment of vertex into a community. Therefore, a crucial question about the variability in community assignment remains unanswered—*can we automatically detect the invariant substructure of a network and if so, what do they signify?*

In this paper, we address the issue of variability of the algorithms in detecting communities from the networks. We show that despite such variability, there exist invariant some groups of vertices in a network which always remain together. The contributions of the paper are fourfold: (i) we propose a local vertex-centric metric, called *permanence* to define the stability of a vertex in a community; (ii) we propose for the first time an algorithm, called *DIGMaP* which is a greedy agglomerative algorithm that maximizes permanence in order to detect such invariant groups from the network; (iii) we run our algorithm on both artificial and real networks, and observe that it quite accurately detects invariant groups from the networks; (iv) finally we show that the invariant groups are the core modules of a community structure.

# 2 Terminologies

In this section, we outline the definitions of certain terms which we shall repeatedly use throughout the rest of the paper.

(i) **Graphs/Networks** A network $G(V, E)$ is represented by a set of vertices $V$ and a set of edges $E$. An edge $e \in E$ connects two vertices $\{u, v\}$ (where $u, v \in V$). A toy example of a network is shown in Fig. 1a.

(ii) **Clustering Coefficient (CC)** The clustering coefficient of a vertex is computed as the ratio of the actual number of edges between the neighbors of the vertex to the total possible connections among the neighbors. For instance, in Fig. 1a node A has three neighbors, B, C and D; among them maximum three edges are possible, but there exist two edges, $\langle B, C \rangle$ and $\langle C, D \rangle$. Therefore, the CC of A is 2/3.

(iii) **Community** A network is assumed to have a *community structure* if the nodes can be easily clustered such that the intre-cluster edge density is higher than the inter-cluster edge-density [1]. In Fig. 1a, 15 nodes are grounded into three communities, C1, C2 and C3 (shown in three colors) and nodes in a group are assumed to possess homogeneous characteristics. Discovering such community structure from the network is non-trivial due to the lack of predefined threshold based on which one can mark a dense group as community.

## 3 Related Work

Research in the area of community detection encompasses many different approaches including, modularity optimization [3], spectral graph-partitioning algorithm [1], clique percolation [4], local expansion [5], fuzzy clustering [6], link partitioning [7], random-walk based approach [8], information theoretic approach [9], diffusion-based approach [10], significance-based approach [11] and label propagation [12]. Most of these community finding algorithms can produce different community assignments if the algorithm factors such as the order of the vertices to be processed changes. Lancichinetti and Fortunato [13] presented a method for consensus clustering where the edges are reweighted depending on the number of times pairs of vertices were assigned to the same community. Different techniques have been proposed to improve the accuracy of the solution. Delvenne et al. [14] invented a the concept of stability of a community, based on the clustered auto-covariance of a dynamic Markov process observed in the network. Recently, Chakraborty et al. [15] point out the effect of vertex ordering in the community detection algorithms.

## 4 Test Suite of Networks

We examine a set of artificially generated networks and two real-world complex networks whose actual community structures are given. The brief description of the used datasets is mentioned below.

(i) **Synthetic Networks** We select the LFR model [13] to construct artificial networks with a community structure. The model allows controlling directly the following properties: $n$: number of nodes, $k$ and $k_{\max}$ are the average and

**Table 1** Characteristics of real networks

| Network | $N$ | $E$ | $<k>$ | $k_{max}$ | $C$ | $n_c^{min}$ | $n_c^{max}$ |
|---------|-----|-----|-------|-----------|-----|-------------|-------------|
| Football | 115 | 613 | 10.57 | 12 | 12 | 13 | 5 |
| Railway | 301 | 1,224 | 6.36 | 48 | 21 | 46 | 10 |

$N$ number of nodes, $E$ number of edges, $C$ number of communities, $\langle k \rangle$ avg. degree, $k_{max}$ maximum degree, $n_c^{min}$ and $n_c^{max}$ the sizes of its smallest and largest communities

maximum degree respectively, $\gamma$: exponent for the degree distribution, $\beta$: exponent for the community size distribution, and $\tau$: mixing coefficient. The parameter $\tau$ controls the average amount of edges between a node and its neighbors outside the community. Unless otherwise stated, in this paper we generate the LFR network with the number of nodes ($n$) as 1000, and $\tau$ is varied from 0.1 to 0.6.

(ii) **Real-world Networks** We use following two real-world networks: (i) Football network and (ii) Railway network. The properties of these networks are summarized in Table 1. The details of the networks can be found in [5].

# 5 Automatic Detection of Invariant Groups

Here, we start with presenting the notion of invariant community structure in a network. Following this, we define a metric which is used to characterize these invariant groups. Finally, we present our proposed algorithm to detect such groups from the network.

## 5.1 Invariance in Community Structure

All the methods for community detection focus on optimizing certain quality functions in detecting community structure. The output of these algorithms is heavily dependent on the ordering in which the vertices are processed during the execution. For instance, consider a network in Fig. 1b. If we run a community detection algorithm on this network, in one iteration it might produce two communities as shown in Fig. 1b(I). In another iteration, it might produce different community structure as shown in Fig. 1b(II). However, a careful inspection reveals that there exist two invariant groups, {A, B, C} and {F, G, H} whose constituent vertices always remain together despite any iteration (see Fig. 1b(III)). A major question about the fluctuation of results remains unanswered in the existing literature—what do the invariance of the results indicates about the network structure? *In this paper, we propose for the first time an automatic way of discovering such invariant groups from a network without using any prior knowledge of network communities.*

## 5.2 Defining Stability of a Node

Chakraborty et al. [5] mentioned that the reason behind the variation of the community structure for different vertex ordering is the instability of the nodes within a community. For instance, in Fig. 1b node D is loosely connected with A and E; similarly node E is loosely connected with D and G. Therefore, when a community finding algorithm is run on this network, it easily combines the well-connected nodes such as A, B, C together (similarly F, G and H are combined together). However, the algorithm gets confused in assigning D and E in the appropriate community and often randomly assigns them in either group {A, B, C} or group {F, G, H}. Therefore, the extent of belongingness of D in group {A, B, C} is lower than that of A. We capture the stability of a node in a group using a metric called *permanence* as follows.

$$Perm(v) = \left[\frac{I(v)}{D(v)} \times \frac{1}{E_{max}(v)}\right] - [1 - C_{in}(v)] \tag{1}$$

where $I(v)$ indicates the number of neighbors of $v$ insider its community, $E_{max}(v)$ is the maximum number of connections of $v$ to any one of the external groups, $D(v)$ indicates the degree of $v$ and $C_{in}(v)$ is the clustering coefficient among the internal neighbors of $v$ in its own group. For the weighted network, $I(v)$, $D(v)$ and $E_{max}(v)$ are calculated by aggregating the edge weights. Figure 1a presents a toy example to calculate the permanence of a vertex.

## 5.3 DIGMaP Algorithm

We propose an algorithm, called DIGMaP (Discovering Invariant Groups using Maximizing Permanence) to detect invariant groups from the networks.

Our algorithm is a greedy agglomerative algorithm, which aims at maximizing permanence. DIGMaP is composed of two steps that are repeated iteratively. First, we initialize every edge (i.e. two end vertices of the edge) to a group. Then, for each vertex $i$ we take each neighbor $j$ of $i$ and check the increase of permanence that would happen by removing $i$ from its group and by assigning it into $j$'s group. Then, the node $i$ is assigned in the group where the increase is maximum, but only if this gain is positive. Otherwise, $i$ is assigned in its original group. The entire steps are iterated repeatedly, and for all nodes until no improvement can be obtained and the first step is then complete. This first step is terminated when permanence reaches to a local maxima.

The second step is started by constricting a super network whose nodes correspond to the groups obtained from the first phase. For that, the edge weight between two new nodes is calculated by adding of the weight of the edges between nodes in the corresponding two groups. Once this second step is over, we keep on repeating

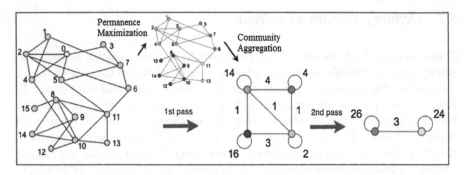

**Fig. 2** Schematic diagram of the steps of DIGMaP algorithm

the first step to the new weighted network and reiterate. The process is repeated until no modification is needed and a maximum of permanence is attained. A schematic diagram of the process is depicted in Fig. 2.

# 6 Experimental Results

We run DIGMaP on LFR networks and two real-world networks mentioned in Sect. 4 and obtain invariant groups. In this section, we present a thorough analysis of these groups from different perspectives.

**Statistical analysis of invariant groups** Table 2 shows the maximum, minimum and average size of the invariant groups for different networks. We observe that with the increase in $\mu$ (which controls the edge density within and across community as mentioned in Sect. 4), the maximum, minimum and the average size decreases. The possible reason could be that when the value of $\mu$ increases, the inter-community edge density becomes higher which results in higher external pull of vertices from outside group. Therefore, DIGMaP gets confused to allocate those nodes into appropriate groups and tends to allocate them separately into singleton groups (i.e. groups consisting of one node).

**Comparison with ground-truth results** One might argue the appropriateness of DIGMaP, i.e. whether the detected groups at all represent the invariant structure of the network. To validate this, we produce ground-truth invariant groups using brute force method invented by Chakraborty et al. [15]. For each network, we permute

**Table 2** Maximum (Max), minimum (Min) and average (Avg.) size of the invariant groups detected by DIGMaP for different networks

| Size | LFR ($\mu = 0.1$) | LFR ($\mu = 0.3$) | LFR ($\mu = 0.6$) | Football | Railway |
|------|-------------------|-------------------|-------------------|----------|---------|
| Max  | 15                | 10                | 7                 | 8        | 12      |
| Min  | 7                 | 4                 | 1                 | 3        | 4       |
| Avg. | 10.30             | 4.95              | 1.92              | 3.02     | 4.56    |

**Table 3** Comparison of the groups detected by DIGMaP with the ground-truth results in terms of three validation measures for all the networks

| Measures | LFR ($\mu = 0.1$) | LFR ($\mu = 0.3$) | LFR ($\mu = 0.6$) | Football | Railway |
|---|---|---|---|---|---|
| NMI | 0.82 | 0.75 | 0.68 | 0.82 | 0.72 |
| ARI | 0.76 | 0.71 | 0.58 | 0.79 | 0.65 |
| PU | 0.84 | 0.70 | 0.61 | 0.78 | 0.72 |

the vertex order, and apply Louvain algorithm [3] to each permuted order. The final community structure can vary across different permutations. We consider the groups whose constituent vertices always remain together across all the permutations. In this way, we construct the ground-truth groups.

To measure the correspondence between ground-truth invariant groups and the groups detected by our algorithm, we use three validation measures [5]: Normalized Mutual Information (NMI), Adjusted Rand Index (ARI) and Purity (PU). For each metric, the values range between 0 and 1, where 0 refers to no match with the ground truth and 1 refers to a perfect match.

Table 3 shows to what extent the output of DIGMaP matches with the ground-truth groups. We notice that the values of NMI, ARI and PU are significantly high which indicates a high similarity of the detected groups with the ground-truth. However, the accuracy decreases with the increase in $\mu$ for LFR network. The possible reason could be that with the increase of $\mu$, the network tends to lose its modular structure resulting in intermingled community structure [13]. Therefore, DIGMaP tends to assign most of the nodes into small size invariant groups. However, the results for Football and Railway networks are reasonably well.

**Coreness of the invariant groups** We hypothesize that the invariant groups detected by DIGMaP are the core of the community structure, i.e. each community is composed of multiple such invariant groups. To validate this hypothesis, we use the actual community structure of each network as mentioned in Sect. 4 and check whether nodes in a certain invariant group are also part of a single community. We use *entropy* measure to check the *coreness* of a detected group. For example, consider a detected group consisting of five vertices {A, B, C, D, E}. Now let us assume two cases: (i) all five nodes are part of a single community (say, Community 1), and (ii) three nodes are part of one community (Community 1) and rest of the two nodes are in another community (Community 2). Therefore, for case (i) the membership vector is {1, 1, 1, 1, 1} which results in zero entropy, and for case (ii) the membership vector is {1, 1, 1, 2, 2} which results in entropy of 0.97. Lower the value of entropy, higher the coreness of the detected invariant groups.

Table 4 presents average entropy of the detected groups for each network. We notice that for most of the networks entropy tends to be zero, which indeed corroborates our hypothesis that DIGMaP is capable of detecting the core of the community structure. We also measure the edge-density of the detected groups and observe that the groups are quite dense, which is an essential criteria to characterize the community cores of a network [15].

**Table 4** The mean coreness (based on the entropy) and the mean edge-density of the detected invariant groups for each network (standard deviation is reported within parenthesis)

| Measures | LFR ($\mu$ = 0.1) | LFR ($\mu$ = 0.3) | LFR ($\mu$ = 0.6) | Football | Railway |
|---|---|---|---|---|---|
| Entropy | 0 (0) | 0.05 (0.01) | 0.25 (0.05) | 0.18 (0.08) | 0.12 (0.09) |
| Edge density | 0.91 (2.48) | 0.94 (3.89) | 0.58 (0.96) | 0.84 (1.28) | 0.81 (0.87) |

# 7  Conclusion and Future Work

In this paper, we examined the invariant groups in different complex networks, which are regions within a network where the community structures are invariant under different vertex orderings. The presence of different results for community detection is well studied; however, this paper is perhaps the one of the first studies of detection of the invariant graphs from different networks. Here, we proposed an algorithm, DIGMaP for automatically discovering such invariant groups and showed that the detected groups quite strikingly match with the ground-truth results generated by the brute force method. We also showed that the invariant groups are the core of the communities in a network. In future, we would like to unfold the dynamics controlling the evolution of such groups in a network. Moreover, we shall investigate the functional significance of these groups in different networks.

# References

1. Newman, M.E.J.: Community detection and graph partitioning. CoRR abs/1305.4974 (2013)
2. Newman, M.E.J.: Modularity and community structure in networks. PNAS **103**(23), 8577–8582 (2006)
3. Blondel, V.D., Guillaume, J.L., Lambiotte, R., Mech. E.L.: Fast unfolding of communities in large networks. J. Stat. Mech. **P10008** (2008)
4. Palla, G., Dernyi, I., Farkas, I., Vicsek, T.: Uncovering the overlapping community structure of complex networks in nature and society. Nature **435**(7043), 814–818 (2005)
5. Chakraborty, T., Srinivasan, S., Ganguly, N., Bhowmick, S., Mukherjee, A.: On the permanence of vertices in network communities. In: 20th ACM SIGKDD Conference on Knowledge Discovery and Data Mining, New York city, pp. 1396–1405 (2014)
6. Psorakis, I., Roberts, S., Ebden, M., Sheldon, B.: Overlapping community detection using Bayesian non-negative matrix factorization. Phys. Rev. E **83**(6), 066114 (2011)
7. Ahn, Y.-Y., Bagrow, J.-P., Lehmann, S.: Link communities reveal multi-scale complexity in networks. Nature **466**, 761–764 (2010)
8. Pons, P., Latapy, M.: Computing communities in large networks using random walks. J. Graph Algorithms Appl. **10**(2), 191–218 (2006)
9. Rosvall, M., Bergstrom, C.T.: An information-theoretic framework for resolving community structure in complex networks. PNAS **104**(18), 7327 (2007)
10. Raghavan, U.N., Albert, R., Kumara, S.: Near linear time algorithm to detect community structures in large-scale networks. Phys. Rev. E **76**, 036106 (2007)
11. Lancichinetti, A., Radicchi, F., Ramasco, J.J., Fortunato, S.: Finding statistically significant communities in networks. CoRR abs/1012.2363 (2010)

12. Xie, J., Szymanski, B.K.: Community Detection Using A Neighborhood Strength Driven Label Propagation Algorithm. CoRR abs/1105.3264 (2011)
13. Lancichinetti, A., Fortunato, A.: Benchmarks for testing community detection algorithms on directed and weighted graphs with overlapping communities. Phys. Rev. E **80**(1), 016118 (2009)
14. Delvenne, J.C., Yaliraki, S.N., Barahona, M.: Stability of graph communities across time scales. Proc. Natl. Acad. Sci. U.S.A. **107**, 12755–12760 (2010)
15. Chakraborty, T., Srinivasan, S., Ganguly, N., Bhowmick, S., Mukherjee, A.: Constant communities in complex networks. Nat. Sci. Reports **3**(1825), (2013)

# Automatic Mucosa Detection in Video Capsule Endoscopy with Adaptive Thresholding

V.B. Surya Prasath and Radhakrishnan Delhibabu

**Abstract** Video capsule endoscopy (VCE) is a revolutionary imaging technique widely used in visualizing the gastrointestinal tract. The amount of big data generated by VCE necessitates automatic computed aided-diagnosis (CAD) systems to aid the experts in making clinically relevant decisions. In this work, we consider an automatic tissue detection method that uses an adaptive entropy thresholding for better separation of mucosa which lines the colon wall from lumen, which is the hollowed gastrointestinal tract. Comparison with other thresholding methods such as Niblack, Bernsen, Otsu, and Sauvola as well as active contour is undertaken. Experimental results indicate that our method performs better than other methods in terms of segmentation accuracy in various VCE videos.

**Keywords** Segmentation · Video capsule · Endoscopy · Mucosa detection · Thresholding

## 1 Introduction

Video capsule endoscopy (VCE) was introduced at the turn of the century for imaging the gastrointestinal tract in a painless way [1]. A tiny pill-sized capsule is ingested by a patient that contains a miniature video camera and a wireless transmitter. The video imagery is sent to a recorder, which is attached to the body, which

V.B. Surya Prasath (✉)
Department of Computer Science, University of Missouri-Columbia, Columbia,
MO 65211, USA
e-mail: prasaths@missouri.edu
URL: http://goo.gl/66YUb

R. Delhibabu
Knowledge Based System Group, Higher Institute for Information Technology
and Information Systems, Kazan Federal University, Kazan, Russia

© Springer India 2016
H.S. Behera and D.P. Mohapatra (eds.), *Computational Intelligence
in Data Mining—Volume 1*, Advances in Intelligent Systems
and Computing 410, DOI 10.1007/978-81-322-2734-2_10

**(a)**                    **(b)**                    **(c)**

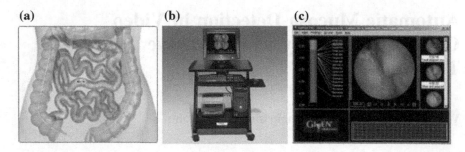

**Fig. 1** Video capsule endoscopy (VCE) system. **a** Capsule endoscopy travels inside gastrointestinal tract and transmits images wirelessly to attached sensors outside the body. **b** The downloaded imagery can be inspected in a workstation where the images are further analyzed for diagnosis. **c** An example frame from a VCE video, showing normal mucosa appearance

is then downloaded to a workstation for further analysis; see Fig. 1a, b. Typically a VCE exam consists of ∼55,000 frames and it requires around 2 h of manual visual inspection by gastroenterologists. To reduce the burden of reading VCE exams, efficient automatic computer aided-diagnosis (CAD) systems are required.

One of the important steps in VCE image analysis is the mucosa tissue detection, which requires a fast and efficient segmentation method for identifying tissues surrounding the gastrointestinal tract. Mucosa tissue lines the gastrointestinal tract and helps in digesting food. Separating mucosal tissue from the intermediate lumen require an adaptive image segmentation method as the appearance of mucosa tissue can vary in color, texture and shape. Active contours can be used for mucosa and lumen segmentation [2, 3], though they suffer from high computational costs. Among a wide variety of segmentation methods available in image processing literature perhaps the simplest but efficient ones are based on intensity thresholding [4]. Thresholding techniques are widely utilized in various image processing pipelines and the methods by Niblack [5], Bernsen [6], Otsu [7], and Sauvola [8] are some of the well-known thresholding approaches.

In this paper, we explore an adaptive fuzzy entropy based thresholding approach for obtaining faster and efficient mucosa segmentation from VCE videos. Our aim here is to study the feasibility of such an intensity thresholding and check its accuracy against other thresholding and active contour methods. Our experimental results on various VCE exams indicate that we obtain better segmentations. Moreover, when compared to other segmentation methods from the literature, in terms of ground-truth comparison, our final segmentations agree strongly with expert marked gold standard.

The rest of the paper is organized as follows. Section 2 introduces our approach for mucosa segmentation with adaptive fuzzy entropy thresholding method. Section 3 provides experimental results and comparisons with other segmentations methods. Section 4 concludes the paper.

## 2  Adaptive Thresholding for Mucosa Segmentation

One of the classical thresholding approaches is based on image entropy calculation. We describe the entropy thresholding briefly here and refer to [9] for more technical details about its statistical interpretations.

### 2.1  Drawbacks of Previous Thresholding Methods for VCE Imagery

The primary aim of image segmentation via thresholding is to obtain an image intensity value and use that as a barrier to define object boundaries, that is to obtain a partition of the given image $\Omega = \Omega_F \cup \Omega_B$ using an 'optimal' thresholding value. To achieve this various methods have been considered in the past. For example, Niblack [5] used a local mean values and variance to determine the threshold. A similar approach was undertaken by Bernsen [6] to determine the dynamic thresholding unique to each image. Perhaps the most widely used thresholding method is due to Otsu [7], who used minimal intra-class variance as a requirement along with bi-modal histogram assumption.

Figure 2 shows the limitations of the traditional thresholding methods on different representative frames from a normal VCE imagery obtained using Pillcam® (Given Imaging Inc, Yoqneam, Israel) capsule endoscopes. The following drawbacks are observed:

1. The Niblack [5] method obtains under segmentation (crucial mucosal regions) since the local mean values are not distributed evenly across the image.
2. The Bernsen [6] method obtains non-optimal segmentation and includes the lumen region as mucosa.
3. The Otsu [7] method obtains spurious regions, as minuscule textural regions present can confuse the local variance computations.
4. The Suvola [8] method obtains a gross under segmentation and fail to separate lumen from mucosa.

In this paper, we consider an adaptive thresholding model utilizing the updated fuzzy entropy based method for obtaining improved segmentation results. We see in Fig. 2 (bottom row) that our method consistently obtained better results.

### 2.2  Adaptive Fuzzy Entropy Thresholding

We use an entropy based adaptive fuzzy thresholding which avoids the problems associated intensity based thresholding methods from the past. Entropy is a measure

**(a)** Input images

**(b)** Niblack [5], Bernsen [6], Otsu [7], and Sauvola [8] results

**(c)** Niblack [5], Bernsen [6], Otsu [7], and Sauvola [8] results

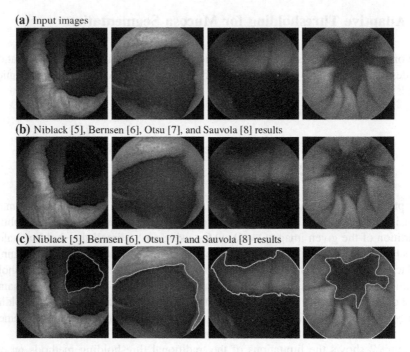

**Fig. 2** VCE segmentations by different thresholding methods. *Top row* Different VCE frames indicating the diversity of appearances of the mucosal tissue in different exams. *Middle row* Niblack [5], Bernsen [6], Otsu [7], and Sauvola [8] thresholding based results, given *left* to *right* respectively. *Bottom row* Results based on fuzzy entropic thresholding used in our work. The binarization results are given as contours to highlight the separation boundaries of mucosa and lumen.

of uncertainty and can be described formally for a finite discrete probability distribution $P = (p_1, p_2, \ldots p_n)$ as,

$$E(P) = -\sum_{k=1}^{n} p_k \log_2 p_k. \tag{1}$$

Tsallis [9] extended this classical entropy definition for non-extensive systems as follows,

$$T_q(P) = \frac{1}{q-1}\left(1 - \sum_{k=1}^{n} p_k^q\right). \tag{2}$$

where $q \geq 0$ is known as entropic order and $q \neq 1$. In our work, we consider Tsallis entropy of generalized distributions corresponding to foreground $P_F$ and background $P_B$ probability distributions from a given image distribution $P$.

That is we let,

$$P = (p_1, p_2, \ldots, p_n) = P_F \cup P_B = \{p_i\}_{i=1}^{h} + \{p_i\}_{i=h+1}^{n} \tag{3}$$

We next use a fuzzy membership function into the Tsallis entropy to indicate which pixel values belong to foreground or background. For this purpose, the following membership functions are used,

$$\mu_F(h - i) = 0.5 + \frac{\sum_{k=0}^{i} p(h-k)}{2p(h)}, \ \mu_B(h - i) = 0.5 + \frac{\sum_{k=1}^{i} p(h+k)}{2(1 - p(h))} \tag{4}$$

Thus, the fuzzy Tsallis entropy equations are given as,

$$T_q^F(h) = \frac{1}{q-1}\left(1 - \frac{\sum_{k=1}^{h} \mu_F(k)^q}{\sum_{k=1}^{h} \mu_F(k)}\right), \ T_q^B(h) = \frac{1}{q-1}\left(1 - \frac{\sum_{k=h+1}^{n} \mu_B(k)^q}{\sum_{k=h+1}^{n} \mu_B(k)}\right) \tag{5}$$

The final thresholding is determined by the maximum of the entropy value,

$$h^* = \mathrm{argmax}\left(T_q^F(h) + T_q^B(h) + (1-q) \cdot T_q^F(h) \cdot T_q^B(h)\right) \tag{6}$$

Note that the above threshold is dynamically set and does not involve manual tuning, this in turn makes the fuzzy Tsallis entropy thresholding an attractive option for obtaining fast VCE video segmentations.

We first utilize an illumination correction as a pre-processing step; this was done using the method described in [10]. Note this method considers La*b* space and performs a contrast enhancement based on spectral optimal contrast-tone mapping. We next apply the adaptive fuzzy Tsallis entropy thresholding using Eq. (6) on the L channel alone. The final binary image shows the segmentation with mucosa as the foreground and lumen the background. The overall pipeline is given in Fig. 3.

## 3 Experimental Results

### 3.1 Setup and Parameters

The overall scheme is implemented using non-optimized MATLAB pipeline and takes on average 0.5 s on a Mac Book Pro Laptop with Processor 2.3 GHz Intel Core i7 and 8 GB 1600 MHz DDR3 RAM memory. It takes on average less than 0.08 s for a RGB $512 \times 512$ VCE frame to obtain thresholding and the remaining time is utilized in illumination correction and for visualizing the results.

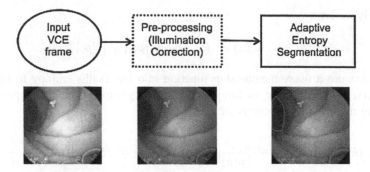

**Fig. 3** Overall pipeline of the VCE mucosa segmentation. For pre-processing we utilize the illumination correction scheme [10]. The final segmentation result is given as a contour overlaid on the color illumination corrected image. The same illumination correction pre-processing step is applied before applying other segmentations schemes considered here. The final lumen region is given by a *red* contour (color online)

## 3.2   Comparisons

Figure 4 shows a comparison of segmentations using Niblack [5], Bernsen [6], Otsu [7], Sauvola [8], methods that are based on intensity thresholdings, and active contour with edges model of [2] implemented using a fast finite difference scheme

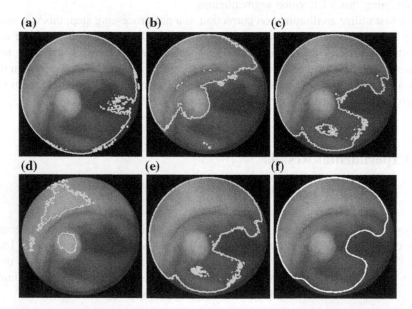

**Fig. 4** Comparison of different segmentation schemes for a normal VCE image. **a** Niblack [5], **b** Bernsen [6], **c** Otsu [7], **d** Sauvola [8], **e** active contour [3], **f** our result. The binarization results are given as contours to highlight the separation boundaries of mucosa and lumen

[3], with our adaptive thresholding technique described in Sect. 2.2. As can be seen, the active contour method [3] obtains spurious segments whereas other thresholding methods obtain either under or over segmentations, see also Fig. 2 for more examples. In contrast, our proposed approach obtains a coherent structure separation without any artifacts. Note that, to make a fair comparison we applied the same contrast enhancement [10] before applying other segmentation methods from the literature as well.

To quantitatively compare the results we utilize the Dice similarity coefficient as an error metric, which is given as follows:

$$D(A, B) = \frac{2|A \cap B|}{|A| + |B|} \tag{7}$$

where $A$, $B$ are two binary segmentations. Here one of the binary segmentation is computed using the automatic segmentation methods and the other using the ground-truth (expert marked). We evaluated the methods with two experienced gastroenterologists marked ground truth boundaries on 5 short videos of 100 frames each (totally 500 frames). Dice values closer to 1 indicate optimal result for an automatic algorithm and imply agreement with manual marked segmentations. Table 1 shows the average Dice metric for different schemes on the 5 videos (averaged across 100 frames) when compared to ground-truth (which is known as gold standard) segmentations. Overall average for our proposed approach is higher when compared to other segmentation methods. This shows that our adaptive thresholding performs better among different videos when compared to other thresholding and active contour methods. The computational time (for the thresholding step alone) is given for all the schemes considered here and it can be seen that our method is comparable to Suvola's method and much faster than the active contour method.

**Table 1** Dice similarity coefficient (D, Eq. 7) values for automatic segmentations when compared with manual ground truth for different schemes

| VCE video | Niblack [5] | Bernsen [6] | Otsu [7] | Sauvola [8] | AC [2] | Our result |
|-----------|-------------|-------------|----------|-------------|--------|------------|
| #1 | 0.6582 | 0.7023 | 0.7332 | 0.7091 | 0.8127 | 0.8725 |
| #2 | 0.6828 | 0.6821 | 0.6619 | 0.6239 | 0.5478 | 0.7040 |
| #3 | 0.7473 | 0.5391 | 0.7683 | 0.5884 | 0.6786 | 0.7902 |
| #4 | 0.6329 | 0.6972 | 0.6289 | 0.6310 | 0.6317 | 0.7321 |
| #5 | 0.6831 | 0.6245 | 0.7934 | 0.6984 | 0.7240 | 0.8242 |

These are average Dice values for different schemes compared with 5 videos of 100 frames each (totally 500 images). Values closer to 1 indicate better segmentation accuracy compared to manual segmentation

# 4 Conclusion

Automatic segmentation for video capsule endoscopy with multilevel adaptive thresholding is considered. By utilizing Tsallis entropy within an automatic fuzzy thresholding model, we devised an efficient mucosa segmentation method. Experimental results on different VCE videos with ground truth data indicate that our method obtains better segmentation accuracy when compared related thresholding methods and active contour method from the past. Note a further segmentation of chromaticity channels may improve the results though in turbid, trash conditions. Though we observed chromaticity can provide wrong information thereby reducing the final segmentation quality. Thus, a balanced and adaptive approach is required for handling the color information and defines our future work.

**Acknowledgments** This work was funded by the subsidy of the Russian Government to support the Program of competitive growth of Kazan Federal University among world class academic centers and universities.

# References

1. Iddan, G., Meron, G., Glukhovsky, A., Swain, F.: Wireless capsule endoscopy. Nature **405**, 417 (2000)
2. Chan, T., Vese, L.: Active contours without edges. IEEE Trans. Image Process. **10**, 266–277 (2001)
3. Prasath, V.B.S., Delhibabu, R.: Automatic image segmentation for video capsule endoscopy. In: Muppalaneni, N.B., Gunjan, V.K. (eds.) Computational Intelligence in Medical Informatics. Springer Briefs in Applied Sciences and Technology, Visakhapatnam, India, pp. 73–80. Springer CIMI (2015)
4. Sezgin, M., Sankur, B.: Survey over image thresholding techniques and quantitative performance evaluation. J. Elect. Imag. **13**, 146–165 (2004)
5. Niblack, W.: An introduction to digital image processing. Prentice Hall, Denmark (1986)
6. Bernsen, J.: Dynamic thresholding of gray-level images. In: International Conference on Pattern Recognition, pp. 1251–1255, Paris, France, October 1986
7. Otsu, N.: A threshold selection method from gray level histograms. IEEE Trans. Syst. Man Cybern. **9**, 62–66 (1979)
8. Sauvola, J., Pietikainen, M.: Adaptive document image binarization. Pattern Recog. **33**, 225–236 (2000)
9. Tsallis, C.: Possible generalization of boltzmann-gibbs statistics. J. Stat. Phy. **52**, 479–487 (1988)
10. Prasath, V.B.S., Delhibabu, R.: Automatic contrast enhancement for wireless capsule endoscopy videos with spectral optimal contrast-tone mapping. In: Jain, L., Behera, H.S., Mandal, J.K., Mohapatra, D.P. (eds.) Computational Intelligence in Data Mining, vol. 1. Odisha, India, pp. 243–250. Springer SIST, December 2015

# Optimization Approach for Feature Selection and Classification with Support Vector Machine

S. Chidambaram and K.G. Srinivasagan

**Abstract** The support vector machine (SVM) is a most popular tool to resolve the issues related to classification. It prepares a classifier by resolving an optimization problem to make a decision which instances of the training data set are support vectors. Feature selection is also important for selecting the optimum features. Data mining performance gets reduced by Irrelevant and redundant features. Feature selection used to choose a small quantity of related attributes to achieve good classification routine than applying all the attributes. Two major purposes are improving the classification functionalities and reducing the number of features. Moreover, the existing subset selection algorithms consider the work as a particular purpose issue. Selecting attributes are made out by the combination of attribute evaluator and search method using the WEKA Machine Learning Tool. In the proposed work, the SVM classification algorithm is applied by the classifier subset evaluator to automatically separate the standard information set.

**Keywords** Data mining · Kernel methods · Support vector machine · Classification

## 1 Introduction

A Support Vector Machine is an overture which is considered to categorize vectors of input features into single of the two categories. By constructing SVMs for dissimilar class pairs, it is possible to reach a classification into various categories. The procedures of the decision function are trained on a set of construction of

S. Chidambaram (✉)
Department of IT, National Engineering College, Kovilpatti, India
e-mail: chidambaramraj1@gmail.com

K.G. Srinivasagan
Department of CSE, National Engineering College, Kovilpatti, India
e-mail: kgsnec@rediffmail.com

© Springer India 2016                                  103
H.S. Behera and D.P. Mohapatra (eds.), *Computational Intelligence in Data Mining—Volume 1*, Advances in Intelligent Systems and Computing 410, DOI 10.1007/978-81-322-2734-2_11

patterns which contains the class labels. The major learning procedures are those that guide to the SVM with the minimum classification error on feature vectors. SVMs with a Gaussian kernel and the learning parameters are interpreted as follows: (1) the kernel parameter $\gamma$, measuring the distribution of the essence in the input space; and (2) the cost parameter C, measuring the comparative meaning of the error phrase in the cost procedure being reduced during SVM training. For that, one or more training and validation cycles are required. Each cycle contains one or more assessment procedures in which an SVM is used as a part of the construction of a model called the training set. The remaining portion is applied for the validation set.

## 2 Related Work

Huang et al. [1] compared neural networks with genetic programming and decision tree classifiers. For some set of input features, SVM classifier has accomplished the maximum classification accuracy. In the proposed hybrid GA-SVM strategy, it is possible to execute feature selection and parameters optimization process. Yang et al. [2] expressed the method which can extract three important attributes from user's network behavior. The extracted feature is categorized into various types, such as browsing news and downloading shared resources etc. Support vector machine is suitable for performing clustering process. It also provides rapid and valid execution, particularly for the tiny datasets.

Nematzadeh Balagatabi et al. [3] described the support vector machine which depends on cognitive style factors. A significant focus of this research is to categorize the researchers with respect to decision tree and Support Vector Machine methods.

## 3 Feature Selection

This procedure starts with subset creation that travels along a search policy to generate feature subsets. Each subset is estimated based on an evaluation condition and make a comparison with the previous solution. If the evaluated subset produces a better result, then it should be replaced with the old one, else remains the same. The method of subset creation and evaluation gets replicated until it satisfies the certain criteria. Figure 1 shows the Feature Selection process. At last, the validation applies to the selected best subset by using the prior data or a few test data. Search method and evaluation criteria are the two important points in feature selection.

Subset creation starts from the collection of attributes in the data set. It can be of any one of the following such as an empty set, the full set or a randomly formed subset. The feature subsets are searched in special ways, such as forward, backward, and random.

**Fig. 1** Feature selection
process

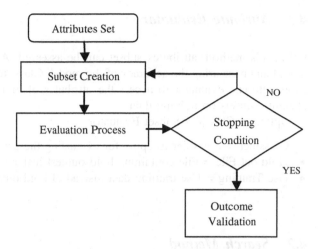

During evolutionary process, after the generation of feature subsets, the results are measured by some criteria to determine their performance. Normally, the main functionality of feature subsets is the discriminating capability of subsets that differentiate between or within various classes. With respect to the dependency of learning algorithms, it can be mostly classified into Wrapper Method, Filter Method. Wrapper method is used to determine attribute subsets with maximum accuracy because the features are coordinated exactly with the learning algorithms.

An attribute selection approach gets completed in any one of the observing conditions:

(i)   If the search process is finished.
(ii)  If a predefined size of feature subsets are finalized.
(iii) If a predefined number of iterations are completed.
(iv)  If an optimal feature subset along with the evaluation process is accomplished.
(v)   If the modification (addition or deletion) of feature subsets does not create a better subset.

In the outcome validation, related features are known in advance. Then it is possible to validate the feature selection outcome by comparing the previous facts. Though, in the real time applications, it is very hard to find which features are mostly related. Feature selection engages searching activity through all potential combinations of attributes in the data to discover which subset of attribute work well for improving performance of classification.

## 4 Proposed Method

In the proposed method, the classifier subset evaluator is combined with the search method which produces the maximum classification accuracy by compared to performing an individual function. In this way optimization can be achieved.

## 4.1 Attribute Evaluator

Using this method, attributes subset can be assessed. Another way is, creating a model and measuring the accuracy of the model. Classifier subset evaluator is used as an attribute evaluator to access the attributes of giving data sets. It Evaluates attribute subsets on training data.

Option in Classifier Subset Evaluator

- Classifier = Classifier to apply for evaluating the accuracy of subsets
- Hold out File = File containing hold out/test instances.
- Use Training = Use training data instead of hold out/test instances.

## 4.2 Search Method

It is the structured representation in which the possible attributes subsets are navigated with respect to the subset evaluation. Best first search may begin with the empty set of attributes and proceed forward, or begin with the complete set of attributes and proceed in the reverse direction, or begin at any position and proceed in both the direction. The combination of best first search method and Classifier Subset Evaluator will give the best feature selection in result which is further used for improving classification accuracy.

Some of the options available in Best First Search method are,

- Direction = Set the direction in which the search process carried out.
- Lookup Cache Size = Position the highest size of the lookup cache of estimated subsets. This can be represented as multiplying the given number of features in the specified dataset.
- Search Termination = Set the quantity of backtracking.

## 5 C4.5 Classification

C4.5 algorithm is an evolution of ID3. It uses Gain Ratio as splitting criteria. But in ID3 considers, gain as splitting criteria in tree growth phase. This algorithm considers both continuous and discrete attributes. In order to work with the continuous attributes, C4.5 sets a threshold for an attribute value and then splits the list of the given attribute value with respect to threshold. C4.5 can operate with the case of attributes with continuous domains by discretization. Missing attribute values are not used for calculating gains and entropy. In the tree pruning those misclassification errors are decreased. The steps of C4.5 algorithm in decision tree construction are specified below:

- Select the attribute of the root node
- Construct branch based on each attribute value and split conditions.
- Replicate the steps for each branch until all instances in the branch come under the single class.

The root node is selected based on the feature whose gain ratio is greater. Gain ratio is measured by Eq. (1).

$$GainRatio(D, S) = \frac{Gain(D, S)}{H\left[\frac{|D_1|}{|D|}, \ldots, \frac{|D_s|}{|D|}\right]} \tag{1}$$

where S = {$D_1$, $D_2$...$D_s$} denotes the new states.

# 6 SVM Classification

Support vector machine is a very efficient and suitable approach for regression, classification and pattern recognition. The main idea of SVM is to calculate the exact classification procedure to differentiate the instances of the two different classes in the training data. The classification parameters can be calculated geometrically.

The margin is the quantity of space or act as a separator between the two different classes by using the hyper plane. To substantiate the maximum margin hyper planes, an SVM classifier attempt to improve the procedure based on W and b.

$$L_p = \frac{1}{2}||W|| - \sum_{i=1}^{t} \alpha_i y_i (w \cdot x_i + b) + \sum_{i=1}^{t} \alpha_i \tag{2}$$

In Eq. (2), 't' is the number of training samples and $\alpha_i$, i = 1,..., t are positive numbers such that the derivatives of Lp based on $\alpha_i$ are zero. $\alpha_i$ is the Lagrange multipliers and Lp is called the Lagrangian. In this equation, 'w' is the vectors and constant 'b' used to identify the hyper plane. A machine learning approach such as the SVM can be designed as a specific task class based on few parameters '$\alpha$'.

## 6.1 Linear Kernel

This type of kernel function is represented by the inner product <x, y> plus an optional constant 'c'. A kernel algorithm uses a linear kernel which is same as the non kernel counterparts. It is calculated by Eq. (3).

$$k(x, y) = x^T y + c \qquad (3)$$

## 6.2  Polynomial Kernel

This type of kernel is a non-stationary kernel. Polynomial kernels are appropriate for the issues where all the training data are normalized.

$$k(x, y) = (\alpha x^T y + c)^d \qquad (4)$$

In Eq. (4), 'α' is the slope and constant term 'c', polynomial degree 'd' and 'x' 'y' are the feature vectors.

## 6.3  Radial Basis Kernel Function

By this type of kernel function, the transformation is denoted by Eq. (5),

$$k(x, y) = \exp\left\{ \frac{\|x - y\|^2}{2\sigma^2} \right\} \qquad (5)$$

where 'σ' is the variance to measure the kernel output. If it is overestimated, the exponential will perform linearly. If it is underestimated, then the procedure will require regularization and decision boundary. It is extremely sensitive to noise during training stage.

## 6.4  Sigmoid Kernel Function

This type, also called as a hyperbolic Tangent kernel method. It is similar to a two layer perceptron neural network. It is denoted by Eq. (6),

$$k(x, y) = \tanh(\alpha x^T y + c) \qquad (6)$$

where 'α' and constant 'c' are the adjustable parameters in this kernel and 'x' and 'y' are the feature vectors. The general value for alpha is 1/N, where N is the data dimension.

# 7   Experimental Results

The proposed method is implemented to the diabetes data set. It consists of eight numbers of numerical attributes. And ten fold crossover validation is applied to validate the results of classifiers.

The following performance metrics such as accuracy, error rate, precision, recall, are measured to find the efficiency of the proposed method. Accuracy measures the proportion of correct predictions considering the positive and the negative inputs. The error rate is the average loss over the test set. Precision is the positive predictions proportions that are correct, and recall is the positive sample proportion that is correctly positively predicted. From the Table 1, the proposed method produced the maximum accuracy (96.5 %) by compared to other classifiers performance. Figure 2 shows the performance comparison of various classifiers.

**Table 1**  Performance comparison of various classifiers with the proposed method

| Algorithms | Acc (%) | ER (%) | TP | FP | TN | FN | Pre | Rec | Sen | Spe | TT (s) |
|---|---|---|---|---|---|---|---|---|---|---|---|
| C4.5 | 68.1 | 11.5 | 0.3 | 0.21 | 0.65 | 0.79 | 0.5 | 0.4 | 0.5 | 0.5 | 1698.8 |
| Linear SVM | 70.6 | 29.3 | 0.7 | 0.53 | 0.30 | 0.47 | 0.6 | 0.7 | 0.7 | 0.6 | 1655.2 |
| Polynomial SVM | 69.9 | 30.6 | 0.6 | 0.47 | 0.31 | 0.53 | 0.4 | 0.7 | 0.5 | 0.7 | 1856.6 |
| RBF SVM | 71.6 | 28.3 | 0.7 | 0.58 | 0.29 | 0.42 | 0.6 | 0.7 | 0.6 | 0.7 | 1620.6 |
| Sigmoidal SVM | 69.9 | 30.6 | 0.6 | 0.69 | 0.31 | 0.31 | 0.5 | 0.6 | 0.6 | 0.7 | 1900.5 |
| Proposed method (RBF SVM+CSE) | 96.5 | 3.49 | 0.9 | 0.04 | 0.51 | 0.55 | 0.9 | 0.8 | 0.9 | 0.8 | 1422.8 |

*RBF* Radial Basis Function, *CSE* Classifier Subset Evaluation, *Acc* Accuracy, *ER* Error Rate, *TP* True Positive, *FP* False Positive, *TN* True Negative, *FN* False Negative, *Pre* Precision, *Rec* Recall, *Sen* Sensitivity, *Spc* Specificity, *TT* Time Taken

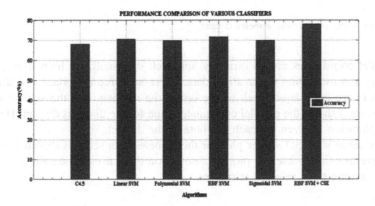

**Fig. 2**  Performance measures of various classifiers

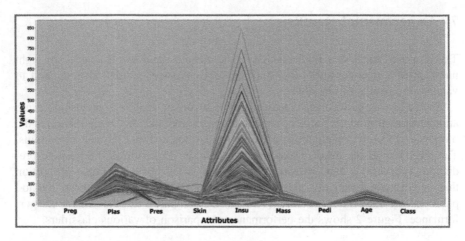

**Fig. 3** Parallel coordinates plot for the diabetes dataset

**Fig. 4** 3D scatter plot for the diabetes dataset

Parallel coordinates are an ordinary approach of visualizing high dimensional geometry and evaluating multivariate data. There are three important visualization parameters are considered, such as the order, the rotation, and the scaling of the axes. Figure 3 shows the parallel coordinate graph for the diabetes dataset. Here the data exploration is given for each attribute in the dataset.

In the same way, 3D scatter plot for the diabetes data set also shown in Fig. 4. This representation is used to plot data points on three axes which show the relationship among three selected attributes.

# 8 Conclusion

In this paper, proposed a SVM based optimization algorithm, which can used to optimize the parameter of SVM values. It is also used for retrieving the features of optimal subset by applying the SVM method to remove insignificant features and effectively find best parameter values. The main purpose of this work is to combine the SVM classifiers along with the attribute evaluators for achieving maximum accuracy in classification. And also from the experimental results, observed that the classifier subset evaluator is best suitable for optimization.

# References

1. Huang, C.-L., Chen, M.-C., Wang, C.-J.: Credit scoring with a data mining approach based on support vector machines. Expert Syst. Appl. **33**, 847–856 (2007)
2. Yang, Z., Su, X.: Customer behavior clustering using SVM. Int. Conf. Med. Phys. Biomed. Eng. Elsevier Physics Proc. **33**, 1489–1496 (2012)
3. Nematzadeh Balagatabi, Z.: Comparison of decision tree and SVM methods in classification of researcher's cognitive styles in academic environment. Int. J. Autom. Artif. Intell. **1**(1) (2013). ISSN:2320–4001
4. Qi, Z., Tian, Y., Shi, Y.: Structural twin support vector machine for classification. Knowl.-Based Syst. **43**, 74–81 (2013). doi:10.1016/j.knosys.2013.01.008
5. Maldonado, S., Weber, R., Basak, J.: Simultaneous feature selection and classification using Kernel-penalized SVM for feature selection. Inform. Sci. **181**(1), 115–128 (2011). doi:10.1016/j.ins.2010.08.047
6. Song, L., Smola, A., Gretton, A., Bedo, J., Borgwardt, K.: Feature selection via dependence maximization. J. Mach. Learn. Res. **13**(1), 1393–1434 (2012)
7. Yu, H., Kim, J., Kim, Y., Hwang, S., Lee, Y.H.: An efficient method for learning nonlinear ranking SVM functions. Inform. Sci. **209**, 37–48 (2012). doi:10.1016/j.ins.2012.03.022
8. Maldonado, S., Lopez, J.: Imbalanced data classification using second-order cone programming support vector machines. Pattern Recogn. **47**(5), 2070–2079 (2014). doi:10.1016/j.patcog.2013.11.021
9. Carrizosa, E., Martín-Barragán, B., Romero-Morales, D.: Detecting relevant variables and interactions in supervised classification. Eur. J. Oper. Res. **213**(16), 260–269 (2014). doi:10.1016/j.ejor.2010.03.020
10. Hassan, R., Othman, R.M., Saad, P., Kasim, S.: A compact hybrid feature vector for an accurate secondary structure prediction. Inform. Sci. **181**, 5267–5277 (2011). doi:10.1016/j.ins.2011.07.019
11. Patel, K., Vala, J., Pandya, J.: Comparison of various classification algorithms on iris datasets using WEKA. Int. J. Adv. Eng. Res. Dev. (IJAERD) **1**(1) (2014). ISSN:2348–4470
12. Tian, Y., Shi, Y., Liu, X.: Recent advances on support vector machines research. Technol. Econ. Dev. Econ. **18**(1), 5–33 (2012)
13. Victo Sudha George, G., Cyril Raj, V.: Review on feature selection techniques and the impact of SVM for cancer classification using gene expression profile. Int. J. Comput. Sci. Eng. Surv. **2**(3), 16–27 (2011)
14. Qi, Z.Q., Tian, Y.J., Shi, Y.: Robust twin support vector machine for pattern classification. Pattern Recogn. 305–316 (2013)

# Design and Development of Controller Software for MAS Receiver Using Socket Programming

K. Haritha, M. Jawaharlal, P.M.K. Prasad
and P. Kalyan Chakravathi

**Abstract** This paper deals with software development for controlling EW (Electronic warfare) subsystem through Embedded system using socket programming. MAS receiver is used to monitor, analyze the communication intelligence signals in the scenario which will be used for Defense purpose. The TCP\IP based socket Application Programming Interface (API's) will be used to realize this application. Vxworks real-time operating system is essential for meeting the time critical application and performance requirement of any system. The RTOS system will also address computational, efficiency and preemptive priority scheduling that is demanded by the user.

**Keywords** Vxworks · Champ AV IV · Monitoring receiver · Windriver

## 1 Introduction

In Communications, there are so many signals received at receiver end, the required signal can be monitored at receiver end by using a MAS receiver. With this, the data can be monitor, analyze and intercepted. In this paper controller software is

K. Haritha (✉) · P.M.K. Prasad · P. Kalyan Chakravathi
GMR Institute of Technology, Rajam, India
e-mail: harithamadhuri6@gmail.com

P.M.K. Prasad
e-mail: mkprasad.p@gmrit.org

P. Kalyan Chakravathi
e-mail: kalyanecebujji@gmail.com

M. Jawaharlal
Defence Electronics Research Laboratory, Hyderabad, India
e-mail: jawaharlam@gmail.com

© Springer India 2016
H.S. Behera and D.P. Mohapatra (eds.), *Computational Intelligence in Data Mining—Volume 1*, Advances in Intelligent Systems and Computing 410, DOI 10.1007/978-81-322-2734-2_12

developed for controlling EW (Electronic warfare) system. The system consisting of CHAMP AV IV board and VxWorks Real time operating system. The IDE Tool wind river workbench from wind river systems is used for development of this application. VxWorks Real time operating system is used to prioritize task to attend the higher priority task in shortest possible time and resume the activity where the present task was interrupted. This application is implemented using Sockets which is used to create a Client-Server environment. This application will interact with user and should be able to send the commands and receive the data.

An Electronic warfare receiver is needed to detect the existence of hostile radar signals [1]. MAS (Monitoring and Analysis Subsystem) receiver will monitor the signals ranging from 20 to 1000 MHz and analyze the signal, it determines the corrected frequency, bandwidth and type of modulation. The type of modulation can be AM, FM, FSK, BPSK, QPSK. The receiver is connected through a LAN and controlled by real time application [2].

Three systems are used for this application, they are MAS (Monitoring and Analysis subsystem) receiver, VXWORKS system, Command control system. The Receiver monitors and analyze the signals from 20 to 1000 MHz [3]. The VXWORKS system is developed on CHAMP AV IV board to control the receiver and it is controlled by the commands given at the command control system. The user gives the commands from the command control system. All the three systems are connected using a Ethernet cable [4]. The data transfer is done using socket programming.

## 2    Socket Programming

Sockets provide the communication between two computers using TCP protocol. Socket is a combination of port numbers and IP Address [5]. For socket programming client and server program should be developed for communication. A server provides service to client; it will never initiate the service unless the client sends the request. A client sends the request to the server. Several numbers of clients can send request to the server, when the service is complete it terminates the communication.

## 2.1    Client Process Involves the Following Steps

- First the socket should be created using TCP() call.
- The address of the server should be connected to the socket using connect() call.
- The data is written on the socket using write() call and the data(send by the server) is read from the socket using read() call.

## 2.2 Server Process Involves the Following Steps

- Socket is create using TCP() call.
- The address of the server should bind to the socket using the bind() call.
- The connections are listened with the listen() call.
- Accept () call is used to accept a connection from the client.
- The data send by the client is read from the socket using read() call and the data is written on the socket using write() call.

# 3 Functional Block Diagram

See Fig. 1.

## 3.1 Command Control System

The command control system is a system where the user gives the command to the receiver through power pc board. According to the command the receiver will complete its works and the data will be sent back to the command control system for display.

## 3.2 Embedded System

The system uses the information of the previous signals, combined with on-board real time information obtained during the monitoring. The receiver is controlled by RTOS and the RTOS connected to Champ AV IV board through LAN and RS232. An VxWorks image is created in the RTOS while building and the image is dumped in the champ V IV board.

**Fig. 1** Block diagram

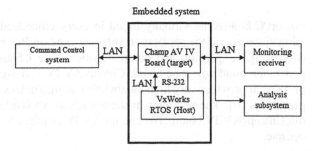

## 3.3 MAS (Monitoring and Analysis Subsystem) Receiver

MAS (Monitoring and Analysis Subsystem) receiver will monitor and analyze the signal and determines the corrected frequency, Bandwidth, Modulation classification [1]. The data is send to the control system based on the request of user. If the signal is AM or FM signal the data related to the signal such as modulation index, modulating signal and frequency deviation is analyzed. If the signal is FSK, BPSK, QPSK signal the data related to the signal such as phase shift, modulating signal, space frequency and mark frequency is analyzed.

# 4 Algorithm for Application Level

- Set the configuration parameters of MAS subsystem as per interface requirements software.
- The Embedded system will forward the commands to the MAS subsystem.
- The MAS subsystem will send the analyzed data to the embedded system.
- The Embedded system will forward the data to command control system for the display of received data.

# 5 Embedded System

An embedded system is a system which has software implanted into hardware that makes a system committed to an application. A real-time operating system (RTOS) has well-defined, fixed time constraints. Processing must be done within the defined constraints, or the system will fail. Real time systems are those systems in which the correctness of the system depends not only on the logical result of computation but also on the time at which the results are produced.

## 5.1 Hardware Development

PowerPC boards are virtually found in every embedded system. Real-Time operating systems and the application software runs on PowerPC because of its system control, I/O expansion and communication. The Champ-AV IV board is fourth generation quad power pc board. Champ-AV IV is designed for signal intelligence purpose. It provides very high bandwidth that maximizes the processing potential of processors [6]. The operating systems such as VxWorks, Linux are supported by the Champ-AV IV board. The champ AV IV requires 5 and 3.3 V power supply to operate.

The champ-AV IV base card is inserted into the chassis. Ethernet LAN and serial cable RS-232 are connected between the RTOS and board. Champ-AV IV provides four LED's on bottom side of board that provide link status and activity information about Ethernet port. Run the terminal and set the serial port for 57600 Baud Rate, 8 Data Bits, 1 Stop Bit, no Parity, no Flow Control. To access the features of embedded firmware on champ AV IV, a terminal must be attached to serial port on the card. The card contains four LED's represents the working of four processors and another one LED that represents the working/failure of board are situated on the front panel of the board. Power is supplied to the chassis and the board will display the configuration information and LED will be flash red. The LEDs will flash green and turn off to indicate a successful load of the On-Board Interrupt and Control. If the LED flash continuously red but information is not displayed, the serial port configuration parameters and connection should be checked. The flashing indicates the board has completed initialization and is in Boot Inhibit mode.

## 5.2 Software Development

VxWorks is a real-time operating system. VxWorks is used for hard real-time application. A Hard real-time application is hardware or software that should be operated with restrict deadline. To make easy and efficient use VxWorks supports powerful developments tools [7].

Multitasking and inter task communications are complementary concepts of Modern real-time systems. In real-time application a multitasking environment is allowed as independent tasks, each one has its own execution. Synchronization and communicating to coordinate the activity are the facilities are allowed in Inter tasking [8]. VxWorks is essential for meeting time critical application and performance requirement of any system. The RTOS system will also address computational, efficiency and preemptive priority scheduling that is demanded by the user [9].

Wind River Workbench 2.2 is a development suite for VxWorks. This provides the best support for cross development. Cross development is to write a process in a system called as Host and that process will run on another system called as target. A target has the resources for testing and debugging. The target is connected to the host using a Ethernet or a LAN. For this application cross development write the code in one system which runs on VxWorks called host system and then runs the code on a target board called Champ AV IV. The host is connected to the target using a LAN.

# 6 Results

The screenshots as in Figs. 2, 3, 4 and 5 indicate the results of the data that is monitored and analyzed. The data from Subsystem is send to the vxworks RTOS (target) for further process and it will be forwarded to command control system for display and for the use of information at the higher echelons. After the Command controller get acceptance from the embedded system the setting of parameters is displayed on the screenshot, when the start analysis command is given by the

```
agnitha_bj@ubuntu:~$ cd Desktop
agnitha_bj@ubuntu:~/Desktop$ cd mtech8
agnitha_bj@ubuntu:~/Desktop/mtech8$ gcc -o stc stc.c
agnitha_bj@ubuntu:~/Desktop/mtech8$ ./stc
connection is sucess
center frequency in MHz:21.4
frequency in mHz: 25
bandwidth in kHz: 6
analysis1
start analysis
msg = 2,length= 10,ack= 1
corrected freq=56
 estimated bw=20
mod=AM
modulating signal:voice
 modulating index 50
agnitha_bj@ubuntu:~/Desktop/mtech8$
```

**Fig. 2** Output of command control system

```
agnitha_bj@ubuntu:~$ cd Desktop
agnitha_bj@ubuntu:~/Desktop$ cd mtech8
agnitha_bj@ubuntu:~/Desktop/mtech8$ gcc -o sts sts.c
agnitha_bj@ubuntu:~/Desktop/mtech8$ ./sts
binding is done
accept is done
center frequency: 21.400000
freq 25
bandwidth 6
msg =1,length= 14
ack
1
cmd recevied
state:start analysis
corrected frequency value in MHz:
56
estimated bw value in kHz:
 20
mod
1
moulation type :AM
modulating signal:voice
modulating index:
50
agnitha_bj@ubuntu:~/Desktop/mtech8$
```

**Fig. 3** Output of monitoring and analysis receiver

```
agnitha_bj@ubuntu:~$ cd Desktop
agnitha_bj@ubuntu:~/Desktop$ cd mtech8
agnitha_bj@ubuntu:~/Desktop/mtech8$ ./stc
connection is sucess
center frequency in MHz:21.4
frequency in mHz: 25
bandwidth in kHz: 6
analysis1
start analysis
msg = 2,length= 10,ack= 1
corrected freq=72
 estimated bw=15
mod=FSK
modulating signal:data
 space freq in mhz: 32
  mark freq in mhz: 56
agnitha_bj@ubuntu:~/Desktop/mtech8$ ▮
```

**Fig. 4** Output of command control system

**Fig. 5** Output of monitoring
and analysis receiver

```
agnitha_bj@ubuntu:~$ cd Desktop
agnitha_bj@ubuntu:~/Desktop$ cd mtech8
agnitha_bj@ubuntu:~/Desktop/mtech8$ ./stc
connection is sucess
center frequency in MHz:21.4
frequency in mHz: 25
bandwidth in kHz: 6
analysis1
start analysis
msg = 2,length= 10,ack= 1
corrected freq=72
 estimated bw=15
mod=FSK
modulating signal:data
 space freq in mhz: 32
  mark freq in mhz: 56
agnitha_bj@ubuntu:~/Desktop/mtech8$ ▮
```

control system the simulated result from the receiver will be displayed on the screen of Command control are shown in Figs. 2 and 4. The command is accepted by the receiver, the start analysis command is read by the receiver and start the analysis of the signal i.e. corrected frequency, estimated bandwidth and type of modulation and that data is send to command control system, the process is shown in Figs. 3 and 5.

# 7 Conclusion

The proposed work is a part of software development for an Electronic warfare system. The main functionality of this software is to command and control various receiver subsystems. The software acts as server for various subsystems and provides necessary data for display. It is to keep track of the enemy even in the peacetime because unfortunately very important signals may not be able to track this will lead to war. The software for commanding and controlling the receiver is developed on TCP/IP protocol stack and has been tested with the simulated receivers on other computers.

# References

1. Tsui, J. B. Y.: Microwave Receivers with Electronic warfare Applications, 1 March 1986
2. Kennedy, G.: Electronic Communication System. Tata McGraw-Hill, (1999)
3. Poisel, R.: Introduction to Communication Electronic Warfare Systems, 1 Feb 2002
4. Stevens, R.: UNIX Network programming, vol. 1, third edition
5. Forouzan, Behrouz A.: TCP/IP Protocol Suite. Mc Graw-Hill, (2010)
6. Compact CHAMP-AV IV QUAD POWERPCTM (SCP-424) User Manual
7. Windriver System Inc, Vx works Programming Guide. Windriver systems, Inc, Alameda, CA (1999)
8. Peng, H., Shen, W., Song, Z.: A real-time management system of flight simulation based on VxWorks. J. Syst. Simul., Jul 2003
9. Windriver systems Inc, Tornado User's Guide. Windriver Systems, Inc, Alameda, CA (1999)

# Inference of Replanting in Forest Fire Affected Land Using Data Mining Technique

T.L. Divya and M.N. Vijayalakshmi

**Abstract** Forest fire is one of the natural calamities. The impact of fire on forest soil depends on its intensity and it also affects soil fertility, nutrients, and properties of soil like texture, color, and moisture content. Fire is beneficial and dangerous for soil nutrients based on its strength and it's duration. In less severity fire, fermentation of soil matter increases nutrients, which results in rapid growth of plants. The natural/artificial biodegradation of soil can rebuild the forest. In the proposed work the effort is made in prediction of possibility of re-plantation in previously forest fire affected area. The analysis and prediction of soil nutrients is done using Naive Bayesian classification.

**Keywords** Forest fire · Soil fertility · Nutrients · Replantation · Naive Bayes classification

## 1 Introduction

Nearly 65 % of world's geographical area is considered as forest. Every year due to forest fire 2 % of forest is under degradation which is of major concern. Occurrence of fire may be due to increase in population and urbanization, meteorological parameters like increase in temperature, Chemicals in air, Dry vegetation and so on. In India nearly 19 % of land is considered as forest and it is also under degradation. In order to protect and save the Nature an effort is made to analyze the fire affected land can be re-used for re-plantation. Table 1 shows forest degradation in India.

T.L. Divya (✉) · M.N. Vijayalakshmi
Department of MCA, R.V. College of Engineering, Bengaluru, India
e-mail: tl.divya@gmail.com

M.N. Vijayalakshmi
e-mail: mnviju74@gmail.com

© Springer India 2016
H.S. Behera and D.P. Mohapatra (eds.), *Computational Intelligence in Data Mining—Volume 1*, Advances in Intelligent Systems and Computing 410, DOI 10.1007/978-81-322-2734-2_13

**Table 1** Statistical information about Indian forest fire using advanced along track scanning radiometer (AATSR) between 2000 and 2010 [6] in Ha

| S. no | State | Jan | Feb | March | April | May | June | Total |
|-------|-------|-----|-----|-------|-------|-----|------|-------|
| 1 | Madhya Pradesh | 2 | 11 | 180 | 527 | 184 | 6 | 910 |
| 2 | Chhattisgarh | – | 205 | 318 | 363 | 52 | 1 | 759 |
| 3 | Maharashtra | 5 | 36 | 325 | 283 | 90 | – | 739 |
| 4 | Karnataka | 3 | 21 | 21 | 10 | 2 | – | 57 |
| 5 | Kerala | 3 | 14 | 21 | 2 | – | 1 | 41 |
| 6 | Andhra Pradesh | – | 42 | 149 | 56 | 32 | 4 | 283 |

## 1.1 Analysis of Soil for Productivity

Increase in intensity of fire affects nutrients contained in the soil, especially Nitrogen. The high temperature (more than 1000 °C) of fire will affect more than 10 cm of soil layer. Poor fire heat (with in 200 °C) results in loss of nutrients within 5 cm of the soil surface. The forest soil contains a lot of elements like nitrogen (N), phosphorous (p), sulphur (S), potassium (K), calcium (Ca), magnesium (Mg), etc. The soil properties that are located on the upper layer are likely to change by fire because they are in upper layers of soil surface with in the range of 5 cm. Nutrients, like Nitrogen (N), Phosphorus (P), and Sulphur (S), which have low temperature thresholds and evaporates [1].

Phosphorus will not appear in the soil profile like Nitrogen compounds. Hence, Phosphorous increases mainly in the ash and on, or near, the soil surface and in air. These Nitrogen and Phosphorous changes in soil are mainly analyzed during pre and post forest fire. The forest fire effects on Soil Organic Material (SOM) by reducing the amount of Nitrogen and Phosphorus. Other nutrients are less affected in comparison to nitrogen [2, 3]. From these chemicals Nitrogen can be recycled in the ecosystem for the growth of plants [3].

## 2 Related Work

Change in climatic conditions helps in decay of soil organic matter. Increase in soil decomposability, helps improving soil nutrients like Nitrogen, Phosphorus, and Zinc. As these nutrients are required for the healthy growth of plants, their data can be collected and prediction of replantation using data mining techniques [1].

Nitrogen is major nutrient that losses by volatilization. The process of growth can be done by fixing the nitrogen in the soil which is the main ingredients for the growth of plant [2, 3]. This fixing of nitrogen can be done in two ways, natural and synthetic manner. The synthetic manner uses microorganism for Nitrogen fixation.

Microorganisms that can fix nitrogen are called prokaryotes, which can be divided into two domain bacteria and archaea, called as diazotrophs. So by placing these microorganisms on the burned soils in the forests, fixing of nitrogen can be done which will help in growing of new plants in the forest areas [2, 3].

The growth of tree will depend on many factors like nutrient availability, the temperature, atmospheric Corban dioxide ($CO_2$). Usually tree growth are often found increased during summer seasons [4, 5]. The post forest fire environmental changes plays role in plants growth by changing soil organic matter [6]. Nitrogen enrichment effects plant productivity and related tree community composition [7, 8]. Soil nutrient and fertilizers requirement for the crops growth is analyzed by classification algorithm [9, 10].

The data mining classification algorithms helps in prediction of plants in a specific land based on the nutrition availability of the soil. The change in soil decomposability can be predicted using supervised algorithm [11, 12].

In the proposed work an effort is made to analyze minimum soil nutrients contents for plants growth in previously fire affected region. The soil organic matter like Dissolved Phosphorus Nitrogen, Ph, Carbon, and Temperature are used as attribute set. For the obtained dataset, Probabilistic classifier i.e. Naïve Bayes classification is applied in order to predict fire affected region eligibility for plants growth.

## 2.1 Naive Bayes Classification Approach

A Naive Bayesian algorithm is based on Bayesian theorem. It has independent assumptions for the occurrence of event. This classification uses conditional probabilities for the each class. This requires prior probability of each attribute. Classification produces posterior probability for the possible categories [3]. The patterns which required calculating maximum likely hood of replantation are detected by dataset. Classifier model depends on accurate class conditional probabilities. Most of the estimates are approximate which further biases the samples search space. Bayesian methods helps in estimation compared to maximum likelihood method. The internal representation of the classifiers data model uses a weighted term and best evaluate the purity of the labeled class.

For data set DS, posteriori probability of a hypothesis H, P(H|DS) follows the Bayes theorem.

Bayesian formula can be written as

$$P(H|DS) = P(DS|H)P(H)/P(DS)$$

The proposed model is Plant replantation is event predictor from the data DS.

## 3 Experiments with Results

Figure 1 shows proposed working model for classification of forest soil data set. The data is collected from the fire affected soil. The data will be sent for the prepossessing for data transformation. Bayesian classification is applied for analysis of pre-processed data.

Table 2 contains the required meteorological data for replantation. Table contains Ph level in the soil with available Phosphorous, Nitrogen, Soil moisture content, Dissolved nitrogen and carbon in the soil, temperature in the fire occurred region with target class for prediction i.e. Replantation (Binary value Yes/No).

Here it is assumed that the Bayesian learner has the sequence of training measurements from data set D. The attribute set considered are, Ph is greater than 0.1, Available Phosphorus is greater > 0.2, Nitrogen > 0.05, Moisture content > 0.1,

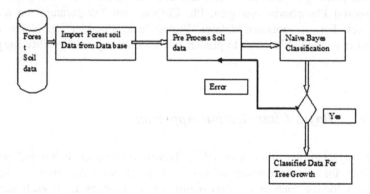

**Fig. 1** Block diagram of forest soil data classification

**Table 2** Sample soil data for the classification and prediction for 0–10 cm mineral soil [2]

| Ph | Available P (mg/kg) | Nitrogen (mg/kg) | Soil moisture content | Dissolved nitrogen (mg/kg) | Dissolved carbon (mg/kg) | Temp (°C) | Replantation |
|---|---|---|---|---|---|---|---|
| 0.48 | 0.292 | 0.08 | 0.35 | 0.248 | 0.66 | 25 | y |
| 0.016 | 0.667 | 0.09 | 0.116 | 0.547 | 0.672 | 28 | y |
| 0.56 | 0.137 | 0.01 | 0.672 | 0.185 | 0.0325 | 30 | n |
| 0.35 | 0.123 | 0.07 | 0.0122 | 0.45 | 0.475 | 31 | n |
| 0.17 | 0.145 | 0.079 | 0.572 | 0.447 | 0.354 | 36 | y |
| 0.18 | 0.15 | 0.081 | 0.475 | 0.123 | 0.3654 | 35 | y |
| 0.45 | 0.22 | 0.067 | 0.445 | 0.25 | 0.335 | 38 | y |
| 0.35 | 0.25 | 0.01 | 0.458 | 0.457 | 0.45 | 40 | n |
| 0.56 | 0.21 | 0.085 | 0.234 | 0.254 | 0.556 | 41 | n |
| 0.39 | 0.15 | 0.084 | 0.354 | 0.456 | 0.254 | 39 | y |

**Table 3** Prior probability of two target classes Cyes and Cno

| Prior probability class yes | Prior probability class no |
|---|---|
| P(Cyes) = P(Replantation = yes)/Total Instances | P(Cno) = P(Replantation = No)/Total Instances |
| 6/10 = 0.6 | 4/10 = 0.4 |

Dissolved Nitrogen in soil is greater than 0.2 and dissolved C > 0.3 and temperature > 15 °C.

For example, consider first record from the sample data set. Prior Probability for Class Yes and Class No is calculated and shown in Table 3.

Prior Probability of the remaining attributes are calculated as follows and shown in Table 4.

- Ph > 0.1/Replantation = Yes = 5/10 = 0.5 and Ph > 0.1/Replantation = No = 5/10 = 0.5
- Available P > 0.2/Replantation = Yes = 4/10 = 0.4 and Available P > Replantation = no = 6/10 = 0.6.
- N > 0.05/Replantation = Yes = 7/10 = 0.7 and N > 0.05/Replantation = No = 3/10 = 0.3
- Soil Moisture > 0.1/Replantation = Yes = 6/10 = 0.6 and Soil Moisture > 0.1/ Replantation = No = 4/10 = 0.4
- Dissolved N > 0.2/Replantation = Yes = 6/10 = 0.6 and Dissolved N > 0.2/ Replantation = No = 4/10 = 0.4
- Dissolved C > 0.3/Replantation = Yes = 5/10 = 0.5 and Dissolved C > 0.3/ Replantation = No = 5/10 = 0.5
- Temperature > 15/Replantation = Yes = 6/10 = 0.6 and Temperature > 15/ Replantation = No = 4/10 = 0.4
- Naive Bayes predicts whether tree re-growth is possible or not.

Maximum likely hood Prediction with Posterior Probability,

(Product of Target class = Yes)/P(CYes) That is, = 0.009/0.6 = 0.015.
(Product of Target class = No)/P(CNo) That is, = 0.001152/0.4 = 0.002.
ClassYes > ClassNo Hence Replantation is possible.

Figure 2 shows loading of soil data for prediction purpose. For each attribute the categorization is done, in order to classify data for conditional probabilities.

Figure 3 shows few attributes categorization for prediction. The data set contains 9 attributes. The internal representation of patterns with individual precision value.

Figure 4 shows the result of analysis. Here total 297 records were considered for experimental purpose.

Table 5 shows confusion matrix for Replantation possibility with Yes/No classification. The data set has 297 instances. The main diagonal values are 180 + 66 = 246 are correct classified data. Hence the success rate for classification is 246/297 = 0.82 that is 82 %.

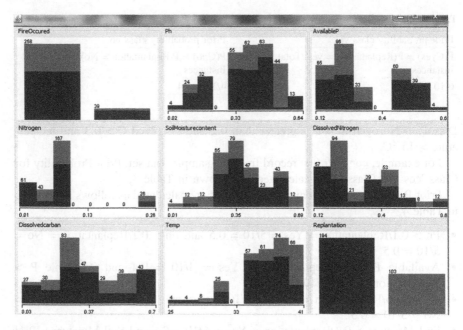

**Fig. 2** Visualization of imported data from database

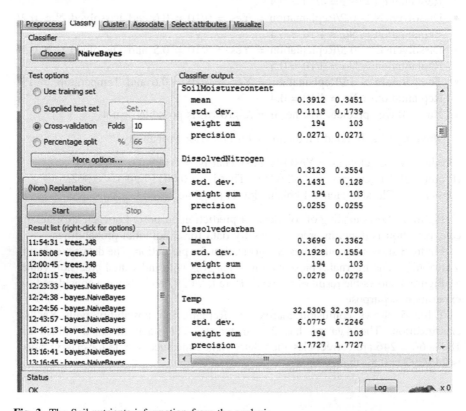

**Fig. 3** The Soil nutrients information from the analysis

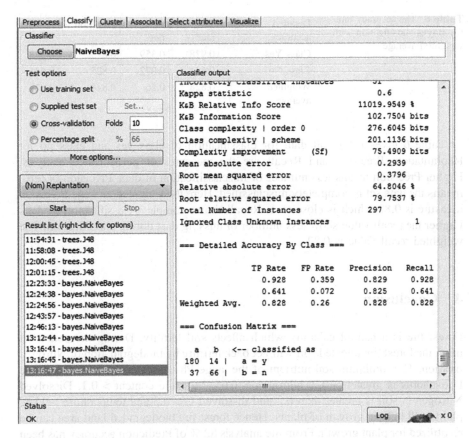

**Fig. 4** Cross validation by Naive Bayes classification

**Table 4** Calculation given in table format

| | Ph | Available P | Nitrogen (mg/kg) | Soil moisture content | Dissolved nitrogen (mg/kg) | Dissolved carbo n (mg/kg) | Temp (°C) | Replantation |
|---|---|---|---|---|---|---|---|---|
| 1 | 0.48 | 0.292 | 0.08 | 0.35 | 0.248 | 0.66 | 25 | Y |
| Class Yes | 0.5 | 0.4 | 0.7 | 0.6 | 0.6 | 0.5 | 0.6 | |
| Class No | 0.5 | 0.6 | 0.3 | 0.4 | 0.4 | 0.5 | 0.4 | |

**Table 5** The confusion matrix for replantation possibility

| Classification | Yes | No |
|---|---|---|
| Yes | 180 | 14 |
| No | 37 | 66 |

**Table 6** The accuracy rate for Bayes classification with weighted average

| Detailed accuracy | TP rate | FP rate | Precision | Recall |
|---|---|---|---|---|
| Class Yes | 0.928 | 0.359 | 0.829 | 0.928 |
| Class No | 0.641 | 0.072 | 0.825 | 0.641 |
| Weighted average | 0.828 | 0.26 | 0.828 | 0.828 |

Table 6 gives the result data for classification of class yes and class no for Replantation. Precision and Recall are the measures of performance analysis. Higher Precision means less number of misclassification of data. Precision 0(Zero) means the data set is completely misclassified. Here the weighted average Precision measure is 0.82 which is close to 1. Recall shows complete classification of data. Higher the recall value show less number of missing data during classification. The weighted recall value is 0.82.

# 4 Conclusion

Forest fire is a natural calamity which affects soil fertility. During the course of time, the forest fire affected soil will get back fertility by biodegradation or Nitrogen fixation. The minimum soil nutrient of the range Ph is greater than 0.1, Available Phosphorus is greater > 0.2, Nitrogen > 0.05, Moisture content > 0.1, Dissolved Nitrogen in soil is greater than 0.2 and dissolved C > 0.3 and temperature > 15°C plants will help in growth of plants. Hence forest fire biodegraded land area can be re-utilized for plant growth. From the analysis 82 % of Prediction accuracy has been achieved from the Bayes algorithm. The results obtained are efficient and acceptable.

# References

1. Xiang, X., Shi, Y., Yang, J., Kong, J., Lin, X., Zhang, H., Zeng, J., Chu, H.: Rapid recovery of soil bacterial communities after wild fire in a Chinese boreal forest, Scientific reports 4, Article Number 3829, doi: 1010381
2. Bowen, J.L., Ward, B.B., Morrison, H.G., Hobbie, J.E., Valiela, I.: Microbial community composition in sediments resists perturbation by nutrient enrichment. ISME **5**, 1540–1548 (2011)
3. Coolon, J.D., Jones, K.L., Todd, T.C., Blair, J.M., Herman, M.A.: Long-term nitrogen amendment alters the diversity and assemblage of soil bacterial communities in tall grass prairie (2013). doi: 10.137
4. Bbhalerao, O., Lobo, L.M.R.J.: Agricultural recommender using data mining techniques. Indian J. Appl. Res. **5**(4):849–851. ISSN: 2249-555X

5. Oren, R.: Differential responses to changes in growth temperature between trees from different functional groups and biomes: a review and synthesis of data. Tree Physiol. **30**:669–688 (2010). doi:10.1093/treephys/tpq015
6. Kitchen, D.J., Blair, J.M., Callaham, M.A. Jr: Annual fire and mowing alter biomass, depth distribution, and C and N content of roots and soil in tall grass prairie. Plant Soil 235–247 (2009)
7. Clark, C.M., Cleland, E.E., Collins, S.L., Fargione, J.E., Gough, L.: Environmental and plant community determinants of species loss following nitrogen enrichment. Ecol. Lett. **10**, 596–606 (2007)
8. Bunn, A.G., Graumlich, L.J., Urban, D.L.: Trends in twentieth-century tree growth at high elevations in the Sierra Nevada and White Mountains, USA. Holocene **15**, 481–488 (2005)
9. Bhuyar, V.: Comparative Analysis of classification techniques on soil data to predict fertility rate for Aurangabad district. Int. J. Emerg. Trends Technol. Comput. Sci. 3(2):200–203 (2014). ISSN-2278-6856
10. Veenadhari, S. Dr., Mishra, B. Dr., Singh, C.D.: Soybean productivity modeling using decision tree algorithms. Int. J. Comput. Appl. **27**(7) (2011)
11. Gholap, J.: Performance tuning of J48 algorithm for soil fertility. Asian J. Comput. Sci. Inf. Technol. **2**:251– 252 (2012)
12. Bharghavi, P., Jyothi, S.: Applying Naive Bayes data mining technique for classification of agricultural land soils. IJCSNS Int. J. Comput. Sci. Network Security **9**(8), 117–122 (2009)

# A New Approach on Color Image Encryption Using Arnold 4D Cat Map

**Bidyut Jyoti Saha, Kunal Kumar Kabi and Chittaranjan Pradhan**

**Abstract** To secure multimedia content such as digital color images, chaotic map can be used to shuffle or encrypt the original pixel positions of an image. A chaotic map such as Arnold cat map is used to encrypt images as the behavior is periodic and deterministic given the correct key values. Even a small change also cannot predict the behavior and properties of image. To secure digital color images, Arnold cat map is being used to encrypt and send it across public network. Various dimensions of Arnold cat map can be used to achieve greater level of secrecy and confusion. In this paper, our work has been extended to encrypt digital color images using Arnold 4D cat map. Here, the color image has been decomposed to its respective RGB planes. The plane's pixel positions and gray scale values are then shuffled using Arnold 3D cat map. Then the resulting shuffled encrypted planes (red, green and blue) are divided into 3 parts vertically respectively. The parts of each plane are shuffled in a specific proposed manner with the parts of other planes to achieve greater level of diffusion and secrecy. Each plane holds information of two other planes. The result and analysis shows better result than Arnold 3D cat map used in image encryption.

**Keywords** Encryption · Decryption · Chaotic map · Arnold 2D cat map · Arnold 3D cat map · Arnold 4D cat map · Image security

The original version of this chapter was revised: The erratum to this chapter is available at DOI 10.1007/978-81-322-2734-2_49

B.J. Saha (✉)
Industrial and Financial Systems, Capgemini, Mumbai, India
e-mail: bidyutjyotisaha@gmail.com

K.K. Kabi
Guidewire Practice, CGI, Bhubaneswar, India
e-mail: kunal.kabi90@gmail.com

C. Pradhan
School of Computer Engineering, KIIT University, Bhubaneshwar, India
e-mail: chitaprakash@gmail.com

© Springer India 2016                                                                                                131
H.S. Behera and D.P. Mohapatra (eds.), *Computational Intelligence in Data Mining—Volume 1*, Advances in Intelligent Systems and Computing 410, DOI 10.1007/978-81-322-2734-2_14

# 1 Introduction

In real time, multimedia content is being propagated and shared as a piece of information across multiple untrusted and unsecured public network. As a result of huge transfer of multimedia content across these networks, it is attracting the attention of intended attackers. The attacker most of the time tries to compromise the content and hence the originality and authenticity is on sake. Image encryption techniques have been proposed across last few decades to avoid unauthorized reception and interruption to the content. Even after the presence of such complex algorithms and encryption techniques, attacker tries to alter the original contents which communicate through network data packets. This interception and alteration of content incurs loss to authorized source and reception. Thus, researchers since last few decades are trying to discover new ways to bypass their alteration or to introduce certain methods to recover the actual multimedia content by the means of certain algorithm.

Arnold cat map is one of the shuffling and encryption algorithm, which either shuffles the pixel positions or the intensity value of a given image. It is one of the popular mechanisms to produce chaotic behavior among all other existing techniques [1].

Arnold cat map has been extended for image encryption to operate on various dimensions to increase the robustness, strength and recoverability [2, 3]. Recently, Arnold 3D map has been introduced to encrypt color images in order to make the transformation mechanism more robust against various differential attacks [4].

In 2010, an improved image encryption algorithm has been proposed by Ding et al. using improved Arnold transform [5]. In 2011, Jin et al. has proposed a digital watermarking algorithm based on Arnold transform [6]. In 2012, Ye et al. proposed an improved technique using generalized Arnold for the image encryption [7]. Similarly, in 2012 Pradhan et al. proposed an imperceptible digital watermarking using 2D Arnold map as well as cross chaos map [8]. Kabi et al. improved the existing Arnold 3D cat map algorithm to encrypt images and showed the differential attack analysis which shows better results than its variation [4].

Our proposed algorithm shows higher level of secrecy in terms of parameters and intensity values. It's nearly impossible to recover the image unless actual keys supplied. Here, we have extended Arnold 3D cat map encryption technique to Arnold 4D cat map. Section 2 describes Arnold 3D cat map. Section 3 shows enhanced Arnold 4D cat map technique and the result analysis of the image encryption using Arnold 4D cat map. Conclusion and future work has been presented in Sect. 4.

## 2 Arnold 3D Cat Map

Arnold 3D cat map is based on its 2D variation. 3D cat map is used to encrypt image which is deterministic and iterates over certain period to reproduce the original image. The earlier approach has also been used to encrypt gray scale image. The following equation is used for Arnold 3D cat map [4]:

$$\begin{cases} \begin{pmatrix} x' \\ y' \end{pmatrix} = \begin{pmatrix} 1 & a \\ b & ab+1 \end{pmatrix} * \begin{pmatrix} x \\ y \end{pmatrix} mod\, N \\ f' = (cx + dy + f) mod\, M \end{cases} \tag{1}$$

Here, $(x, y)$ is the original image and $(x', y')$ is the resulting image after 2D transformation. 'a' and 'b' are two parameters whose values are supplied by authorized user to encrypt. The same parameters are used to decrypt the scrambled image back to original image. After 2D shuffling of images, the intensity values are changed by passing another couple of user supplied parameters 'c' and 'd'. Parameter 'f' is the resultant intensity value depending upon the original intensity value. It combines both the technique of shuffling and substituting.

## 3 Proposed Approach

### 3.1 Arnold 4D Cat Map

The proposal of this paper is Arnold 4D cat map to encrypt digital color images. The Arnold 4D map is an enhancement over Arnold 3D cat map in order to encrypt the color image in a more secure and efficient way. Similar to Arnold 3D map, Arnold 4D cat map also adopt its shuffling and substituting phases. Apart from that, one more dimension has been added to increase the robustness and authenticity. The fourth dimension is focused on diffusion mechanism of different RGB planes. The Arnold 4D cat map follows the same encryption technique as Arnold 3D map on each plane and the equation which represents Arnold 4D cat map on each plane is also same as Arnold 3D cat map as shown in Eq. (1), where, 'N' is the size of image and 'M' is the color level.

Here, the Arnold cat map shuffles the pixel positions and it substitutes gray values depending on the positions and original gray values of pixels with 3D Arnold cat map. The color image is separated into its respective RGB planes ($R$, $G$, and $B$). Each plane is divided vertically into 3 parts such as ($R1$, $R2$, $R3$), ($G1$, $G2$, $G3$), ($B1$, $B2$, $B3$) so that ($R1 = G1 = B1$), ($R2 = G2 = B2$) and ($R3 = G3 = B3$) (*in terms of dimension*).

These planes undergoes shuffling and substitution phases as per Eq. (1) and gives the result as ($R1'$, $R2'$, $R3'$), ($G1'$, $G2'$, $G3'$), ($B1'$, $B2'$, $B3'$). The resulting planes after Arnold 3D cat map encryption, are permuted and gives result as ($R1'$, $G2'$, $B3'$), ($G1'$, $B2'$, $R3'$), ($B1'$, $R2'$, $G3'$). The resultant planes so formed are combined to generate final encrypted image.

## 3.2 Steps for Encryption

The different steps of our encryption process are:

1. The original color image has been taken as input and divided into its respective RGB planes.
2. These planes undergoes encryption process of shuffling and substitution as per pixel information and gives ($R1'$, $R2'$, $R3'$), ($G1'$, $G2'$, $G3'$), ($B1'$, $B2'$, $B3'$) as per Eq. (1).
3. The parts of each plane is shuffled into alternate parts of other planes. The resulting planes are then permuted in a given manner ($R1'$, $G2'$, $B3'$), ($G1'$, $B2'$, $R3'$), ($B1'$, $R2'$, $G3'$).
4. The final resulting encrypted planes are combined to form final encrypted color image, which is nearly impossible to perceive.

The schematic diagram is shown in Fig. 1. Figure 2 shows the proposed encryption process.

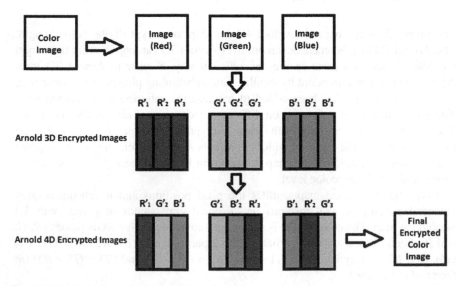

**Fig. 1** Schematic diagram of Arnold 4D cat map

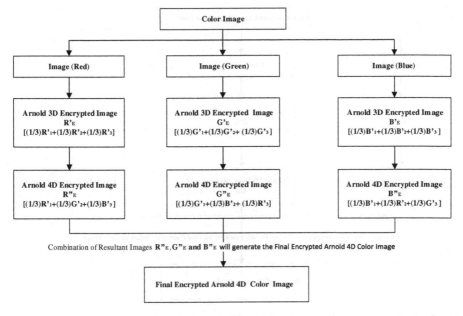

**Fig. 2** Block diagram of encryption process of Arnold 4D cat map


## 3.3 Steps for Decryption

The different steps involved in Arnold 4D cat map decryption are given below. Note that the same encryption keys and initial parameters should be used to encrypt and decrypt back to original image.

1. The final encrypted color image is taken as input to decrypt it back to original image.
2. The encrypted image is again divided into its respective RGB planes $(R, G, B)$.
3. Each of these planes are divided vertically into 3 parts resulting in $(R1', G2', B3'), (G1', B2', R3'), (B1', R2', G3')$.
4. These planes are shuffled and permuted back to $(R1', R2', R3'), (G1', G2', G3'), (B1', B2', B3')$.
5. These planes undergoes Arnold 3D cat map decryption process $(R1, R2, R3), (G1, G2, G3), (B1, B2, B3)$.
6. The decrypted planes are then combined to form original image. The decryption process is shown in Fig. 3.

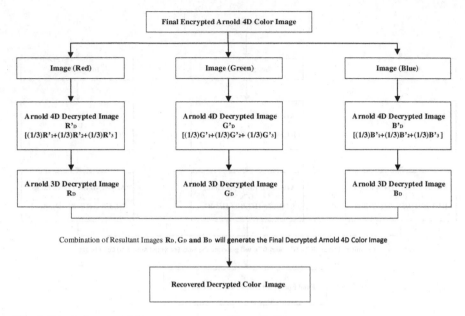

**Fig. 3** Block diagram of decryption process of Arnold 4D cat map

## 3.4 Result Analysis

For simulation and experiments, MATLAB 2014a has been used. 'Lena.png' of size $512 \times 512$ has been taken as color image and shown in Fig. 4a. Figure 4b shows the individual RGB components of color image. The individual components of color image are encrypted using 3D Arnold map with different secret keys as shown in Fig. 4c. Figure 4d shows the resultant images after Arnold 4D. The resulting encrypted images are further concatenated to produce final encrypted Arnold 4D colored image and shown in Fig. 4e.

Similarly, for decryption process the final encrypted colored image is decomposed into its respective RGB components shown in Fig. 4f. The Arnold 4D Decrypted images are shown in Fig. 4g. Figure 4h shows the results of different components of image after Arnold 3D decryption process. The combined image of different RGB components is shown in Fig. 4i, which gives the final color image.

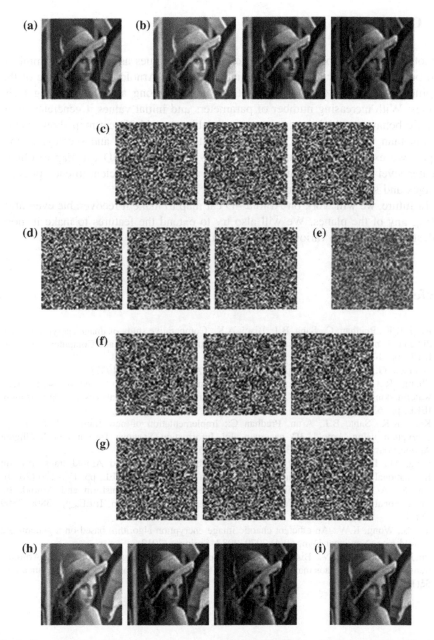

**Fig. 4** Results of Arnold 4D cat map. **a** Color image. **b** *Red*, *Green* and *Blue* component of color image. **c** Arnold's 3D encrypted image (R′$_E$, G′$_E$ and B′$_E$). **d** Resultant Arnold's 4D encrypted image (R″$_E$, G″$_E$ and B″$_E$). **e** Final encrypted Arnold 4D color image. **f** *Red*, *Green* and *Blue* component of color image (R′$_E$, G′$_E$ and B′$_E$). **g** Arnold's 4D decrypted image (R″$_E$, G″$_E$ and B″$_E$). **h** Arnold's 3D decrypted image (R″$_E$, G″$_E$ and B″$_E$). **i** Recovered color image (color online)

## 4 Conclusion

Chaotic sequence is being used in encryption techniques as it poses deterministic and periodic behavior for specific initial parameters. Arnold cat map is one of the popular mechanism to produce chaotic behavior among all other existing techniques. With increasing number of parameters and initial values it generates more chaotic behavior, which is almost unpredictable. Arnold 3D cat map shows better results than its variations in terms of robustness, efficiency and secrecy. In this paper, we extended and enhanced the behavior of Arnold 3D cat map to obtain greater level of robustness and secrecy. This algorithm is efficient to encrypt color images and shows better results than Arnold 3D cat map.

In future, we will work on Arnold 4D cat map to make it recoverable even after losing any of the planes. We will also try to extend the features to make it more indestructible against various differential attacks.

## References

1. Kabi, K.K., Pradhan, C., Saha, B.J., Bisoi, A.K.: Comparative study of image encryption using 2D chaotic map. In: International Conference on Information Systems and Computer Networks, IEEE, pp. 105–108 (2014)
2. Peterson, G.: Arnold's Cat Map, Math 45-Linear Algebra, pp. 1–7 (1997)
3. Zhang, R., Wang, H., Wang, Y.: A novel image authentication based on semi-fragile watermarking. In: International Joint Conference on Computational Sciences and Optimization, IEEE, pp. 631–634 (2012)
4. Kabi, K.K., Saha, B.J., Arun, Pradhan C.: Implementation of new framework for image encryption using Arnold 3D cat map. In: Information Systems Design and Intelligent Applications, Vol. 1, pp. 379–384. Springer (2015)
5. Ding, M.: Digital image encryption algorithm based on improved Arnold transform. In: International Forum on Information technology and Applications, IEEE, pp. 174–176 (2010)
6. Jin, X.: A digital watermarking algorithm based on wavelet transform and Arnold. In: International Conference on Computer Science and Service System, IEEE, pp. 3806–3809 (2011)
7. Ye, G., Wong, K.W.: An efficient chaotic image encryption algorithm based on a generalized Arnold map. Nonlinear Dyn. 69(4):2079–2087. Springer (2012)
8. Pradhan, C., Saxena, V., Bisoi, A.K.: Imperceptible watermarking technique using Arnold's transform and cross chaos map in DCT domain. International Journal of Computer Applications 55(15), 50–53 (2012)

# Neutrosophic Logic Based New Methodology to Handle Indeterminacy Data for Taking Accurate Decision

**Soumitra De and Jaydev Mishra**

**Abstract** We proposed a method using Neutrosophic set and data to take the most suitable decision with the help of three different members truth, indeterminacy and falsity. Neutrosophic logic is capable of handling indeterministic and inconsistent information. We have focused to draw a meaning full outcome using neutrosophic concept about the illness of a patient who is suffering from a disease. Fuzzy logic can only handle incomplete information using truth membership value. Vague logic also can handle incomplete information using truth and false membership values. It is an upgrade part of fuzzy and vague concept.

**Keywords** Neutrosophic logic · Neutrosophic set · Neutrosophic operation · New method

## 1 Introduction

Some of the problems including uncertainties are a major issue in different fields of real life such as engineering, social sciences, medical sciences and business management. Uncertain data in these fields could be caused by difficulties in classical mathematical modeling. The fuzzy sets and vague set are applied in various real problems in uncertain and incomplete information based environment. The fuzzy sets [1–4] are signature by the truth membership value. Few cases it may be very difficult to assign the membership value for fuzzy sets. In some of our daily

S. De (✉) · J. Mishra
Computer Science & Engineering Department, College of Engineering
and Management, Kolaghat, 721171 Purba Medinipur, West Bengal, India
e-mail: soumitra@cemk.ac.in

J. Mishra
e-mail: jsm03@cemk.ac.in

© Springer India 2016
H.S. Behera and D.P. Mohapatra (eds.), *Computational Intelligence in Data Mining—Volume 1*, Advances in Intelligent Systems and Computing 410, DOI 10.1007/978-81-322-2734-2_15

problems in different area, we have used the truth and false membership for proper explanation of an entity in unpredictable and incomplete information based environment. The vague set [5–10] is only used for incomplete information based operation considering the truth and false membership values. It does not handle the indeterminacy oriented information which presents in our daily life. Smarandache [11–13] first proposed the neutrosophic set and data which are used for dealing problems involving imprecise data. Recent research works on neutrosophic set theory and its applications in various fields are progressing rapidly. A lot of literature can be found in this regard in [14–17]. Decision making can be more accurate in neutrosophic logic due to handling indeterminacy membership. We are introduced a new method using neutrosophic logic to take better decision rather than fuzzy or vague based system. Here we are reducing more confusion by using indeterminacy data and we got an optimal solution. We are basically applying the concept of newly introduced neutrosophic set to solve the pathological problem by making a new methodology.

## 2 Basic Concepts

The truth, indeterminacy and falseness values were applied to investigate the benefits of one person affected by another. The combination of evidence to find a probabilistic model. Any function which is based on belief is a link between incomplete reasoning and probability. Probability uses the mathematics in a simple way, and is based on evidence which is probabilistic in nature.

This kind of model usually applicable for data which are defined very clearly. So, when we discuss regarding the indeterminacy data, it will not give any answer. Therefore, to signify and control different types of imperfect sequence of data, this new methodology will help us to draw a decision.

### 2.1 Example

"Today is sunny day" is a proposition which is represents by neutrosophic data such as at time $t_1$ truth, indeterminacy and false percentages are 30, 55 and 40 %, at time $t_2$ these percentages will alter to 57, 42 and 34 % accordingly. So, the truth, indeterminacy and false values are changing from time to time in dynamic nature.

# 3 Basic Definitions of Neutrosophic Set

Here we are focusing on some few definitions.

## 3.1 Definition

A neutrosophic set X on U is defined as $X = \{\langle a_1, T_X(a_1), I_X(a_1), F_X(a_1)\rangle, a_1 \in U\}$, where $T_X(a_1) \to [0,1]$; $I_X(a_1) \to [0,1]$; $F_X(a_1) \to [0,1]$ and $T_X(a_1) + I_X(a_1) + F_X(a_1) \le 2$ and U is the discourser of world.

From realistic observation, the neutrosophic set will consider the data from the closed interval of $[0, 1]$. Truth or False membership value cannot exceed 1 and if anyone is zero (0) other may reach maximum one (1). Here separately indeterminacy membership value also can reach maximum 1, which will play crucial role for taking proper decision.

## 3.2 Definition

X and Y both are neutrosophic sets and $X \subseteq Y$ if $a \in U$ for all cases, $T_X(a) \le T_Y(a)$, $I_X(a) \le I_Y(a)$, $F_X(a) \ge F_Y(a)$.

### 3.2.1 Example

Considering $U_1$ as the universe of discourse which is represents as $U_1 = \{u_1, u_2, u_3\}$, where $u_1$ is signifies the capability, $u_2$ is signifies the reliability and $u_3$ is signifies availability. We are assuming that the values of $m_1$, $m_2$ and $m_3$ are obtained from some judges. The judges may impose their opinion using truth, indeterminacy and false values of neutrosophic representation which will explain the nature of the entities. Consider $B_1$ is a Neutrosophic Set (N S) of U1, which is shown in Table 1.

### 3.2.2 Example

Now consider books $(U_1)$ and parameters (P) two sets. Consider $P = \{$physics, chemistry, biology$\}$. Suppose that, there are three books in $U_1$ which is given by,

**Table 1** Neutrosophic set of universe of discourse with T, I, F values

| Neutrosophic set name | Neutrosophic data of $U_1$ | Inference of capability |
|---|---|---|
| $B_1$ | $\langle u_1, 0.6, 0.5, 0.3\rangle$, $\langle u_2, 0.8, 0.4, 0.2\rangle$, $\langle u_3, 0.4, 0.5, 0.6\rangle$ | Truth, indeterminacy and false membership values are 0.6, 0.5 and 0.3 |

**Table 2** Neutrosophic set with parameter names which are containing neutrosophic values (T, I, F) of three existing books

| Neutrosophic set with parameter name | Neutrosophic representation of three books |
|---|---|
| B(physics) | $\langle b, 0.5, 0.7, 0.3\rangle$, $\langle c, 0.3, 0.7, 0.5\rangle$, $\langle d, 0.5, 0.2, 0.3\rangle$ |
| B(chemistry) | $\langle b, 0.8, 0.4, 02\rangle$, $\langle c, 0.7, 0.4, 0.3\rangle$, $\langle d, 0.6, 0.5, 0.3\rangle$ |
| B(biology) | $\langle b, 0.5, 0.4, 0.3\rangle$, $\langle c, 0.3, 0.5, 0.6\rangle$, $\langle d, 0.6, 0.7, 0.3\rangle$ |

$U_1 = \{b, c, d\}$ and the set of parameters $P = \{p_1, p_2, p_3\}$, where $p_1$ stands for the parameter 'physics', $p_2$ stands for the parameter 'chemistry', $p_3$ stands for the parameter 'biology'. We are expressing the knowledge from Table 2.

The neutrosophic set (NS) (B, P) is a parameterized family $\{B\ (p_i)$; i should vary from 1 to 10$\}$ of all neutrosophic sets of $U_1$. Here B is book and $(\cdot)$ is to be filled up by a parameter $p \in P$. So B $(p_3)$ means Book (biology) whose functional value is the neutrosophic set $\{\langle b, 0.5, 0.4, 0.3\rangle, \langle c, 0.3, 0.5, 0.6\rangle, \langle d, 0.6, 0.7, 0.3\rangle\}$.

## 3.3   Definition

Let (C, X) and (D, Y) are the NSs in $U_1$ and the union of these two sets represents as $(C, X) \cup (D, Y) = (S, Z)$, where $Z = X \cup Y$ and the membership of truth, indeterminacy and falseness of (S, Z) given below

$T_S(m) = \text{maximum } \{T_C(m), T_D(m)\}$,
$F_S(m) = \text{minimum } \{T_C(m), T_D(m)\}$,
$I_S(m) = $ If max of common column indeterminacy value within the range of true and false value in that (T, I, F) value set then consider it finally. Otherwise final indeterminacy value would be $\{(I_C(m) + I_D(m))/2\}$.

## 3.4   Example

(See Tables 3, 4 and 5).

**Table 3** For NS of (C, X)

| Book | Physics | Chemistry | Biology |
|---|---|---|---|
| $b_1$ | (0.4,0.3,0.6) | (0.5,0.4,0.4) | (0.6,0.2,0.2) |
| $b_2$ | (0.7,0.4,0.3) | (0.3,0.8,0.6) | (0.5,0.7,0.4) |
| $b_3$ | (0.6,0.5,0.4) | (0.4,0.6,0.3) | (0.3,0.5,0.6) |

**Table 4** For NS of (D, Y)

| Book | Chemistry | Mathematics |
|------|-----------|-------------|
| $b_1$ | (0.8,0.4,0.2) | (0.7,0.6,0.2) |
| $b_2$ | (0.4,0.4,0.5) | (0.5,0.3,0.4) |
| $b_3$ | (0.5,0.7,0.4) | (0.3,0.4,0.7) |

**Table 5** For NS of (S, Z)

| Book | Physics | Chemistry | Biology | Mathematics |
|------|---------|-----------|---------|-------------|
| $b_1$ | (0.4,0.3,0.6) | (0.8,0.4,0.2) | (0.6,0.2,0.2) | (0.7,0.6,0.2) |
| $b_2$ | (0.7,0.4,0.3) | (0.4,0.6,0.5) | (0.5,0.7,0.4) | (0.5,0.3,0.4) |
| $b_3$ | (0.6,0.5,0.4) | (0.5,0.65,0.3) | (0.3,0.5,0.6) | (0.3,0.4,0.7) |

**Table 6** For NS of (S, Z)

| Book | Chemistry |
|------|-----------|
| $b_1$ | (0.5,0.4,0.4) |
| $b_2$ | (0.3,0.6,0.6) |
| $b_3$ | (0.4,0.65,0.4) |

## 3.5  Definition

Let (C, X) and (D, Y) are the NSs in $U_1$ and the intersection of these two sets represents as (C, X) ∩ (D, Y) = (S, Z), where Z = X ∩ Y and the membership of truth, indeterminacy and falseness of (S, Z) given below

$T_S(m) = \text{minimum}\{T_C(m), T_D(m)\}$,
$F_S(m) = \text{maximum}\{T_C(m), T_D(m)\}$,
$I_S(m) = \text{maximum}\{\text{minimum } (I_C(m), I_D(m)), (I_C(m) + I_D(m))/2\}$.

## 3.6  Example

Using the Tables 3 and 4 we will get the Table 6 of NS (S, Z).

## 4  New Method of Neutrosophic Logic for Taking Accurate Decision

### 4.1  Method

Step 1   If total number of '+ve' results more than the total number of '−ve' results then we can say that patient is suffering from cancer disease.
If step 1 fails then step 2 will consider

Step 2   First calculate 'T' value greater equal to 'I' value and 'T 'value greater than 'F' value for each laboratory data.

If the above condition is true at least 50 % then we can say that patient is suffering from cancer disease.

If step 2 fails then step 3 will consider

Step 3   If total number of '+ve' differences between (T, I) greater than total number of '+ve' differences between (F, I) then we can say that patient is suffering from cancer disease.

If step 3 fails then step 4 will consider

Step 4   If summation of T values + summation of I values greater than summation of F values + summation of I values then we can say that patient is suffering from cancer disease.

If any step is true then we can conclude that patient is definitely suffering from cancer disease then we have to show the accuracy of decision percentage (%) of affected patient.

If all steps are true the n we can say patient is 100 % affected by cancer.

If three steps are true the n we can say patient is 75 % affected by cancer.

If two steps are true the n we can say patient is 50 % affected by cancer.

If only step is true the n we can say patient is 25 % affected by cancer.

Otherwise we can say that patient is totally free from cancer disease.

### 4.1.1   Problem

We have taken a pathological problem to take the decision whether the patient is suffering from cancer or not. We have taken different laboratory based data in neutrosophic format (T, I, F) and analyzing these data with the above method to confirm the status of cancer patient (Table 7).

**Table 7**  Problem data

| Patient | Laboratory data (T, I, F) |
|---------|---------------------------|
| $P_1(Lab_1)$ | (0.5,0.6,0.4) |
| $P_1(Lab_2)$ | (0.8,0.3,0.2) |
| $P_1(Lab_3)$ | (0.3,0.4,0.6) |
| $P_1(Lab_4)$ | (0.6,0.5,0.4) |
| $P_1(Lab_5)$ | (0.5,0.7,0.5) |

**Solution:**

$P_1(Lab_1) = \max(0.5, 0.6, 0.4) = 0.6$ (Indeterminacy) (+ve/−ve)
$P_1(Lab_2) = \max(0.8, 0.3, 0.2) = 0.8$ (+ve)
$P_1(Lab_3) = \max(0.3, 0.4, 0.6) = 0.6$ (−ve)
$P_1(Lab_4) = \max(0.6, 0.5, 0.4) = 0.6$ (+ve)
$P_1(Lab_5) = \max(0.5, 0.7, 0.5) = 0.6$ (Indeterminacy) (+ve/−ve)

Total number of Indeterminacy (I) = 2
Total number of Truthness (T) = 2
Total number of Falseness (F) = 1

Step 1   Total number of Truthness (T) > Total number of Falseness (F) that is
2 > 1 is true
So patient is suffering from cancer disease.

Step 2   First calculate 'T' value ≥ 'I' value and 'T 'value > 'F' value for each
laboratory data.
If the above condition is true at least 50 % then we can say patient is
suffering from cancer disease.

$P_1(Lab_1)$ = (0.5, 0.6, 0.4); 0.5 ≥ 0.6(f) and 0.5 > 0.4(t)
$P_1(Lab_2)$ = (0.8, 0.3, 0.2); 0.8 ≥ 0.3(t) and 0.8 > 0.2(t)
$P_1(Lab_3)$ = (0.3, 0.4, 0.6); 0.3 ≥ 0.4(f) and 0.3 > 0.6(f)
$P_1(Lab_4)$ = (0.6, 0.5, 0.4); 0.6 ≥ 0.5(t) and 0.6 > 0.4(t)
$P_1(Lab_5)$ = (0.5, 0.7, 0.5); 0.5 ≥ 0.7(f) and 0.5 > 0.5(f)

Here above condition is not true for at least 50 %.
So we can say patient is not suffering from cancer disease for step 2
process.

Step 3   If summation of '+ve' differences between (T, I) > summation of '+ve'
differences between (F, I) then we can say that patient is suffering from
cancer disease.

$P_1(Lab_1)$ = (0.5, 0.6, 0.4); T − I = 0.5 − 0.6 (−ve),   F − I = 0.4 − 0.6 (−ve)
$P_1(Lab_2)$ = (0.8, 0.3, 0.2); T − I = 0.8 − 0.3 = 0.5 (+ve),   F − I = 0.2 − 0.3 (−ve)
$P_1(Lab_3)$ = (0.3, 0.4, 0.6); T − I = 0.3 − 0.4 (−ve),   F − I = 0.6 − 0.4 = 0.2 (+ve)
$P_1(Lab_4)$ = (0.6, 0.5, 0.4); T − I = 0.6 − 0.5 = 0.1 (+ve),   F − I = 0.4 − 0.5 (−ve)
$P_1(Lab_5)$ = (0.5, 0.7, 0.5); T − I = 0.5 − 0.7 (−ve),   F − I = 0.5 − 0.7 (−ve)

Summation of +ve differences between (T, I) = 0.5 + 0.1 = 0.6
Summation of +ve differences between (F, I) = 0.2 that is, 0.6 > 0.2
So we can say that patient is suffering from cancer disease.

Step 4   If summation of T values + summation of I values > summation of F
values + summation of I values then we can say that patient is suffering
from cancer disease.
From the Table 1 we can calculate these values.

Total values of T (including all lab data) = 0.5 + 0.8 + 0.3 + 0.6 + 0.5 = 2.7
Total values of I (including all lab data) = 0.6 + 0.3 + 0.4 + 0.5 + 0.7 = 2.5
Total values of F (including all lab data) = 0.4 + 0.2 + 0.6 + 0.4 + 0.5 = 2.1
From the step 4 condition we can write that, 2.7 + 2.5 > 2.5 +
2.1 ≡ 5.2 > 4.6

So we can say that patient is suffering from cancer disease.

At last we can take the decision that the patient has been suffering from cancer disease (75 %).

# 5 Conclusion

A new method using neutrosophic data has been introduced and we applied this method to solve the medical problem and to take appropriate decision about the suffering patient, with the neutrosophic membership values of truth, indeterminacy and falseness. In different field of application the neutrosophic data will be used in a database framework for query based operation on which our next research work has been focused. This newly method is only used for neutrosophic based problems and this method cannot be applicable on fuzzy and vague related problems due to presence of indeterminacy data.

# References

1. Zadeh, L.A.: Fuzzy sets. Inf. Control 8(3), 338–353 (1965)
2. Raju, K.V.S.V.N., Majumdar, A.K.: Fuzzy functional dependencies and lossless join decomposition of fuzzy relational database system. ACM Trans. Database Syst. 13(2), 129–166 (1988)
3. Sözat, M.I., Yazici, A.: A complete axiomatization for fuzzy functional and multivalued dependencies in fuzzy database relations. Fuzzy Sets Syst. 117(2), 161–181 (2001)
4. Bahar, O., Yazici, A.: Normalization and lossless join decomposition of similarity-based fuzzy relational databases. Int. J. Intell. Syst. 19(10), 885–917 (2004)
5. Gau, W.L., Buehrer, D.J.: Vague Sets. IEEE Trans. Syst. Man Cybern. 23(2), 610–614 (1993)
6. Zhao, F., Ma, Z.M., Yan, L.: A vague relational model and algebra. In: Fourth International Conference on Fuzzy Systems and Knowledge Discovery (FSKD 2007), 1:81–85 (2007)
7. Zhao, F., Ma, Z.M.: Vague query based on vague relational model. In: AISC 61, pp. 229–238, Springer, Berlin (2009)
8. Mishra, J., Ghosh, S.: A new functional dependency in a vague relational database model. Int. J. Comput. Appl. (IJCA) 39(8), 29–33 (2012)
9. Mishra, J., Ghosh, S.: A vague multivalued data dependency. Int. J. Fuzzy Inform. Eng. 5(4), 459–473 (2013). ISSN: 1616-8666 (Springer)
10. Mishra, J., Ghosh, S.: Uncertain query processing using vague sets or fuzzy set: which one is better? Int. J. Comput. Commun. Control (IJCCC) 9(6), 730–740 (2014)
11. Smarandache, F.: First International Conference on Neutrosophy, Neutrosophic Probability, Set, and Logic, University of New Mexico, vol. 1, no. 3 (2001)
12. Smarandache, F.: A unifying field in logics: neutrosophic logic, in multiple-valued logic. Int. J. 8(3), 385–438 (2002)
13. Smarandache, F.: Definitions derived from neutrosophics, multiple-valued logic. Int. J. 8(5–6), 591–604 (2002)
14. Arora, M., Biswas, R.: Deployment of neutrosophic technology to retrieve answers for queries posed in natural language. In: 3rd International Conference on Computer Science and Information Technology ICCSIT 2010, vol. 3, pp. 435–439 (2010)

15. Arora, M., Biswas, R., Pandy, U.S.: Neutrosophic relational database decomposition. Int. J. Adv. Comput. Sci. Appl. **2**(8), 121–125 (2011)
16. Broumi, S.: Generalized neutrosophic soft set. Int. J. Comput. Sci. Eng. Inform. Technol. **3**(2), 17–30 (2013)
17. Deli, I., Broumi, S.: Neutrosophic soft relations and some properties. Ann. Fuzzy Math. Inform. **9**(1), 169–182 (2015)

# Prognostication of Student's Performance: An Hierarchical Clustering Strategy for Educational Dataset

Parag Bhalchandra, Aniket Muley, Mahesh Joshi,
Santosh Khamitkar, Nitin Darkunde, Sakharam Lokhande
and Pawan Wasnik

**Abstract** The emerging field of educational data mining gives us better perspectives for insights in educational data. This is done by extracting hidden patterns in educational databases. In these lines, the objective of this research work is to introduce hierarchical clustering models for student's collected data. The ultimate goal is to find attributes in terms of set of clusters which severely affect the student's performance. Here clustering is intentionally used as the most common causes affecting performance within the database which cannot be seen normally. The results enable us to use discovered characteristic or patterns in palpating student's learning outcomes. These patterns can be useful for teachers to identify effective prognostication strategies for students.

**Keywords** Educational data mining · Clustering · Performance analysis · Pattern discovery

P. Bhalchandra (✉) · S. Khamitkar · S. Lokhande · P. Wasnik
School of Computational Sciences, S.R.T.M. University, Nanded 431606, MS, India
e-mail: srtmun.parag@gmail.com

S. Khamitkar
e-mail: s.khamitkar@gmail.com

S. Lokhande
e-mail: sana_lokhande@rediff.com

P. Wasnik
e-mail: pawan_wasnik@yahoo.com

A. Muley · N. Darkunde
School of Mathematical Sciences, S.R.T.M. University, Nanded 431606, MS, India
e-mail: aniket.muley@gmail.com

N. Darkunde
e-mail: darkundenitin@gmail.com

M. Joshi
School of Educational Sciences, S.R.T.M. University, Nanded 431606, MS, India
e-mail: maheshmj25@gmail.com

© Springer India 2016                                                      149
H.S. Behera and D.P. Mohapatra (eds.), *Computational Intelligence in Data Mining—Volume 1*, Advances in Intelligent Systems and Computing 410, DOI 10.1007/978-81-322-2734-2_16

# 1  Introduction

Student's data is a vital resource for all higher educational organizations. In today's era, Computer/Electronic/IT/ICT devices are used for data collection which helps for collected student's data to be available in digital format. Due to such digitization, data can be processed easily leading to faster decision processes [1]. The function of data oriented intelligence is becoming core application in educational organizations. Since all educational organizations have been computerized, they have updated student's databases with them. These databases have information about all required data pertaining to students, learning processes and other essential things. Even though, higher educational institutes face very hard for predicting accurate knowledge from databases. Thus Data mining finds applications in educational industry [1, 2] mainly to explore hidden knowledge in crucial aspects including student's performance, outcome of teaching processes, result analysis, course information, etc. We can apply Data Mining to traditional processes also. Such applications are very challenging [3]. In Educational data mining, we primarily use existing data mining algorithms over educational databases to give us good insights [1, 4, 5]. These insights are in terms of patterns. The discovered patterns can be used for characterizations of models for prediction, association, classification and better resource management [3]. These models are also known as Academic analytics. Despite the Government, UGC, etc. are pouring lot of funds for improving education standards, academic performance of students from rural Universities of India is not improving [2, 5]. Over the years, University like Swami Ramanand Teerth Marathwada University, Nanded have taken attempts to create large amount of databases from the data pertaining to academic activities, educational activities and administration contexts. If we apply educational data mining algorithms then it will be very interesting to see hidden patterns in these databases. Our work is the first most attempts. This work is an example of a joint interdisciplinary work undertaken by three University Departments, viz, School of Computational Sciences, School of Mathematical Sciences and School of Educational Sciences. The objective of this study is to use data mining processes over collected educational datasets. Using these results, a role model could be set for academic institutions to data mine their data for accurate insight. Thus the research objective of proposed work is to propose a strategy for automatic gaining of intelligence from vast educational data. In this paper we suggest approaches for finding patterns in student's achievements and performances. Some real word data related to performance of University students was collected using questionnaire and progress reports. A student's database was created with 360 records and 46 fields using closed questionnaire method. This includes student's personal details like social, intellectual, demographic, habitual, health and economical data. The core attributes where we witness downfall in performance of students were found out and were modeled using Hierarchical clustering models. Hierarchical clustering algorithms were implemented using SPSS Ver. 22 software [6].

## 2 Research Methodology

In India, the educational system consists of $10 + 2 + 3$ pattern of undergraduate education for no-professional streams and $10 + 2 + 4$ pattern for undergraduate professional streams. The students admitted to School of Computational Sciences are mainly from the first pattern. Most of the students are under graduate from the public, grant-in-aids sanctioned and nominal expensive education system. There are several courses like BCS, BCA, B.Sc (CS or IT or CM) under science and technology faculty from where the students mainly come. The data is collected for the duration 2009–2012. The database was built from two sources: previous examination's progress reports and our questionnaires. We have also collected information of the fresher students. However their progress reports have not yet displayed. We will show that the performance degrading pattern we have discovered will be helpful to predict some students which will also have failure. Our questionnaire has questions with predefined options [7] related to student's attributes as defined in Pritchard and Wilson [7, 8]. The attributes in questionnaire have relations with the performance of students. The questionnaire was reviewed by experts from School of Educational Sciences. There was some trial testing of questionnaire on some small groups of students. These trials helped us to understand over all feedback as such. Two revisions with subsequent trial testing have been made to devise out the final version of questionnaire. The final questionnaire consisted of 43 closed type questions [7]. The questionnaire were distributed to students and demonstrated for feedback. After this, a dataset/database was integrated with 360 student records and each record consists of 46 fields. Originally, there were 43 fields, equal to the total number of questions in questionnaire. Three additional fields were added for seeking information of students. Microsoft Excel 2007 software is used to record the dataset. Data set values like Yes/No were converted into numeric values like 1 or 0. Other numerical codes in the range 0, 1, 2, 3, 4 …6 were also given depending upon the number of possible answers a question can have. Likewise other answers are also converted into numeric values. During the pre-processing of data, we found some ambiguities and false information in student's data. For example, all students live at their homes with their parents. This cannot be true as we know that many students live in hostels or at relative's home or even they share rooms. We got around 18 % confusing data or falsely filled data or partly filled questionnaire in first step. In order to correct them, the students were called on one to one basis and convinced to fill the missing data. There were some confidential issues discovered after going through the filled questionnaire. Appropriate actions were taken time to time. For example, students shied to disclose their personal mobile numbers. This was very redundant with girl students. Significant number of students got convinced and for remaining students, we put our departmental number as their contact number. A snapshot of questionnaire and the dataset is as given in Fig. 1. Since we aim for discovery of attributes which affects performance of students, we primarily investigated literature across the globe to see what other people have done. To understand performance of a student, we underwent discussions with educationalist.

| 1 | Course code | MSc (5) , MCA (6) | | | | |
|---|---|---|---|---|---|---|
| 2 | Your name | | | | | |
| 3 | Gender (sex) | Male (1) | | Female (0) | | |
| 4 | Marital status | Married (2) | | unmarried(3) | | |
| 5 | Age | | | | | |
| 6 | Home address | Urban(1) | | rural (2)   foreign(3) | | |
| 7 | Mobile no. | | | | | |
| 8 | Personal email id | | | | | |
| 9 | Degree passer and percentage | General B.Sc. /   B.Sc.(computer CS) /   BCA / BCS/   Other / <br> (1)           (2)                   (3)      (4)        (5) | | | | |
| | Percentage | | | | | |
| 10 | Degree collage name | | | | | |
| 11 | Father's Education | Below or SSC/   HSC/   Graduate/   Post Graduate/   other <br> (1)         (2)        (3)           (4)            (5) | | | | |
| 12 | Fathers job and annual income | Service /   Business/   Agriculture/   In house/   Other/ <br> (1)        (2)           (3)             (4)         (5) | | | | |
| | Income | 0-1 lakh (1)  , 1.1-2 lakh(2), 2.1-5 lakh(3) , 5lakh - above (4) | | | | |
| 13 | Mothers education | Below or SSC/   HSC/   Graduate/   Post Graduate/   other <br> (1)         (2)        (3)           (4)            (5) | | | | |
| 14 | Mothers job and annual income | Service /   Business/   Agriculture/   In house/   Other/ <br> (1)        (2)           (3)             (4)         (5) | | | | |
| | Income | 0-1 lakh (1)  , 1.1-2 lakh(2), 2.1-5 lakh(3) , 5lakh - above (4) | | | | |
| 15 | Family size | | | | | |
| 16 | Family relationship | Excellent /   Good/   Satisfactory/   Bad/   Very Bad <br> (1)        (2)        (3)            (4)        (5) | | | | |
| 17 | Family support to your education | Excellent /   Good/   Satisfactory/   Bad/   Very Bad <br> (1)        (2)        (3)            (4)        (5) | | | | |
| 18 | Reason to choose this course | Career in IT/   Near to Home/   Reputation of course /   Blind Decision/   Parents wish: <br> (1)          (2)                      (3)                   (4)      (5) | | | | |
| 19 | Travel mode and time needed | Bus/   Railway/   City Bus/   Rickshaw/   Self Vehicle /   walking (6) <br> (1)     (2)      (3 but taken as 1)      (4)                (5) | | | | |

**Fig. 1** Sample questionnaire

The faculties from School of Educational Sciences of our University had given us orientation on the same. We finally understood that mere marks in final examination cannot be taken as main indicator of performance. The performance in broader sense is how well a student does in over all courses. For proper understanding the performance terminology, we primarily relied on the work of Ali et al. [9]. This work is a lucid discussion on performance analysis of students. This work is then taken as main base for our analysis and some terminology is also borrowed from the same. As per above refereed studies, there are numerous factors which affect student's academic growth and learning outcome. This consideration matches with few attributes of our dataset. In the literature review Ali et al. [9] cites other references for proper understanding of factors important for performance analysis. A similar works of Graetz et al. [10] suggested that the social status of parents and performance of students are reliant with each other. This also matches with our understanding of proportionate relationship of performance with social and economical conditions of students. The Ali et al. [9] also cites that Considine and Zappala [11] have noticed that parent's economical conditions have positive effects on the performance. The work of Staffolani and Bratti [12] highlights that the future achievements of students have correlation with previous year's scores.

The combined finding that affects performance as per [9] highlights many social, economical and psychological factors affecting performance. We felt that, once we discover these hidden truths, we can undertake appropriate, corrective actions to improve overall performance. This is the motto behind our prognostication work. Further, it was matter of curiosity that whether such investigations were made in past or not. So we started investigating profiling studies for student's performances. In effect, several studies have been found which have addressed related issues. Some of them includes, Ma et al. [13] which used Association Rules [7], the Minaei-Bidgoli et al. [14] used Classification technique, the Kotsiantis et al. [ 15] used Prediction approach, and more recently, Pardos et al. [7] used Regression technique for prognostication purpose. Keeping the knowledge earned after careful review of above primary sources, in this work, we aim to predict attributes which are directly related to student's performance. We are of the opinion that not only productive teaching/learning but also other social and economical aspects will hamper the performance. Our analysis was carried out with some thoughts in mind including social aspects of students, economical favorableness of students to peruse education, personal desires and ambitions and finally resources available with students which can help them to get success. Since we started with unsupervised mode, we stick to clustering methods. Several clustering techniques are available. Similar attempts were made in past whose review gave us detailed insights [2, 16]. We stick up to hierarchical clustering [2, 17, 18]. A tree in hierarchical clustering is known as Dendogram [18]. In Hierarchical clustering, a binary tree is build by merging similar groups of points. It can be Divisive or Agglomerative [17, 18]. Agglomerative clustering is bottom up and Divisive clustering is top down. These two approaches are most commonly used because their underlying representation of clusters in hierarchy resembles their application domain. Hierarchical agglomerative clustering algorithms merge pair of clusters at each step based on linkage metrics for measuring proximity between clusters (Fig. 2).

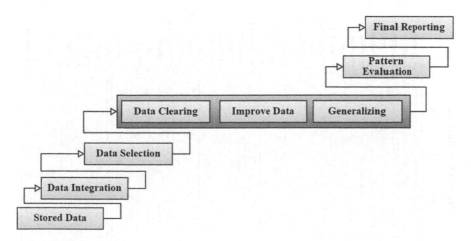

**Fig. 2** Research methodology

## 3 Experimentations, Results and Discussions

We used legal SPSS version 22 for the implementation of our idea [6]. Our student's dataset in MS-Excel is exported in SPSS and following procedure is performed. Once the dataset is fetched, the clustering process is initiated using Ward's method [17]. Below are some of the screenshots during experimentations. From experimental analysis, we found negative as well as positive insights that affect the performance of students. It is also appeared that, the UG percentage is the highest cluster which contains number of small sub clusters which affect student's performance. While analyzing the performance of students, we found that, at the first level of clustering, student's family environment including family income, family support and family relations matters a lot. These attributes have direct affect on the performance (Figs. 3 and 4).

Further, the use of internet, the available study material, own notes and additional tutorials also have corresponding positive effect on the performance. The personal computer and personal book library also shows self help seeking attitude of the student and have the closure association with performance. Students residential accommodation (place of living), travel time needed to reach University have direct impacts on the performance. If they are worst, there will be negative performance. The reason to join educational course that is ambitions or interest of student in a particular course also shows positive association with the performance. The hobbies like watching more movies per week, other aspect like less privileged background from which student has came have negative impact on the performance. The primary reason for joining to the course shows closure association with pattern of family size. Hence if family size is large, students prefer the same places

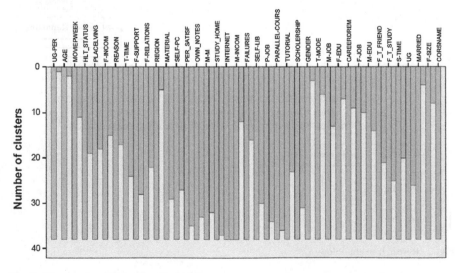

**Fig. 3** The number of clusters of variable cases

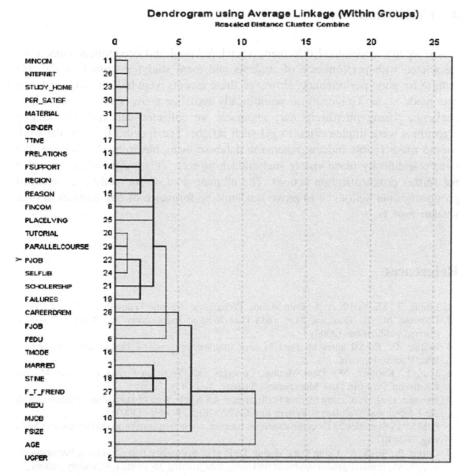

**Fig. 4** The resultant dendogram

for education as of his brothers or sisters. The attributes like parent's education, part time job, travelling mode, self study time, marital status, and time to spare with friends, also affects his performance. The UG level education is closely associated with family educational background and shows closure association with student's career dreams. The poor performance or failure is associated with reason to choose this course and free time spent with his friends. These highlight prognostications avenues for increasing performance of the students.

# 4 Conclusion

It was openly understood that many social, habitual and economical aspects are associated with performance of students and mere studying cannot be the sole criteria for good performance. However, these aspects were hidden and no attempt was made at our University to scientifically visualize them. The study took it as challenge. Using interdisciplinary approach, we collected data and data mining algorithms were implemented to get such insight. The specific focus of the work was to make visible hidden patterns in database using hierarchical clustering. We have scientifically made visible such hidden aspects. This insight is then recorded for further prognostication actions. The ultimate goal of our study is in terms of prognostication actions to improve academic performance of our students in their weaker aspects.

# References

1. Linoff, G., Michael J, et al.: Data Mining Techniques, 3rd edn. Wiley Publications (2011)
2. Dunham, M.: In: Dunham, M.H. (ed.) Data Mining: Introductory and Advanced Topics. Pearson publications (2002)
3. Indrato, E.: Edited notes on Data Mining. http://recommender-systems.readthedocs.org/en/latest/datamining.html
4. Han, J., Kamber, M.: Data Mining: Concepts and Techniques, 2nd edn. The Morgan Kaufmann Series in Data Management Systems, Jim Gray (2006)
5. Behrouz, et al.: Predicting Student Performance: An Application of Data Mining Methods with the Educational Web-Based System Lon-CAPA. IEEE, Boulder (2003)
6. IBM SPSS Statistics 22 Documentation on Internet. www.ibm.com/support/docview.wss?uid=swg27038407
7. Cortez, P., Silva, A.: Using Data Mining To Predict Secondary School Student Performance. http://www.researchgate.net/publication/Using_data_mining_to_predict_secondary_school_student_performance
8. Pritchard, M.E., Wilson, G.S.: Using emotional and social factors to predict student success. J. Coll. Student Dev. **44**(1), 18–28 (2003)
9. Ali, S., et al.: Factors contributing to the students academic performance: a case study of Islamia University Sub-Campus. Am. J. Educ. Res. **1**(8), 283–289 (2013)
10. Graetz, B.: Socio-economic status in education research and policy in John Ainley et al., Socio-economic Status and School Education DEET/ACER Canberra. J. Pediatr. Psychol. **20** (2), 205–216 (1995)
11. Considine, G., Zappala, G.: Influence of social and economic disadvantage in the academic performance of school students in Australia. J. Sociol. **38**, 129–148 (2002)
12. Bratti, M., Staffolani, S.: Student Time Allocation and Educational Production Functions. University of Ancona Department of Economics Working Paper No. 170 (2002)
13. Ma, Y., Liu, B., Wong, C.K., Yu, P.S., Lee, S.M.: Targeting the right Students using data mining. In: Sixth ACM SIGKDD International Conference, Boston, MA (Conference Proceedings) pp. 457–464 (2000)
14. Minaei-Bidgoli, B., Kashy, D.A., Kortemeyer, G., Punch, W.F.: Predicting student performance: an application of data mining methods with the educational web-based system

LON-CAPA. In: Proceedings of ASEE/IEEE Frontiers in Education Conference. IEEE, Boulder, CO (2003)

15. Kotsiantis, S.: Educational data mining: a case study for predicting dropout—prone students. Int. J. Knowl. Eng. Soft Data Paradigms **1**(2), 101–111 (2009)

16. Berkhin, P.: Survey of Clustering Data Mining Techniques, Accrue Software. www.cc.gatech.edu/ ~ isbell/reading/papers/berkhin02survey.pdf

17. Sasirekha, K., Baby, P.: Agglomerative hierarchical clustering algorithm—a review. Int. J. Sci. Res. Publ. **3**(3) (2013). ISSN 2250-3153

18. Murugesan, K., Zhang, J.: Hybrid hierarchical clustering: an experimental analysis. Technical Report: CMIDA-hipsccs #001-11. www.cs.uky.edu/ ~ jzhang/pub/techrep.html

# Design and Implementation ACO Based Edge Detection on the Fusion of Hue and PCA

Kavita Sharma and Vinay Chopra

**Abstract** The edge detection has become very popular due to its use in various vision applications. Edge detections produces a black and white image where objects are distinguished by lines (either objects boundary comes in black color or white color) depends upon where sharp changes exist. Many techniques have been proposed so far for improving the accuracy of the edge detection techniques. The Fusion of PCA and HUE based edge detector has shown quite better results over the available techniques. But still fusion technique forms unwanted edges so this paper has proposed a new Color and ACO based edge detection technique. The MATLAB tool is used to design and implement the proposed edge detection. Various kinds of images has been considered to evaluate the effectiveness of the proposed technique. Pratt figure of merit and F-measure parameters has been used to evaluate the effectiveness of the available edge detectors.

## 1 Introduction

Edge is a boundary among two homogeneous regions [1]. Image edge is the most essential characteristic of image. The edge is the set of pixels that has step changes in pixel gray value. Image edge reveals most of the information of image [2].

Edge detection is to find object boundaries, which is done by locating sharp variations in intensity values in an image [3]. In image processing and vision applications edge detection is an essential task. Edge detection also has wide utilization in segmentation of image, recognition of an object, image coding etc. Lately, the chief attention has been focused in the direction of algorithms utilized for colour images and multispectral images. Colour images as well as multispectral

K. Sharma (✉) · V. Chopra
DAVIET, Jalandhar, India
e-mail: kavitasharma147@yahoo.com

V. Chopra
e-mail: vinaychopra222@yahoo.co.in

© Springer India 2016                                                              159
H.S. Behera and D.P. Mohapatra (eds.), *Computational Intelligence in Data Mining—Volume 1*, Advances in Intelligent Systems and Computing 410, DOI 10.1007/978-81-322-2734-2_17

images enclose larger information as compared to the grey-scale images. The benefit of edge detection methods of colour image over edge detection methods of gray image is simply shown by believing the reality that those edges that lie at the boundary among areas of dissimilar colours cannot be found in grey images if there is nil variation in intensity. The studies illustrate that transforming colour images to gray-scale images will cause 10 % loss of edges [2].

## 2 Improved Edge Detection Using ACO and Color Gradients on Fusion of Hue and PCA

A new approach using ant colony optimization [4–6] and color gradients for color image edge detection is proposed. The methodology is shown in Fig. 1.

Step 1 **Input Image**: Digital image will be given as input to the system.
Step 2 **Check color or not**: Then it is checked whether the input image is colored or not.

**If (image is colored)**

Step 3 **Convert RGB to HSV**: Here input image is converted from RGB to HSV [2] image. Applying edge detection on hue component causes better edge detection as it is consistent with the human perception of colors also.

**Fig. 1** Proposed methodology

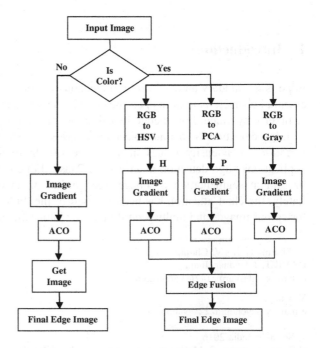

Step 4 **Apply image gradients Hue Image**: HSV image is converted to gradient [7] image. The pixels where large gradients are coming are more probable to be the edges in an image.

Step 5 **Apply ACO edge detection**: Ant colony optimization is applied to detect the edges of hue component. Call procedure ACO edge detection.

Step 6 **Convert RGB to PCA image**: PCA [8, 9, 16] is an arithmetic tool which changes a quantity of related variables into a quantity of unrelated variables. It makes complex object uncorrelated. It makes color edge detection simple and consistent to real time.

Step 7 **Apply image gradients On PCA image**: Image gradient is used to extract important information from image.

Step 8 **Apply ACO edge detection**: Ant colony optimization is applied to detect the edges of hue component. Call procedure ACO edge detection

Step 9 **Apply RGB to GRAY**: Now the RGB image is converted to Gray image because gray images also include some information which will be useful for final result.

Step 10 **Apply image gradients on Gray image**: Image gradient is used to extract important information from image.

Step 11 **Apply ACO edge detection**: Ant colony optimization is applied to detect the edges of hue component. Call procedure ACO edge detection

Step 12 **Apply edge fusion**: Fusion is the process of combine significant information from two or more images into an image. By integrating all the edge detected images of hue, PCA and Gray image we get more edge results.

Step 13 **Final edge detected image**: The output is edge detected image.

**Else**

Step 3 **Apply image gradient on input image**: Image gradient is used to extract important information from image.

Step 4 **Apply ACO edge detection**: Ant colony optimization is applied to detect the edges of hue component. Call procedure ACO edge detection

Step 5 **Final edge detected image**: The output is edge detected image.

**ACO Procedure for Edge Detection** [10–12]—(1) Initialize the locations of every ant and pheromone matrix (identical size as image). (2) Next comes the construction step here ant shifts from the node (a, b) to its neighbouring node (i, j) according to a transition probability that is defined as

$$P^{(N)}_{(a,b),(i,j)} = \frac{(\tau_{i,j}^{(n-1)})^\alpha (\eta_{i,j})^\beta}{\sum_{(i,j) \in \Omega_{(a,b)}} (\tau_{i,j}^{(n-1)})^\alpha (\eta_{i,j})^\beta} \qquad (1)$$

where $\tau_{i,j}^{(n-1)}$ is the pheromone value of the node (i, j), $\Omega_{(a,b)}$ is the neighbourhood nodes of the node (a, b), $\eta_{i,j}$ corresponds to the heuristic information at the node (i, j). $\alpha$ and $\beta$ are constants. (3) Update the pheromone matrix using (2) where $\rho$ denotes the evaporation rate and $\Delta_{i,j}^{(k)} = \eta_{i,j}$

$$\tau_{i,j}^{(n-1)} = \begin{cases} (1-\rho) \cdot \tau_{i,j}^{(n-1)} + \rho \cdot \Delta_{i,j}^{(k)}, & \text{if } (i,j) \text{ is visited by the current kth ant} \\ \tau_{i,j}^{(n-1)} & \text{otherwise.} \end{cases}$$

(2)

(4) Binary choice is made utilizing threshold value T on the final pheromone matrix to find whether there exists an edge or not.

$$E_{i,j} = \begin{cases} 1, & if \tau_{i,j}^{(N)} \geq T^{(l)}; \\ 0 & \text{otherwise.} \end{cases}$$

(3)

## 3 Experimental Setup and Results

The hue, PCA, integrated PCA and hue edge [13], Fuzzy [14] and proposed edge detectors has been designed and implemented with MATLAB 1.7. Different images are taken for experiment and the results have shown that the proposed edge detectors have more efficient and effective results over the existing techniques. Following section shows the result of proposed algorithm and existing techniques.

Figure 2 has shown the input image. Figure 3 shows the Hue edge detected image of the hue image. Figure 4 shows the PCA edge detected image of the PCA image. Figure 5 shows the Fusion based edge detected image of the input image shown in the Fig. 2. Figure 6 shows Fuzzy based edge detected image of the input image shown in the Fig. 2. Figure 7 shows Proposed edge detected image of the input image shown in the Fig. 2. The results have shown that proposed technique is

**Fig. 2** Input image

**Fig. 3** Hue edge detected image

**Fig. 4** PCA edge detected image

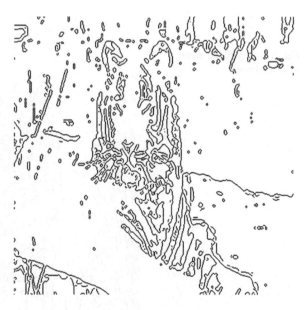

more effective than other techniques because of having maximum edges. The results has exposed that the each methods has pretty effective results above each other but the result of integrated hue and PCA based edge detectors and fuzzy based edge detectors are pretty more effective because of having highest edges. But the outcome of the proposed edge detectors is much more informative and clear than other methods.

**Fig. 5** Fusion based edge
detected image

**Fig. 6** Fuzzy based edge
detected image

**Fig. 7** Proposed method

## 4 Performance Evaluation

Three types of images are taken for experimental purposes like natural images, underwater images and remote sensing images. For each type five images are taken for evaluation and two performance metrics are taken i.e. Pratt's figure of merit and F-Measure which are shown below.

### 4.1 Pratt's Figure of Merit [13]

Pratt introduced function FM for computing quantitatively the performance of different edge detectors.

$$FOM = \frac{1}{\max(I_D, I_I)} \sum_{i=1}^{I_D} \frac{1}{1 + \beta(d_i)^2} \tag{4}$$

Table 1 has shown the comparison among proposed and the existing strategy based on Pratt's figure of merit. The proposed technique has more value of Pratt's figure of merit for all three types of images. The graph Fig. 8 shows how well the proposed technique has performed for Pratt's parameter.

**Table 1** Pratt's figure of merit analysis for existing and proposed techniques

| Pratt's figure of merit | | | | | |
|---|---|---|---|---|---|
| S. no | Hue | PCA | Fusion | Fuzzy | ACO |
| *Natural images* | | | | | |
| 1 | 0.0989 | 0.0946 | 0.5391 | 0.6952 | 0.7326 |
| 2 | 0.0693 | 0.0692 | 0.6511 | 0.8415 | 0.9462 |
| 3 | 0.1053 | 0.1090 | 0.4122 | 0.8067 | 0.9115 |
| 4 | 0.0794 | 0.0787 | 0.6575 | 0.8431 | 0.9834 |
| 5 | 0.0860 | 0.0820 | 0.5908 | 0.7123 | 0.8698 |
| *Underwater images* | | | | | |
| 1 | 0.0899 | 0.0900 | 0.5682 | 0.8387 | 0.9573 |
| 2 | 0.1181 | 0.1142 | 0.5190 | 0.7395 | 0.8801 |
| 3 | 0.1164 | 0.1120 | 0.4494 | 0.5082 | 0.8477 |
| 4 | 0.1537 | 0.1552 | 0.7082 | 0.7633 | 0.8816 |
| 5 | 0.0826 | 0.0826 | 0.4892 | 0.5838 | 0.7974 |
| *Remote sensing images* | | | | | |
| 1 | 0.1462 | 0.1371 | 0.7534 | 0.7805 | 0.8265 |
| 2 | 0.1549 | 0.1553 | 0.5884 | 0.7639 | 0.9240 |
| 3 | 0.0859 | 0.0895 | 0.3803 | 0.7975 | 0.8305 |
| 4 | 0.1537 | 0.1552 | 0.7082 | 0.8654 | 0.9845 |
| 5 | 0.0826 | 0.0826 | 0.4892 | 0.5838 | 0.7974 |

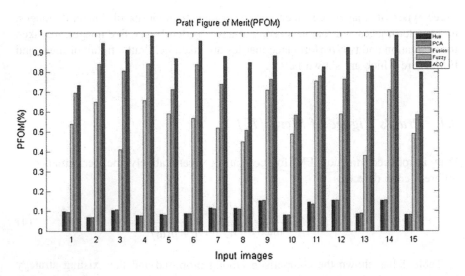

**Fig. 8** Pratt's figure of merit analysis for existing and proposed technique

**Table 2** F-measure for existing and proposed techniques

| S. no | Hue | PCA | Fusion | Fuzzy | ACO |
|---|---|---|---|---|---|
| F-measure | | | | | |
| *Natural images* | | | | | |
| 1 | 11.1870 | 2.7312 | 92.4138 | 94.0153 | 95.8726 |
| 2 | 6.4354 | 6.9012 | 94.3825 | 97.1820 | 98.5830 |
| 3 | 13.0565 | 18.3701 | 88.8177 | 94.9250 | 97.5907 |
| 4 | 7.3846 | 5.6449 | 94.0113 | 96.5605 | 98.2316 |
| 5 | 12.3868 | 3.2602 | 92.6965 | 94.8051 | 97.8792 |
| *Underwater images* | | | | | |
| 1 | 9.5375 | 9.7626 | 92.7305 | 94.5831 | 98.2086 |
| 2 | 7.5963 | 0.2563 | 92.8500 | 92.8670 | 96.4918 |
| 3 | 9.0498 | 1.3006 | 92.2002 | 93.7070 | 96.0455 |
| 4 | 9.9521 | 11.4188 | 88.4015 | 92.1221 | 95.4094 |
| 5 | 9.0787 | 9.1544 | 91.8648 | 94.8115 | 96.6932 |
| *Remote sensing images* | | | | | |
| 1 | 13.3659 | 1.4866 | 90.2046 | 92.5470 | 95.1594 |
| 2 | 16.0382 | 16.4552 | 87.3486 | 93.6316 | 96.5886 |
| 3 | 8.9287 | 16.0499 | 90.4307 | 95.5895 | 97.3718 |
| 4 | 9.5051 | 15.7614 | 87.8580 | 92.8963 | 97.8552 |
| 5 | 10.9371 | 8.7471 | 89.8273 | 93.7146 | 98.2645 |

## 4.2 F-Measure

The average of the precision of information extraction as well as recall metrics known as F-Measure [15]. Higher the value of it better will be edge detection.

$$Fmeasure = \frac{100 \times 2 \times (precision \times recall)}{precision + recall} \qquad (5)$$

Table 2 has shown the comparison among proposed and the existing strategy based on F-measure parameter. The proposed technique has more value of F-Measure for all three types of images The graph Fig. 9 shows how well the proposed technique has performed for F-Measure parameter.

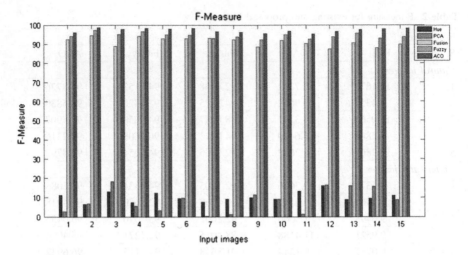

**Fig. 9** F-measure analysis for existing and proposed technique

## 5 Conclusion

This paper has evaluated the performance of ACO based color based edge detection techniques. Studies has shown that most of existing schemes ignored the use of color component in an image for finding edges but in numerous applications a area can be classified based upon the color. This paper has shown the result of different color based edge detectors i.e. hue, PCA, integrated Hue and PCA based methods and Fuzzy methods and compared them with the proposed technique. The results of the proposed edge detectors have shown the effectiveness of the fusion based edge detection. The proposed edge detector outperforms over the available techniques. In this work PFOM and F-Measure is considered for experimental purpose in near future some more performance parameters will also be considered to verify the proposed algorithm.

## References

1. Ju, Z.W., Chen, J.Z., Zhou, J.L.: Image segmentation based on edge detection using K-means and an improved ant colony optimization. In: Proceedings of the International Conference on Machine Learning and Cybernetics, Tianjin, 14–17 July, IEEE (2013)
2. Lei, Tao, Fan, Yangyu, Wang, Yi: Colour edge detection based on the fusion of hue component and principal component analysis. IET Image Proc. **8**(1), 44–55 (2014)
3. Deng, C.X., Wang, G.B., Yang, X.R.: Image edge detection algorithm based on improved canny operator. In: Proceedings of the 2013 International Conference on Wavelet Analysis and Pattern Recognition IEEE, Tianjin, 14–17 July (2013)
4. Manish, T.I., Murugan, D., Ganesh Kumar, T.: Hybrid edge detection using canny and ant colony optimization. Commun. Inf. Sci. Manag. Eng. **3**(8), 402–405 (2013)

5. Gupta1, C., Gupta, S.: Edge detection of an image based on ant colony optimization technique. Int. J. Sci. Res. (IJSR) **2**(6) (2013)
6. Thukaram, P., Saritha, S.J.: Image edge detection using improved ant colony optimization algorithm. Int. J. Res. Comput. Commun. Technol. **2**(11), 1256–1260 (2013)
7. Jin, L., Song, E., Li, L., Li, X.: A quaternion gradiet operator for color image edge detection. ICIP IEEE (2013)
8. Dikbas, S., Arici, T., Altunbasak, Y.: Chrominance edge preserving grayscale transformation with approximate first principal component for color edge detection. Proc. IEEE Conf. Image Process. (ICIP'07), 497–500 (2007)
9. Mahajanl, S., Singh, A.: Integrated PCA & DCT based fusion using consistency verification & non-linear enhancement. IJECS **3**(3), 4030–4039 (2014)
10. Tung, C.H., Han, G.W.: Efficient image edge detection using ant colony optimization
11. Rahebi, J., Tajik, H.R.: Biomedical image edge detection using an ant colony optimization based on artificial neural networks. Int. J. Eng. Sci. Technol. (IJEST) **3**(12), 8211–8218 (2011) (ISSN:0975-5462)
12. Tyagi, Y., Puntambekar, T.A., Sexena, P., Tanwani, S.: A hybrid approach to edge detection using ant colony optimization and fuzzy logic. Int. J. Hybrid Inf. Technol. **5**(1) (2012)
13. Lei, T., Fan, Y., Wang, Y.: Colour edge detection based on the fusion of hue component and principal component analysis. IET Image Proc. **8**(1), 44–55 (2014)
14. Walia, S.S., Singh, G.: Improved color edge detector using fuzzy set theory. Int. J. Adv. Res. Comput. Sci. Electronics Eng. (IJARCSEE) **3**(5) (2014)
15. Powers, D.M.W.: Evaluation: from precision, recall and F-factor to ROC, informedness, markedness and correlation (2007)
16. Mahajan, S., Singh, A.: Evaluated the performance of integrated PCA & DCT based fusion using consistency verification & non-linear enhancement. IJESRT, (2014)

# Effective Sentimental Analysis and Opinion Mining of Web Reviews Using Rule Based Classifiers

Shoiab Ahmed and Ajit Danti

**Abstract** Sentiment Analysis is becoming a promising topic with the strengthening of social media such as blogs, networking sites etc. where people exhibit their views on various topics. In this paper, the focus is to perform effective Sentimental analysis and Opinion mining of Web reviews using various rule based machine learning algorithms. we use SentiWordNet that generates score count words into one of the seven categories like strong-positive, positive, weak-positive, neutral, weak-negative, negative and strong-negative words. The proposed approach is experimented on online books and political reviews and demonstrates the efficacy through Kappa measures, which has a higher accuracy of 97.4 % and lower error rate. Weighted average of different accuracy measures like Precision, Recall, and TP-Rate depicts higher efficiency rate and lower FP-Rate. Comparative experiments on various rule based machine learning algorithms have been performed through a Ten-Fold cross validation training model for sentiment classification.

**Keywords** Sentiment analysis · Opinion mining · Rule based classifier · Score count

## 1 Introduction

Sentiment Analysis and Opinion Mining is an accelerating research area, ranging over assorted disciplines such as data mining, text analysis etc. Sentiment Analysis is a type of natural language processing which aims to determine the attitude of a speaker or a writer with respect to some topic or the overall contextual polarity of a

S. Ahmed (✉) · A. Danti
Department of Computer Applications, Jawaharlal Nehru National
College of Engineering, Shimoga, Karnataka, India
e-mail: shoiabahmed@gmail.com

A. Danti
e-mail: ajitdanti@yahoo.com

© Springer India 2016                                                                                    171
H.S. Behera and D.P. Mohapatra (eds.), *Computational Intelligence in Data Mining—Volume 1*, Advances in Intelligent Systems and Computing 410, DOI 10.1007/978-81-322-2734-2_18

document. The existing works of Sentimental Analysis have used SentiWordNet as a lexical resource. The major drawback of SentiWordNet is that it produces the count of only basic scored words such as positive, negative and neutral. This doesn't provide a strong discrimination for decision.

The popularity and attainability of opinion-rich resources are growing day by day. Such resources like personal blogs, online review sites etc., can be actively used in information technology to understand others opinion. To do this manually analyzing such huge amount of data is the tedious process. For the purpose of processing these types of data, Natural Language Processing (NLP) came into existence. Opinion mining and sentiment analysis are type of NLP. The main focus here is the extraction of people's opinion from the opinionated sites and analyzes those opinions. Liu [1], in his work related sentiment analysis with opinion mining by stating "Sentiment analysis is the field of study that analyses people's opinions, sentiments, evaluations, appraisals, attitudes, and emotions towards entities such as products, services, organizations, individuals, issues, events, topics, and their attributes.

For the task of experimentation, Data mining uses wide variety of datasets. Datasets consist of all of the information gathered during a survey which needs to be analyzed. The data in a dataset can be anything like product, movie, hospital, tourist place, Political reviews etc. The success of sentiment analysis approach depends on the quality, size and domain of the dataset. In this work, online book reviews and online political reviews are used as dataset for experimentation.

## 2   Literature Survey

Several research works has been published in the area of opinion mining and sentimental analysis. Pang et al. [2] have implemented the machine learning techniques to classify movie reviews according to the sentiments using POS Tagger and SentiWordNet for scoring. In this work, reviews are classified into positive, negative and neutral. Dave et al. [3] developed a document level opinion classifier that uses statistical techniques and POS tagging information for sifting through and synthesizing product reviews, essentially automating the sort of work done by aggregation sites or clipping services. "Sentiment Analysis Using Collaborated Opinion Mining" [4] given by the teacher using Thesaurus and Word net-a lexical resource.

Thenmozhi et al. [5] proposed a novel approach for retrieving opinion information from customer reviews. The work showed that the ontology based learning produces better analysis of sentence structure. Emelda [6] work used review analyzer system which works on the basis of sentiment words. Three types of classification methods are used for comparison and concluded that the accuracy is greater than that of the emotions based method.

Routray et al. [7] have made a survey on various approaches followed for sentimental analysis using WordNet, Support Vector Machine, Naive Bayes,

Maximum Entropy, language Model for performing sentimental analysis of data. Rahate and Emmunuel [8] have used SVM for Sentimental analysis of Movie reviews as either positive or negative. Vaidya and Rafi [9] have developed an improved SentiWordNet method for sentimental analysis in which POS tagging and three words scoring technique is used.

# 3 Proposed Methodology

In this paper, a novel approach is proposed to produce the word counts and categorizing them into seven classes as strong-positive, positive, weak-positive, neutral, weak-negative, negative and strong-negative words using SentiWordNet.

Online books and political web reviews are analyzed using the score count of these seven classes of opinions and further classified using different Rule Based machine learning algorithms which have resulted in better results. Therefore, this approach will be a more powerful aid for the web users to take decision based on reviews available on the internet.

For example, the sample output generated for the online book data review by proposed approach using SentiWordNet is shown in Fig. 1. Each line represents a word's score value and its associates score category.

The proposed approach is implemented on online book reviews and online political polls data using the Web Crawler. Figure 2 depicts the proposed architecture of sentimental analysis and opinion.

## 3.1 Data Acquisition

Huge amount of data from web is collected using JSoup Web Crawler for getting the various online reviews. The JSoup produces plain text sentences by parsing each page and eliminates all tag information. The produced reviews are deposited in the form of text documents.

**Fig. 1** Sample output of proposed approach

```
0.12091857471887278 weak_positive
0.02062568512992925 weak_positive
0.6022727272727273 positive
-0.031161089283294946 weak_negative
-0.01037537345531905 weak_negative
0.0 neutral
0.02737226277372263 weak_positive
0.3741496598639456 positive
-0.4583333333333333 negative
```

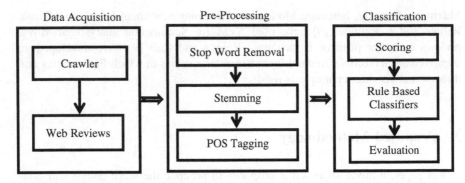

**Fig. 2** Proposed architecture of sentimental analysis and opinion mining

## 3.2 Pre-processing

### 3.2.1 Stop Words Removal

Stop words are cleaned out prior to or after processing of natural language data. These stop words can cause problem while searching for phrases that include them. Following are few common stop words: the, is, at, which, on etc. Thus such words are not essential to carry our pre-processing. Those words are removed. In this work more than 1124 stop words are removed from test document.

### 3.2.2 Stemming

Online reviews are generally used with informal language and they include internet slang and contemporary spellings like use of apostrophes, ing form of words to name a few. So such words must be re-visited and stemmed for correct data retrieval. In this approach Porter Stemming algorithm is employed which utilizes suffix stripping but not addresses prefixes. Steps involved in this algorithm are given below.

Step 1  Gets rid of plurals and -ed or -ing suffixes
Step 2  Turns terminal y to i when there is another vowel in the stem
Step 3  Maps double suffixes to single ones: -ization, -ational, etc.
Step 4  Deals with suffixes, -full, -ness etc.
Step 5  Takes off -ant, -ence, etc. from the suffixes. Step 6: Removes a final -e from given suffixes.

### 3.2.3    Parts of Speech Tagging

The reviews are tagged by their respective parts of speech using POS Tagger. A POS tagger parses a string of words, (e.g., a sentence) and tags each term with its part of speech using the Eqs. (1) and (2). For example, every term has been associated with a relevant tag indicating its role in the sentence, such as VBZ (verb), NN (noun), JJ (adjective) etc.

$$T\left(w_{i,n}\right) = \arg\max \max_{t_i,n} xe^{-x^2} \prod_{i=1}^{n} P\left(w_i | w_{i,i-1}\right) \tag{1}$$

$$T\left(w_{i,n}\right) = \arg\max \max_{t_i,n} xe^{-x^2} \prod_{i=1}^{n} P(t_i | w_i) \tag{2}$$

where 'w' represents the word, '$t_i$' represents the word tag. The tagging problem is defined as the sequence of tags as '$t_i, n$'.

In this work Stanford POS Tagger is employed and its conversion to SentiWordNet tags are shown in Table 1.

## 3.3    Classification

For effective sentimental analysis, proposed approach uses SentiWordNet and produces the count of scored words by classifying them into the seven classes such as strong-positive, positive, weak-positive, neutral, weak-negative, negative and strong-negative words. These score counts are vital to accomplish the task of opinion mining.

The SentiWordNet is a lexical resource, where each WordNet synset 's' is associated to two numerical scores Pos(s) and Neg(s), describing how positive or negative terms contained in the synsets. If the analyzer finds a pre-defined keyword in a sentence of a given blog page for a specific gadget, it looks for the modifiers associated with that keyword. If it finds such a word, it obtains its score from SentiWordNet for further process.

| Table 1 Conversion of POS tags to SentiWordNet tags | SentiWordNet Tag | POS tag |
|---|---|---|
| | a (adjective) | JJ, JJR, JJS |
| | n (noun) | NN, NNS, NNP, NNPS |
| | v (verb) | VB, VBD, VBG, VBN, VBP VPZ |
| | r (adverb) | RB, RBR, RBS |

# 4  Experimental Results

The proposed approach is experimented on various online web datasets ranging from different web domains like book review dataset and Political polling review dataset. Scores for the words and their counts are determined. These counts are used for the classification and sentimental analysis.

In this work, the Online Books Review Dataset is prepared by collecting the reviews of the web users for the book entitled "God of Small Things" written by Arundhati Roy from the website "www.goodreads.com". Reviews of the 200 online users are collected for experimentation. Similarly, the Political Review Dataset is collected from online reviews of 200 web users on the poll "BJP to get majority in Delhi elections Opinion poll" from internet.

Both the datasets undergo pre-processed phase in which stop words are removed and stemming is performed. Further employed with Stanford POS tagger to the dataset along with their parts of speech. The tagged documents are used for generating scores and count of each scored word is obtained using SentiWordNet.

Experimental results reveal the efficiency of the proposed approach. The Table 2 shows the count of scored words for book and political online reviews datasets.

Figure 3 shows the Precision Values for both online book and political datasets obtained by different Rule based machine learning algorithms.

For experimenting, 3496 instances of online book reviews and 2518 online political reviews were collected and the success rate of the proposed approach is as shown in Table 3.

From Table 3 reveals that the Decision Table Naïve Bayes Rule Classifier and PART 4.5 algorithm has a higher classification rate than others.

Kappa Statistic measures the accuracy of the system using Eq. (3).

$$(Kappa)K = \frac{a_0 - a_e}{1 - a_e} \tag{3}$$

where '$a_0$' is observed accuracy and '$a_e$' is expected accuracy.

Fleiss considers kappa (K) > 0.75 as excellent, 0.40–0.75 as fair to good, and <0.40 as poor. In this work, the kappa is 0.98, which indicates the proposed approach is most efficient and excellent.

**Table 2** Score count of online datasets

| Opinions | Book reviews | Political reviews |
|---|---|---|
| Neutral | 1444 | 1143 |
| Positive | 212 | 216 |
| Weak_Positive | 1071 | 690 |
| Strong_Positive | 60 | 70 |
| Negative | 169 | 19 |
| Weak_Negative | 540 | 380 |
| Strong_Negative | 0 | 0 |

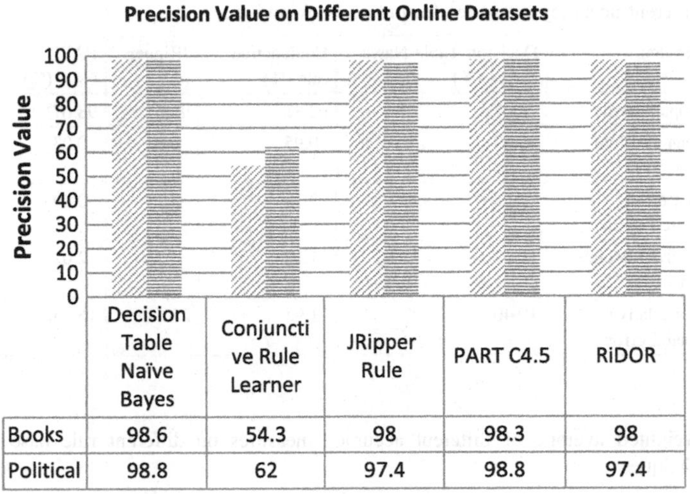

**Precision Value on Different Online Datasets**

| | Decision Table Naïve Bayes | Conjunctive Rule Learner | JRipper Rule | PART C4.5 | RiDOR |
|---|---|---|---|---|---|
| Books | 98.6 | 54.3 | 98 | 98.3 | 98 |
| Political | 98.8 | 62 | 97.4 | 98.8 | 97.4 |

**Fig. 3** Comparision of different rule based algorithms

**Table 3** Success rate on different online reviews

| Success rate | Decision table Naïve Bayes (%) | Conjunctive rule (%) | JRipper (%) | PART C4.5 (%) | RiDOR (%) |
|---|---|---|---|---|---|
| Book reviews | 98.54 | 73.09 | 97.95 | 98.24 | 97.95 |
| Political reviews | 99.14 | 78.54 | 97.85 | 99.14 | 97.85 |

Means Absolute Error (MAE) is the average absolute difference between classifier predicted output and actual output as given in the Eq. (4).

$$\text{MAE} = \frac{1}{N} \sum_{i=1}^{N} (\text{Desired}_i - \text{Actual}_i) \leq \varepsilon \qquad (4)$$

Root Mean Square Error (RMSE) is a frequently used measure of the differences between value predicted by a model and the values actually observed. RMSD is a good measure of accuracy as given by the Eq. (5).

$$\text{RMSE} = \sqrt{\frac{1}{N} \sum_{i=1}^{N} (\text{Desired}_i - \text{Actual}_i)^2} \leq \varepsilon \qquad (5)$$

Accuracy of the proposed approach is evaluated using different measures as given in Table 4.

The Table 5 depicts the comparison of accuracy rate of the proposed approach on different accuracy parameters like True Positive Rate (TP Rate), False Positive Rate (FP Rate), Precision, Recall, F-Measure and ROC Area using various rule based machine learning techniques.

**Table 4** Different accuracy measures for online book reviews

| Sl. no. | Measures | Decision Table Naïve Bayes (%) | Conjunctive rule (%) | JRipper (%) | PART C4.5 (%) | RiDOR |
|---------|----------|-------------------------------|----------------------|-------------|---------------|-------|
| 1 | Kappa statistic | 98.67 | 62.81 | 96.65 | 98.67 | 96.66 |
| 2 | Mean absolute error | 0.19 | 0.95 | 0.08 | 0.03 | 0.06 |
| 3 | Root mean squared error | 0.58 | 2.18 | 0.77 | 0.49 | 0.78 |
| 4 | Relative absolute error | 10.69 | 51.12 | 4.35 | 2.02 | 3.29 |
| 5 | Root relative squared error | 19.40 | 71.92 | 25.64 | 16.15 | 25.80 |

**Table 5** Weighted average of different accuracy measures on different rule based machine learning techniques

| Weighted average | TP rate | FP rate | Precision | Recall | F-measure | ROC area |
|------------------|---------|---------|-----------|--------|-----------|----------|
| Decision Table Naïve Bayes | 0.991 | 0.002 | 0.988 | 0.991 | 0.989 | 0.997 |
| Conjunctive rule learner | 0.785 | 0.18 | 0.62 | 0.785 | 0.692 | 0.847 |
| JRipper rule | 0.979 | 0.014 | 0.974 | 0.979 | 0.976 | 0.983 |
| PART C4.5 | 0.991 | 0.002 | 0.988 | 0.991 | 0.989 | 0.995 |
| RiDOR (ripple down rule learner) | 0.979 | 0.008 | 0.974 | 0.979 | 0.976 | 0.985 |

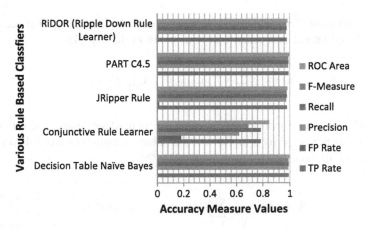

**Fig. 4** Accuracy measures of different rule based algorithms

From Fig. 4 it is inferred that, proposed approach has highest in Precision, Recall, TP Rate and F-Measure Rate and lowest in FP Rate for Decision Table Naïve Bayes, JRipper, PART C4.5, RiDOR and Conjunctive Rule

Algorithms. This analysis determines that proposed approach has a very higher efficiency that can be effectively deployed in Sentimental Analysis and Opinion Mining for the task of decision making in web.

# 5 Conclusion

This work showcases a novel approach which will be useful for the users for taking decisions on the online reviews available in the web. The proposed approach is experimented on Book and Political online reviews and experimental results reveal the efficiency of the proposed approach with highest accuracy. The proposed method can be further improved by different feature selection and evaluation techniques for classification in opinion mining.

# References

1. Liu, B.: Sentiment Analysis and Opinion Mining, Synthesis Lectures on Human Language Technologies. San Rafael. Calif. Morgan Claypool 5(1), 1–167 (2012)
2. Pang B., Lee L., Vaithyanathan S., "Thumbs up?: Sentiment Classification using Machine Learning Techniques", in Proceedings of the ACL-02 Conference on Empirical Methods in Natural Language Processing-Volume 10, Association for Computational Linguistics, pp-79–86, 2002
3. Dave, K., Lawrence, S., Pennock, D.M.: Mining the peanut gallery: opinion extraction and semantic classification of product reviews. In: Proceedings of the 12th International Conference on World Wide Web, ACM, pp. 519–528. (2003)
4. Virmani, D., Malhotra, V., Tyagi, R.: Sentimental analysis using collaborated opinion mining. Int. J. Soft Comput. Eng. 4, Issue-ICCIN-2014. ISSN 2331-2037
5. Thenmozhi, D., Ajay Shrivatsav, V.P., Manahas, B.P.S., Kishore Kumar, P.: Extracting customer needs for electronic gadgets using opinion mining. Int. J. Adv. Eng. Res. Stud., E-ISSB 2249–8974
6. Emelda, C.: A Comparative Study on Sentimental Classification and Ranking Reviews. Int. J. Innovative Res. Adv. Eng., ISSN. 2349–2163, 1(10) 2014
7. Routray, P., Swain, C.K., Mishra, S.P.: A survey on sentiment analysis (0975–8887). Int. J. Comput. Appl. 76(10) (2013)
8. Rahate, R.S., Emmunuel, M.: Feature selection for sentimental analysis using SVM. Int. J. Comput. Appl. (0975–8887) 84, 5
9. Vaidya, S., Rafi, M.: An improved SentiWordNet for opinion mining and sentiment analysis. J. Adv. Database Manage. Syst. 1(2) (2014)
10. Sharma, A., Dey, S.: Performance investigation of feature selection methods and sentiment lexicons for sentiment analysis. Special Issue Int. J. Comput. Appl. (0975–8887) Adv. Comput. Commun. Technol. PC Appl.—ACCTHPCA, (2012)
11. Baccianella, S., Esuli, A., Sebastiani, F.: SentiWordNet 3.0: an enhanced lexical resource for sentiment analysis and opinion mining. LREC 2200–2204 (2010)
12. Ohana, B., Tierney, B.: Sentiment classification of reviews using SentiWordNet. In: 9th IT&T Conference, Dublin Institute of Technology, p. 13, Dublin, Ireland (2009)
13. Chinsha, T.C., Joseph, S.: A syntactic approach for aspect based opinion mining. In: 2015 IEEE International Conference Semantic Computing (ICSC), pp. 24–31, Feb 2015

# Trigonometric Fourier Approximation of the Conjugate Series of a Function of Generalized Lipchitz Class by Product Summability

B.P. Padhy, P.K. Das, M. Misra, P. Samanta and U.K. Misra

**Abstract** Trigonometric Fourier approximation and Lipchitz class of function had been introduced by Zygmund and McFadden respectively. Dealing with degree of approximation of conjugate series of a Fourier series of a function of Lipchitz class Misra et al. have established certain theorems. Extending their results, in this paper a theorem on trigonometric approximation of conjugate series of Fourier series of a function $f \in Lip\,(\alpha, r)$ by product summability $(E,\ s)\ (N,\ p_n,\ q_n)$ has been established.

**Keywords** Fourier approximation $\cdot$ $Lip\,(\alpha, r)$ class of function $\cdot$ $(E,\ q)$ mean $\cdot$ $(N,\ p_n,\ q_n)$ mean $\cdot$ $(E,\ s)\ (N,\ p_n,\ q_n)$ product mean $\cdot$ Conjugate fourier series $\cdot$ Lebesgue integral

**2010-Mathematics Subject Classification:** 42B05 $\cdot$ 42B08

B.P. Padhy (✉)
Department of Mathematics, School of Applied Sciences, KIIT University,
Bhubaneswar, India
e-mail: iraady@gmail.com

P.K. Das
Biswasray Science College, Patapur, Ganjam, Odisha, India

M. Misra
Department of Mathematics, B.A. College, Berhampur, Odisha, India

P. Samanta
Department of Mathematics, Berhampur University, Berhampur, Odisha, India

U.K. Misra
National Institute of Scince and Technology, Ganjam, Odisha, India

© Springer India 2016
H.S. Behera and D.P. Mohapatra (eds.), *Computational Intelligence
in Data Mining—Volume 1*, Advances in Intelligent Systems
and Computing 410, DOI 10.1007/978-81-322-2734-2_19

# 1 Introduction

Let $\sum a_n$ be a given infinite series with sequence of partial sums $\{s_n\}$. Let $\{P_n\}$ and $\{q_n\}$ be sequences of positive real numbers such that

$$P_n = \sum_{v=0}^{n} p_v \quad \text{and} \quad Q_n = \sum_{v=0}^{n} q_v. \tag{1.1}$$

Let

$$t_n = \frac{1}{r_n} \sum_{v=0}^{n} p_{n-v} q_v s_v, \tag{1.2}$$

where $r_n = p_0 q_n + p_1 q_{n-1} + \ldots + p_n q_0 (\neq 0)$, $p_{-1} = q_{-1} = r_{-1} = 0$.
Then $\{t_n\}$ is called the sequence of $(N, p_n, q_n)$ mean of the sequence $\{s_n\}$. If

$$t_n \to s, \text{ as } n \to \infty, \tag{1.3}$$

the series $\sum a_n$ is said to be $(N, p_n, q_n)$ summable to $s$.
The necessary and sufficient conditions for regularity of $(N, p_n, q_n)$ method are [1]:

$$\frac{p_{n-v} q_v}{r_n} \to 0, \tag{1.4}$$

for each integer $v \geq 0$ as $n \to \infty$ and

$$\sum_{v=0}^{n} |p_{n-v} q_v| < H |r_n|, \tag{1.5}$$

where $H$ is a positive number independent of $n$.
The sequence-to-sequence transformation [2],

$$T_n = \frac{1}{(1+q)^n} \sum_{v=0}^{n} \binom{n}{v} q^{n-v} s_v, \tag{1.6}$$

defines the $(E, q)$ mean of the sequence $\{s_n\}$. If

$$T_n \to s, \text{ as } n \to \infty, \tag{1.7}$$

the series $\sum a_n$ is said to be $(E, q)$ summable to $s$. Clearly $(E, q)$ method is regular [2].

Further, the $(E, q)$ transform of the $(N, p_n, q_n)$ transform of $\{s_n\}$ is defined by

$$\tau_n = \frac{1}{(1+q)^n} \sum_{k=0}^{n} \binom{n}{k} q^{n-k} t_k$$

$$= \frac{1}{(1+q)^n} \sum_{k=0}^{n} \binom{n}{k} q^{n-k} \left\{ \frac{1}{r_k} \sum_{v=0}^{k} p_{k-v} q_v s_v \right\} \tag{1.8}$$

If

$$\tau_n \rightarrow s, \text{ as } n \rightarrow \infty, \tag{1.9}$$

$\sum a_n$ is said to be $(E, q)$ $(N, p_n, q_n)$—summable to $s$.

Let $f(t)$ be a periodic function with period $2\pi$ and, L-integrable over $(-\pi, \pi)$. The Fourier series associated with $f$ at any point $x$ is defined by

$$f(x) \sim \frac{a_0}{2} + \sum_{n=1}^{\infty} (a_n \cos nx + b_n \sin nx) \equiv \sum_{n=0}^{\infty} A_n(x). \tag{1.10}$$

and the conjugate series of the Fourier series (1.10) is

$$\sum_{n=1}^{\infty} (b_n \cos nx - a_n \sin nx) \equiv \sum_{n=1}^{\infty} B_n(x). \tag{1.11}$$

The $L_\infty$-norm of a function $f : R \rightarrow R$ is defined by

$$\|f\|_\infty = \sup\{|f(x)| : x \in R \} \tag{1.12}$$

and the $L_v$-norm is defined by

$$\|f\|_v = \left( \int_0^{2\pi} |f(x)|^v \right)^{\frac{1}{v}}, \; v \geq 1. \tag{1.13}$$

The degree of approximation of a function $f : R \rightarrow R$ by a trigonometric polynomial $P_n(x)$ of degree n under norm $\|\cdot\|_\infty$ is defined by [7]

$$\|P_n - f\|_\infty = \sup\{|P_n(x) - f(x)| : x \in R \} \tag{1.14}$$

and the degree of approximation $E_n(f)$ of a function $f \in L_v$ is given by [6]

$$E_n(f) = \min_{P_n} \|P_n - f\|_v.$$  (1.15)

This method of approximation is called Trigonometric Fourier approximation. A function $f \in Lip\alpha$, if [3]

$$|f(x+t) - f(x)| = O(|t|^\alpha), 0 < \alpha \le 1.$$  (1.16)

and $f \in Lip\,(\alpha, r)$, for $0 \le x \le 2\pi$, if [3]

$$\left( \int_0^{2\pi} |f(x+t) - f(x)|^r dx \right)^{\frac{1}{r}} = O(|t|^\alpha),\ 0 < \alpha \le 1, r \ge 1, t > 0.$$  (1.17)

We use the following notation throughout this paper:

$$\psi(t) = \frac{1}{2}\{f(x+t) - f(x-t)\},$$  (1.18)

$$\bar{s}_n(f; x) = \text{nth partial sum of the conjugate Fourier}$$  (1.19)

Series and

$$\overline{K_n}(t) = \frac{1}{\pi(1+s)^n} \sum_{k=0}^{n} \binom{n}{k} s^{n-k} \left\{ \frac{1}{r_k} \sum_{v=0}^{k} p_{k-v} q_v \frac{\cos\frac{t}{2} - \cos\left(v + \frac{1}{2}\right)t}{\sin\frac{t}{2}} \right\}.$$  (1.20)

Further, the method $(E, q)\,(N, p_n, q_n)$ is assumed to be regular and this case is supposed throughout the paper.

## 2  Known Theorems

Dealing with the degree of approximation by product summability, Misra et al. [4] proved the following theorem using $(E, q)\,(\overline{N}, p_n)$ mean of the conjugate series of a Fourier series:

## 2.1  Theorem

If $f$ is a $2\pi-$ periodic function of class $Lip\,\alpha$, then the degree of approximation by the product $(E, q)\,(\overline{N}, p_n)$ summability means of the conjugate series (1.11) of the Fourier series (1.10) is given by $\|\tau_n - f\|_\infty = O\left(\frac{1}{(n+1)^\alpha}\right)$, $0 < \alpha < 1$, where $\tau_n$ is as defined in (1.8).

Recently, Misra et al. [5] established another theorem on degree of approximation by the product mean $(E,q)(\overline{N},p_n)$ of the conjugate series of Fourier series of a function of class $Lip\ (\alpha,r)$. They prove:

## 2.2  Theorem

If $f$ is a $2\pi-$ periodic function of class $Lip(\alpha,r)$, then the degree of approximation by the product $(E,q)(\overline{N},p_n)$ summability means of the conjugate series (1.11) of the Fourier series (1.10) is given by $\|\tau_n-f\|_\infty = O\left(\dfrac{1}{(n+1)^{\alpha+\frac{1}{r}}}\right), 0<\alpha<1, r\geq 1$, where $\tau_n$ is as defined in (1.8).

# 3  Main Theorem

In this paper, we have studied a theorem on the degree of approximation by the product mean $(E,\ s)\ (N,\ p_n,\ q_n)$ of the conjugate series of the Fourier series of a function of class $Lip\ (\alpha,r)$. We prove:

## 3.1  Theorem

If $f$ is a $2\pi-$ Periodic function of the class $Lip(\alpha,l)$, then the degree of approximation by the product $(E,s)\ (N,p_n,q_n)$ summability means on its conjugate Fourier series (1.11) is given by $\|\tau_n-f\|_\infty = O\left(\dfrac{1}{(n+1)^{\alpha-\frac{1}{l}}}\right),\ 0<\alpha<1, l\geq 1$, where $\tau_n$ is as defined in (1.8).

# 4  Required Lemmas

We require the following Lemma for the proof the theorem.

## 4.1  Lemma

$$|\overline{K}_n(t)| = O(n), 0\leq t\leq \frac{1}{n+1}.$$

**Proof of Lemma 4.1:**

For $0 \le t \le \frac{1}{n+1}$, we have $\sin nt \le n \sin t$ then

$$|\overline{K_n}(t)| = \frac{1}{\pi(1+s)^n} \left| \sum_{k=0}^{n} \binom{n}{k} s^{n-k} \left\{ \frac{1}{r_k} \sum_{v=0}^{k} p_{k-v} q_v \frac{\cos \frac{t}{2} - \cos\left(v+\frac{1}{2}\right)}{\sin \frac{t}{2}} \right\} \right|$$

$$\le \frac{1}{\pi(1+s)^n} \left| \sum_{k=0}^{n} \binom{n}{k} s^{n-k} \left\{ \frac{1}{r_k} \sum_{v=0}^{k} p_{k-v} q_v \frac{\cos\frac{t}{2} - \cos vt \cdot \cos\frac{t}{2} + \sin vt \cdot \sin\frac{t}{2}}{\sin\frac{t}{2}} \right\} \right|$$

$$\le \frac{1}{\pi(1+s)^n} \left| \sum_{k=0}^{n} \binom{n}{k} s^{n-k} \left\{ \frac{1}{r_k} \sum_{v=0}^{k} p_{k-v} q_v \left( \frac{\cos\frac{t}{2}\left(2\sin^2 v\frac{t}{2}\right)}{\sin\frac{t}{2}} + \sin vt \right) \right\} \right|$$

$$\le \frac{1}{\pi(1+s)^n} \left| \sum_{k=0}^{n} \binom{n}{k} s^{n-k} \left\{ \frac{1}{r_k} \sum_{v=0}^{k} p_{k-v} q_v \left( O\left(2\sin v\frac{t}{2}\cos v\frac{t}{2}\right) + v\sin t \right) \right\} \right|$$

$$\le \frac{1}{\pi(1+s)^n} \left| \sum_{k=0}^{n} \binom{n}{k} s^{n-k} \left\{ \frac{1}{r_k} \sum_{v=0}^{k} p_{k-v} q_v (O(v) + O(v)) \right\} \right|$$

$$\le \frac{1}{\pi(1+s)^n} \left| \sum_{k=0}^{n} \binom{n}{k} s^{n-k} \left\{ \frac{O(K)}{R_k} \sum_{v=0}^{k} p_{k-v} q_v \right\} \right|$$

$$= O(n).$$

This proves the Lemma.

## 4.2 Lemma

$$|\overline{K_n}(t)| = O\left(\frac{1}{t}\right), \quad for \quad \frac{1}{n+1} \le t \le \pi.$$

**Proof of Lemma 4.2**

By Jordan's Lemma, for $\frac{1}{n+1} \le t \le \pi$, we have $\sin\left(\frac{t}{2}\right) \ge \frac{t}{\pi}$.

Then

$$\left|\overline{K_n}(t)\right| = \frac{1}{\pi(1+s)^n}\left|\sum_{k=0}^{n}\binom{n}{k}s^{n-k}\left\{\frac{1}{r_k}\sum_{v=0}^{k}p_{k-v}q_v\frac{\cos\frac{t}{2}-\cos\left(v+\frac{1}{2}\right)t}{\sin\frac{t}{2}}\right\}\right|$$

$$\leq \frac{1}{\pi(1+s)^n}\left|\sum_{k=0}^{n}\binom{n}{k}s^{n-k}\left\{\frac{1}{r_k}\sum_{v=0}^{k}p_{k-v}q_v\frac{\cos\frac{t}{2}-\cos vt\cdot\cos\frac{t}{2}+\sin vt\cdot\sin\frac{t}{2}}{\sin\frac{t}{2}}\right\}\right|$$

$$\leq \frac{1}{\pi(1+s)^n}\left|\sum_{k=0}^{n}\binom{n}{k}s^{n-k}\left\{\frac{1}{r_k}\sum_{v=0}^{k}\frac{\pi}{2t}p_{k-v}q_v\left(\cos\frac{t}{2}\left(2\sin^2\frac{t}{2}\right)+\sin vt\cdot\sin\frac{t}{2}\right)\right\}\right|$$

$$\leq \frac{\pi}{2\pi(1+s)^n t}\left|\sum_{k=0}^{n}\binom{n}{k}s^{n-k}\left\{\frac{1}{r_k}\sum_{v=0}^{k}p_{k-v}q_v\right\}\right|$$

$$= \frac{1}{2(1+s)^n t}\left|\sum_{k=0}^{n}\binom{n}{k}s^{n-k}\left\{\frac{1}{r_k}\sum_{v=0}^{k}p_{k-v}q_v\right\}\right|$$

$$= \frac{1}{2(1+s)^n t}\left|\sum_{k=0}^{n}\binom{n}{k}s^{n-k}\right|$$

$$= O\left(\frac{1}{t}\right).$$

This proves the Lemma.

## 5 Required Lemmas

Using Riemann–Lebesgue theorem, for the nth partial sum $\bar{s}_n(f;x)$ of the conjugate Fourier series (1.11) of $f(x)$ and following Titchmarch [6], we have

$$\overline{s_n}(f;x) - f(x) = \frac{2}{\pi}\int_0^{\pi}\psi(t)\frac{\cos\frac{t}{2}-\sin\left(n+\frac{1}{2}\right)t}{2\sin\left(\frac{t}{2}\right)}dt.$$

Using (1.2), the $(N,p_n,q_n)$ transform of $\overline{s_n}(f;x)$ is given by

$$t_n - f(x) = \frac{2}{\pi r_n}\int_0^{\pi}\psi(t)\sum_{k=0}^{n}p_{n-k}q_v\frac{\cos\frac{t}{2}-\sin\left(n+\frac{1}{2}\right)t}{2\sin\left(\frac{t}{2}\right)}dt$$

Denoting the $(E,s)$ $(N,p_n,q_n)$ transform of $\overline{s_n}(f;x)$ by $\tau_n$, we have

$$\|\tau_n - f\| = \frac{2}{\pi(1+s)^n} \int\limits_0^\pi \psi(t) \sum_{k=0}^n \binom{n}{k} s^{n-k} \left\{ \frac{1}{r_k} \sum_{v=0}^k p_{n-k} q_v \frac{\cos\frac{t}{2} - \sin\left(n+\frac{1}{2}\right)t}{2\sin\left(\frac{t}{2}\right)} \right\} dt$$

$$= \int\limits_0^\pi \psi(t) \overline{K_n}(t) dt$$

$$= \left\{ \int\limits_0^{\frac{1}{n+1}} + \int\limits_{\frac{1}{n+1}}^\pi \right\} \psi(t) \overline{K_n}(t) dt$$

$$= I_1 + I_2, say \tag{5.1}$$

Now

$$|I_1| = \frac{2}{\pi(1+s)^n} \int\limits_0^{\frac{1}{n+1}} \psi(t) \sum_{k=0}^n \binom{n}{k} s^{n-k} \left\{ \frac{1}{r_k} \sum_{v=0}^k p_{n-k} q_v \frac{\cos\frac{t}{2} - \sin\left(n+\frac{1}{2}\right)t}{2\sin\left(\frac{t}{2}\right)} \right\} dt$$

$$= \left| \int\limits_0^{\frac{1}{n+1}} \psi(t) \overline{K_n}(t) dt \right|$$

$$\leq \left( \int\limits_0^{\frac{1}{n+1}} |\psi(t)|^l dt \right)^{\frac{1}{l}} \left( \int\limits_0^{\frac{1}{n+1}} |K_n(t)|^m dt \right)^{\frac{1}{m}},$$

where $\frac{1}{l} + \frac{1}{m} = 1$, using Holder's inequality

$$= O\left(\frac{1}{(n+1)^\alpha}\right) \left( \int\limits_0^{\frac{1}{n+1}} n^m dt \right)^{\frac{1}{m}}$$

$$= O\left(\frac{1}{(n+1)^\alpha}\right) \left( \frac{n^m}{n+1} \right)^{\frac{1}{m}}$$

$$= O\left(\frac{1}{(n+1)^{\frac{1}{m}-1+\alpha}}\right)$$

$$= O\left(\frac{1}{(n+1)^{\alpha-\frac{1}{l}}}\right) \tag{5.2}$$

Next

$$|I_2| = \frac{2}{\pi(1+s)^n} \int_{\frac{1}{n+1}}^{\pi} \psi(t) \sum_{k=0}^{n} \binom{n}{k} s^{n-k} \left\{ \frac{1}{r_k} \sum_{v=0}^{k} p_{n-k} q_v \frac{\cos\frac{t}{2} - \sin\left(n+\frac{1}{2}\right)t}{2\sin\left(\frac{t}{2}\right)} \right\} dt$$

$$= \left| \int_{\frac{1}{n+1}}^{\pi} \psi(t) \overline{K_n}(t) dt \right|$$

$$\leq \left( \int_{\frac{1}{n+1}}^{\pi} |\psi(t)|^l dt \right)^{\frac{1}{l}} \left( \int_{\frac{1}{n+1}}^{\pi} |\overline{K_n}(t)|^m dt \right)^{\frac{1}{m}},$$

where $\frac{1}{l} + \frac{1}{m} = 1$, using Holder's inequality

$$= O\left( \frac{1}{(n+1)^{\alpha}} \right) \left( \int_{\frac{1}{n+1}}^{\pi} \left( \frac{1}{t} \right)^m dt \right)^{\frac{1}{m}},$$

using Lemma 4.2

$$= O\left( \frac{1}{(n+1)^{\alpha}} \right) \left( \left[ t^{-m+1} \right]_{\frac{1}{n+1}}^{\pi} \right)^{\frac{1}{m}}$$

$$= O\left( \frac{1}{(n+1)^{\alpha}} \right) \left( \frac{1}{n+1} \right)^{\frac{1-m}{m}}$$

$$= O\left( \frac{1}{(n+1)^{\alpha-1+\frac{1}{m}}} \right)$$

$$= O\left( \frac{1}{(n+1)^{\alpha-\frac{1}{l}}} \right). \tag{5.3}$$

Then from (5.2) and (5.3), we have

$$|\tau_n - f(x)| = O\left( \frac{1}{(n+1)^{\alpha-\frac{1}{l}}} \right), \quad \text{for} \quad 0 < \alpha < 1, l \geq 1.$$

Hence

$$\|\tau_n - f(x)\|_\infty = \sup_{-\pi < x < \pi} |\tau_n - f(x)| = O\left(\frac{1}{(n+1)^{\alpha - \frac{1}{l}}}\right) \, , \, 0 < \alpha < 1, \; l \geq 1.$$

This completes the proof of the theorem.

# References

1. Borwein, D.: On product of sequences. J. London Math. Soc. **33**, 352–357 (1958)
2. Hardy, G.H.: Divergent Series (First Edition). Oxford University Press, Oxford (1970)
3. Mcfadden, L.: Absolute Norlund summabilty. Duke Math. J. **9**, 168–207 (1942)
4. Misra, U.K., Misra, M., Padhy, B.P., Buxi, S.K.: On degree of approximation by product means of conjugate series of Fourier series. Int. J. Math. Sci. Eng. Appl. **6**(1), 363–370 (2012). ISSN 0973–9424
5. Misra, U.K., Paikray, S.K., Jati, R.K, Sahoo, N.C.: On degree of Approximation by product means of conjugate series of Fourier series. Bull. Soc. Math. Serv. Stan. **1**(4), 12–20 (2012). ISSN 2277–8020
6. Titchmarch, E.C.: The Theory of Functions. Oxford University Press, Oxford (1939)
7. Zygmund, A.: Trigonometric Series (Second Edition), vol. I. Cambridge University Press, Cambridge (1959)

# Optimizing Energy Efficient Path Selection Using Venus Flytrap Optimization Algorithm in MANET

S. Sivabalan, R. Gowri and R. Rathipriya

**Abstract** As routing protocols very essential for transmission, power consumption is one of the major prevailing issues to find the optimal path. Improve energy performance of MANET Routing protocol by choosing the optimal path which jointly reduces Total Transmission, Power is consuming the packet latency and improves the network lifetime. Propose a new VFO routing based on energy metrics and to reduce the total transmission power, maximize the lifetime of the connection which is implemented using Venus Flytrap Optimization (VFO). The VFO is the novel algorithm devised based on the closure behavior of the Venus Flytrap plant. The VFO algorithm generates the optimal path and energy aware routing.

**Keywords** Energy efficient path · MANET · VFO · Optimal energy

## 1 Introduction

The nodes in MANET are unstructured and dynamic state, so it will take power for making operation. Generate an optimal route using optimal path selection policy between source and destination. Each node of the power consumption must be used efficient manner. We can find optimum energy consumption for network route. Efficient traffic, path, optimal power [1] are major network factors of a node life. Energy model used to control the transmission, receiving power to set the energy limitations of nodes. Some difficulties to optimize the energy [2] in wireless

S. Sivabalan (✉) · R. Gowri · R. Rathipriya
Department of Computer Science, Periyar University, Salem, Tamilnadu, India
e-mail: sivabalan1990s@gmail.com

R. Gowri
e-mail: gowri.candy@gmail.com

R. Rathipriya
e-mail: rathi_priyar@periyaruniversity.ac.in

© Springer India 2016                                                           191
H.S. Behera and D.P. Mohapatra (eds.), *Computational Intelligence in Data Mining—Volume 1*, Advances in Intelligent Systems and Computing 410, DOI 10.1007/978-81-322-2734-2_20

networks. Optimization techniques used to find the best path selection for all possible routes.

AODV [3] and DSR [4] protocols are used to discover the shortest route in perfect manner. In order to solve this problem, energy aware protocols [5, 6] have been heavily studied in recent years. Most of them consider energy related metrics instead of the intermediate node or distance metrics. Routing optimization can be used energy-aware routing protocols, and this scheme to select the Minimum Total Transmission Power Routing (MTPR). The VFO optimize routes and optimal Minimum Total Transmission Power.

## 1.1 Energy-Aware Metric

Energy aware metrics in MANETs in each node at dynamic position with power supplied, and the data forward through of the intermediate nodes. Optimizing energy level means find shortest packet delivery and increasing the lifetime. Energy Optimization for an intermediate node of the network will be calculated and using topology based energy metrics [7]. Intermediate nodes are dynamic position, so can move freely from the network area. Optimization technique can perform a huge number of nodes [8]. Each intermediate node for packet transmission to spent energy, the VFO algorithm's path selection policy used by MANET's lifetime.

The most optimization problem optimize the nodes: there is multicasts relay connection used to the optimization problems. In the multi objective energy aware scheme classified based on the objective of the problem. Following problem can use to solve: (i) optimal path selection; (ii) increases the residual energy of node selection; (iii) less draining for data transmission for intermediate nodes (iv) optimize energy consumption [9].

## 2  Methods and Material

VFO Routing (VFOR) is a routing technique can find an optimal path selection between source to destination and calculate the overall network energy consumption [9]. We can set multiple sources and destination for a particular network. For all possible routes r select the optimal path between our source and destination. Calculate the intermediate nodes and energy value of those nodes. VFOR selects the optimal route r* such that r* = min r ∈ R Pr, where R generic routes. A VFO algorithm can find optimum route. A group of similar nodes considers trap and the source, destination selection may be different traps. Each trap different from the one or more topologies.

## 2.1 Venus Flytrap Inspired Optimization Algorithm (VFO)

The botanical name of Venus Flytrap is Dionaea Muscipula is shown in Fig. 1a. The great scientist Darwin quoted this plant as "one of the most wonderful in the world" [10]. This algorithm is devised on the rapid closure action of its leaves. This trap closure is due to the stimulation of the trigger hairs (Fig. 1c) that present in the two lobes of the leaf by the movements of prey (small insects, small animals, Rain drop, fast blowing wind, etc.) Fig. 1a–d. This plant is an insectivorous plant [10].

If two trigger hairs are stimulated or same trigger hair is stimulated twice within 30 s, then it rapidly initiates the closure of the trap [11–16]. The prey will be stuck inside the trap. The continuous movement of prey inside the trap causes the tightening of the trap. The trap, then sealed to digest the prey (5–7 days) [5] and reopened after digestion. If the prey escapes, then there will be no movements inside then there is no sealing phase and the trap will be opened within 12 h [11]. Similarly, the larger prey will not be caught fully and it cannot escape from the Flytrap so it will hang out of the trap causes trap death Sect. 2.2.

## 2.2 Implementation of Venus Flytrap Optimization (VFO)

VFO is the new algorithm devised first time and used for selecting optimal energy efficient paths in this paper. The rapid closure behavior of the Venus Flytrap leaves (trap) to capture the prey is mimicked in this algorithm. The flow of the Venus Flytrap Algorithm is described in the Fig. 1d.

The number of traps is decided based on the problem objectively. Each trap represents one path. It is represented as binary strings of length n, where n is the maximum possible length of a path, i.e. the active nodes in the network.

The presence of each node in a particular trap (path) is depicted using this binary string. The trap parameters like trigger time (t), action potential (ut), charge accumulated (C), flytrap status ($\delta$(ft)), object status (s(C)) are initialized in the beginning of the algorithm. The trapping process is performed until the optimal solution is attained. Initially, the trap is kept open, seeking for the prey. The nodes of each path are randomly chosen initially. When the prey has arrived, the trigger hair is stimulated first at t = 0, the second stimulation is performed at t < T (T = 30 s) then the flytrap parameter values are updated. The fitness of the flytrap is defined as the total transmission energy utilization of that path. The energy of each node I in the path up is calculated using the Eq. (1). The fitness of each trap is evaluated using the Eq. (2).

$$energy(p_i) = \sum_{j=1}^{n} power(p_{ij})^* time(p_{ij}) \qquad (1)$$

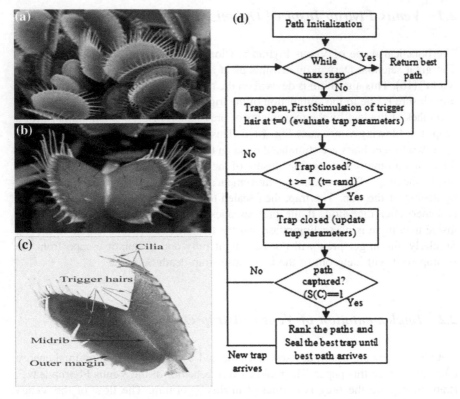

**Fig. 1** a Venus Flytrap (Dionaea Muscipula) plant with many flytraps (leaves). **b** Inner view of a Flytrap leaf shows trigger hairs on the midrib. **c** Parts of Venus Flytrap leaf. **d** Steps involved in VFO

where energy of the ith path is calculated using the Eq. (2), power(pij) is power spend by each node j present in that path, time(pij) represents the time taken for data transmission, n represents number of nodes in the path.

The objective function of this paper is to minimize the energy utilization of the paths by selecting the efficient nodes of the path. At each iteration the energy utilization of the path is evaluated, the presence of which node has a minimal energy requirement is selected.

$$f(p_i) = \begin{cases} energy(p_i), & energy(p) \leq threshold \\ 0, & otherwise \end{cases} \tag{2}$$

So the fitness of the object is necessary to decide the object capture. The flytrap will be sealed is shown in Sect. 2.2 if the trap is closed as well as the object is trapped otherwise the flytrap will be reopened for the next capture. The sealed trap will not be undergone to next iteration until better flytrap than the sealed trap is arrived. The process of capturing the prey will be performed until the maximum snaps of the flytraps. After max_snap iterations the best flytrap is returned which is the required optimal energy efficient path.

## 2.3 Parameter Setup

The action potential [11, 15] required for the closure of the leaf. The potential generated is dissipated at a particular rate and reaches to zero. It's jumped to 0.15 V at 0.001 s and simultaneously dissipated to zero after 0.003 s [11]. The action potential $u_t$ required is described using exponential function is given in the Eq. (4).

$$u_t = \begin{cases} 0.15e^{-2000t}, & t \geq 0 \\ 0, & t < 0 \end{cases} \tag{3}$$

Where, t is the trigger time at which the trigger hairs are stimulated, t < 0, before the first stimulation, t = 0, for the first stimulation, t > 0, for the second stimulation.

---

**Venus Flytrap Algorithm**

```
//Initial phase
Objective function f(x), x = (x₁, ..., xₐ)ˢ          //Energy of Path
Initialize a population of flytraps xᵢ (i = 1, 2, ..., n)  //flytraps- paths
While iter< = S
  For i=1 : n all n flytraps                        //trapping phase
      At t=0,                                        //first stimulation
      Evaluate Action Potential uₜ of flytrap i
      Charge Accumulation C of flytrap i  is determined by f(xᵢ)
      At t=rand()                                    //second stimulation
      if t<=T then
              update uₜ, C of flytrap i
              evaluate the Object Status
      end if
  end for
  Rank the flytraps and find the current best      //sealing phase
  seal the best flytrap until another best flytrap arrives
  end while
Post process results and visualization
```

---

The time between the two successive stimulations will be less than 30 s in order to initiate the flytrap closure. Thus there is no action potential for the flytrap before the first stimulation. The charge accumulation [11] may relate to the stepwise accumulation of a bioactive substance, resulting in ion channel activation by the action potential. The charge accumulation can be described by the following linear dynamic system is given in the Eq. (5).

$$C = -k_cC + k_au_t \tag{4}$$

where, C is the charge accumulated by the lobes for trap shutter, $k_c$ is rate of dissipation of charge between the first and second stimulations and C reaches to zero after 30 s, $k_a$ is rate of accumulation of charge. So it is implied that the second stimulation will be occurred within 30 s to attain the maximum charge of 14 μC to shut the flytrap. In these computational flytraps, the charge is calculated based on the objective function f(x) given in Eq. (2).

The next parameter is the flytrap status, used to know the current status of the flytraps. The status of the flytrap will be either 0 or 1 or 2 (open or close or seal) is evaluated using the Eq. (3).

$$\delta(ft) = \begin{cases} 2, & S(C) = 1, \delta(ft) = 1 \\ 1, & 0 \leq t \leq T \\ 0, & \text{otherwise} \end{cases} \tag{5}$$

The flytrap will be initially in the opened the (0) state. If the trap closure is triggered by the prey (object) then the trap will be in closed (1) state. The time point of the first stimulation is taken as zero, here the t represents the second stimulation time. The T is the time threshold between two stimulations (say 30 s) [11, 15, 16]. All of the closed traps will not be sealed. The flytrap with best fitness will be placed in sealed state. The sealed flytraps will be reopened when the fittest flytrap than it will be achieved.

# 3 Optimal Energy Efficient Path Selection

## 3.1 Experimental Setup

Energy optimization only counts packet transferring different traps. We can set transmission power 0.5 W, receiving power 0.3 W. A packet A with time $t(p_i)$; a node transmits $A_i$, decreasing energy $E_{tx}(A_i)$, where $E_{tx}(A_i) = 0.5 \times t(A_i)$; receiving packet p, decreasing energy $E_{rx}(A_i)$, where $E_{rx}(A_i) = 0.3 \times t(A_i)$. VFO algorithm to take less count of hops for transmission. Each trap will be considering hops.

The Table 1 represents the parameters. A simulated network is shown in the Fig. 2a. This network is simulated using NS2.

A network with mobile nodes is simulated randomly distributed in a region. The mobility model uses the random waypoint model, and node speed is randomly distributed between (0–20) meters per second. Two sets of simulations for mobile and static scenarios are done respectively. For the traffic models, we use CBR sources, but the source-destination pairs are randomly chosen over the network. Network settings are represented in Table 1. The VFO algorithm is used to discover an optimal energy efficient path from the simulated dataset. This novel algorithm is

**Table 1** Simulation parameters setup

| Energy parameter | Value | Network parameter | Value |
|---|---|---|---|
| Initial energy | 30 J | Nodes | 31 |
| Transmission power | 0.5 W | Area | 1327*661 |
| Receiving power | 0.3 W | Packet size | 3000 |
| Sleep power | 0.005 W | Traffic type | CBR |
| Idle power | 0.1 W | Rate | 1.0 MB |

**(a)**

**(b)**

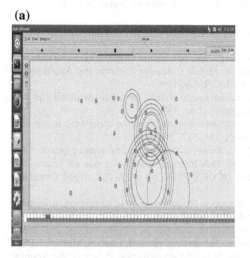
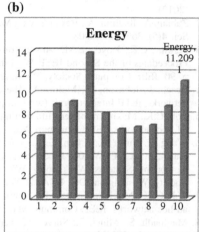

**Fig. 2  a** Simulated network. **b** Energy consumed by top ten paths

executed using the MATLAB. The optimal result is taken out of 50 runs. The energy of the various paths in best run is represented using the graph given in the Fig. 2b.

The Fig. 2 shows energy consumed by top ten paths that are discovered using the VFO algorithm. The optimal energy efficient path is 12-16-17-29 with minimal energy consumption 5.97573 J. The VFO algorithm helps to retain the best optimal path using the sealing phase.

# 4  Conclusion

Finally analyzed that VFO optimize energy efficiency in Mobile Ad hoc Networks. Experimental results have demonstrated to the energy aware techniques. The VFO routing technique based optimal energy consumed. Optimization in terms of area and VFO algorithm for efficient implementation of optimal path selection technique with optimal energy consumption. The optimal energy is marginally reduced due to optimal path selection policy. It is also demonstrated that the path can be easily reconfigured to match a range of performance requirements.

# References

1. Shivashankar, V.G.: Study of Routing Protocols for Minimizing Energy Consumption Using Minimum Hop Strategy in MANETS. Int. J. Comput. Commun. Netw. Res. **1**(3), 10–21 (2012)
2. Zabian, A., Ibrahim, A.: Power Saving Mechanism in Clustered Ad-Hoc Networks. J. Comput. Sci. **4**(5), 366–371 (2008)
3. Perkins, C.E., Royer, E.M.: Ad-hoc on-demand distance vector routing. In: WMCSA '99: Proceedings of the Second IEEE Workshop on Mobile Computer Systems and Applications, p. 90. IEEE Computer Society, Washington, DC, USA (1999)
4. Broch, J., Johnson, D., Maltz, D.: The dynamic source routing protocol for mobile ad hoc networks. iETF Internet Draft (work in progress), (Dec. 1998)
5. Toh, C.K.: Maximum battery life routing to support ubiquitous mobile computing in wireless ad hoc networks. IEEE Commun. Mag. **39**(6), 138–147 (2001)
6. Wang, K. Xu, Y.-L., Chen, G.-L., Wu, Y.-F.: Power-aware on-demand routing protocol for manet. In: ICDCSW '04: Proceedings of the 24th International Conference on Distributed Computing Systems Workshops- W7: EC (ICDCSW'04), pp. 723–728. IEEE Computer Society, Washington, DC, USA (2004)
7. Toh, C.-K.: Maximum battery life routing to support ubiquitous mobile computing in wireless ad hoc networks. IEEE Commun. Mag. **39**(6), 138–147 (2001)
8. Mahfoudh, S., Minet, P.: Survey of energy efficient strategies in wireless ad hoc and sensor networks. In: IEEE International Conference on Networking, pp. 1–7. Cancun, Mexico, (2008)
9. Sivabalan, S., Thangavel, K., Sathish, S.: Analysis of Path Selection Policy of the Routing Protocols Using Energy Efficient Metrics For Mobile Ad-Hoc Networks
10. Darwin, C.: Insectivorous Plants. Murray, London (1875)
11. Yang, Ruoting, Lenaghan, Scott C., Zhang, Mingjun, Xia, Lijin: A mathematical model on the closing and opening mechanism for Venus flytrap. Plant Signal Behav **5**(8), 968–978 (2010)
12. Volkov, A.G., Adesina, T., Jovanov, E.: Closing of Venus flytrap by electrical stimulation of motor cells. Plant Signal Behav **2**, 139–144 (2007)
13. Volkov, A.G., Adesina, T., Jovanov, E.: Closing of venus flytrap by electrical stimulation of motor cells. Plant Signal Behav **2**(3), 139–145 (2007)
14. Hodick, D., Sievers, A.: The action potential of Dionaea muscipula Ellis. Planta **174**, 8–18 (1988)
15. Fagerberg, W.R., Howe, D.G.: A quantitative study of tissue dynamics in Venus's flytrap dionaea muscipula (Droseraceae) II. Trap reopening. Am J Bot **83**, 836–842 (1996)
16. Forterre, Y., Skotheim, J.M., Dumais, J., Mahadevan, L.: How the Venus flytrap snaps. Nature **433**, 421–425 (2005)

# Biclustering Using Venus Flytrap Optimization Algorithm

R. Gowri, S. Sivabalan and R. Rathipriya

**Abstract** Digging up the coregulated gene biclusters using a novel Nature-inspired Meta-Heuristic algorithm named Venus Flytrap Optimization (VFO). This optimized biclustering approach will yield highly correlated biclusters. This algorithm is based on the rapid closure behavior of the Venus Flytrap (Dionaea Muscipula) leaves. Gene temperament is understood from their exposure under specific conditions. So far, Optimal Biclusters are extracted using various optimization algorithms like PSO, Genetic algorithm, SA, etc., were used for this kind of analysis. In this paper, VFO algorithm is used for extracting optimal biclusters and results are compared with those obtained by applying PSO, SA, PSO-SA Biclustering algorithms.

**Keywords** Venus flytrap optimization · VFO · Biclustering · Gene expression · Nature-Inspired optimization · Green intelligence algorithm

## 1 Introduction

In this modern era, there are numerous algorithms are devised to extract coregulated gene biomarkers. Normally some genes are active under specific set of conditions which are not amiable to other genes. Biclustering technique is an apt data mining technique to mine the local models (bicluster) which considers even a very small pattern. These biclusters are helpful in learning the peculiar nature of the genes. Optimizing the biclustering approach will refine the quality of biclusters further

R. Gowri (✉) · S. Sivabalan · R. Rathipriya
Department of Computer Science, Periyar University, Salem, Tamil Nadu, India
e-mail: gowri.candy@gmail.com

S. Sivabalan
e-mail: sivabalan1990s@gmail.com

R. Rathipriya
e-mail: rathi_priyar@periyaruniversity.ac.in

© Springer India 2016                                                                     199
H.S. Behera and D.P. Mohapatra (eds.), *Computational Intelligence in Data Mining—Volume 1*, Advances in Intelligent Systems and Computing 410, DOI 10.1007/978-81-322-2734-2_21

than that of those extracted using the existing biclustering techniques. Quality of the biclusters is analyzed using the Average Correlation Value (ACV) [1]. Bicluster with higher ACV is the highly correlated group present in the given dataset. The optimization algorithms like Particle Swarm Optimization, Simulated Annealing, Genetic algorithm, etc. were used so far. The novel VFO is devised based on the rapid closure of the Venus flytrap leaves. The trapping behavior of the Venus Flytrap plant is mimicked to formulate this novel algorithm. VFO is newly proposed for the Biclustering Gene expression data. The basic Yeast dataset [2] is chosen to test this VFO. The trapping behavior of the Venus Flytrap and its implementation along with the parameter setup is discussed in this paper. Application of this algorithm for biclustering the coregulated genes and its performance analysis is studied in further sections of this paper.

## 2 Venus Flytrap Behavior

The botanical name of Venus Flytrap is Dionaea Muscipula is shown in Fig. 1a. The great scientist Darwin quoted this plant as "one of the most wonderful in the world" [3]. This algorithm is devised on the rapid closure action of its leaves. This trap closure is due to the stimulation of the trigger hairs (Fig. 1c) that present in the two lobes of the leaf by the movements of prey (small insects, small animals, Rain drops, fast blowing wind, etc.) Fig. 1d–g. This plant is an insectivorous plant [1].

If two trigger hairs are stimulated or same trigger hair is stimulated twice within 30 s, then it rapidly initiates the closure of the trap [4]. The prey will be stuck inside the trap Fig. 1e. The continuous movement of prey inside the trap causes the tightening of the trap. The trap then sealed to digest the prey (5–7 days) [5] and reopened after digestion. If the prey escapes then there will be no movements inside then there is no sealing phase and the trap will be opened within 12 h [6]. Similarly, the larger prey will not be caught fully and it cannot escape from the flytrap so it will hang out of the trap causes trap death 3(F).

## 3 Implementation of Venus Flytrap Optimization (VFO)

VFO is the new algorithm devised first time and used for biclustering purpose in this paper. The rapid closure behavior of the Venus Flytrap leaves (trap) to capture the prey is mimicked in this algorithm. The flow of the Venus Flytrap Algorithm is described in the Fig. 1h.

The number of traps is decided based on the number of biclusters we want to extract. Each trap represents one bicluster. It is represented as binary strings of length $n$ is given in Eq. (1).

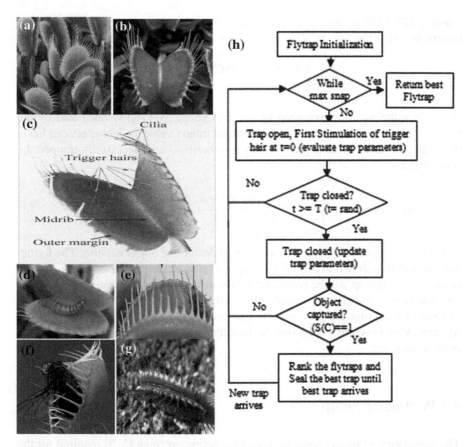

**Fig. 1 a** Venus flytrap (Dionaea Muscipula) plant with many flytraps (leaves). **b** Inner view of a flytrap leaf shows trigger hairs on the midrib. **c** Parts of Venus flytrap leaf. **d** A worm in Venus flytrap bending trigger hairs. **e** Semi closed Venus flytrap with trapped fly using fringed hairs of blade. **f** Flytrap with meal hanging out. **g** Sealed Venus flytrap in prey digestion. **h** Steps involved in VFO

$$n = \text{no. of Genes} + \text{no. of Conditions} \qquad (1)$$

The presence of each gene and each condition in a particular trap (bicluster) is depicted using this binary string. The trap parameters like trigger time ($t$), action potential ($u_t^-$), charge accumulated (C), flytrap status ($\delta(ft)$), object status (s(C)) are initialized in the beginning of the algorithm. The trapping process is performed until the optimal solution is attained. Initially, the trap is kept open, seeking for the prey. The elements of each biclusters are randomly chosen. When the prey has arrived, the trigger hair is stimulated first at $t = 0$, the second stimulation is performed at $t < T$ (T = 30 s) then the flytrap parameter values are updated. The fitness of the flytrap is defined as the maximum ACV of the biclusters. Average Correlation

Value (ACV) [7] is used to evaluate the correlation homogeneity of a bicluster matrix $B = (b_{ij})$ has the ACV. The fitness is calculated using the Eq. (2).

$$f(b_i) = \begin{cases} \text{ACV}(b_i), & \text{ACV}(b_i) \geq \text{threshold} \\ 0, & \text{otherwise} \end{cases} \tag{2}$$

where $b_i$ is the bicluster extracted by the $i$th flytrap, $\text{ACV}(b_i)$ is calculated by the following Eq. (3), the threshold is the nominal fitness value the biclusters can have, the flytraps whose fitness is greater than this threshold are successfully trapped the prey object.

$$\text{ACV}(b_i) = \max\left\{ \frac{\sum_{i=1}^{n}\sum_{j=1}^{n}\left|\text{row}_{ij}\right| - n}{n^2 - n}, \frac{\sum_{k=1}^{m}\sum_{l=1}^{m}\left|\text{col}_{kl}\right| - m}{m^2 - m} \right\} \tag{3}$$

So the fitness of the object is necessary to decide the object capture. The flytrap will be sealed is shown in 3(D) if the trap is closed as well as the object is trapped otherwise the flytrap will be reopened for the next capture. The sealed trap will not be undergone to next iteration until better flytrap than the sealed trap is arrived. The process of capturing the prey will be performed until the maximum snaps of the flytraps. After max_snap iterations the best flytrap is returned which is the required optimal bicluster.

## 4  Parameter Setup

The stimulation of trigger hairs generates the action potential [5, 8] required for the closure of the leaf. The potential generated is dissipated at a particular rate and reaches to zero. The action potential jumped to 0.15 V at 0.001 s and rapidly dissipated to zero after 0.003 s [3]. The action potential $u_t$ required is described using exponential function is given in the Eq. (4).

$$u_t = \begin{cases} 0.15e^{-2000t}, & t \geq 0 \\ 0, & t < 0 \end{cases} \tag{4}$$

where, $t$ is the trigger time at which the trigger hairs are stimulated, $t < 0$, before the first stimulation, $t = 0$, for the first stimulation, $t > 0$, for the second stimulation. The time between the two successive stimulations will be less than 30 s in order to initiate the flytrap closure. Thus there is no action potential for the flytrap before the first stimulation. The charge accumulation [5] may relate to the stepwise accumulation of a bioactive substance, resulting in ion channel activation by the action potential. The charge accumulation can be described by the following linear dynamic system is given in the Eq. (5).

$$C = -k_cC + k_au_t \tag{5}$$

where, $C$ is the charge accumulated by the lobes for trap shutter, $k_c$ is rate of dissipation of charge between the first and second stimulations and $C$ reaches to zero after 30 s, $k_a$ is rate of accumulation of charge. So it is implied that the second stimulation will be occurred within 30 s to attain the maximum charge of 14 μC to shut the flytrap. In these computational flytraps, the charge is calculated based on the objective function $f(x)$ given in Eq. (2).

---

**Venus Flytrap Algorithm**

---

//Initial phase
Objective function f(x), x = (x₁, ..., x_d)ˢ          //ACV of biclusters
Initialize a population of flytraps xᵢ (i = 1, 2, ..., n)  //flytraps- biclusters
While iter< = S
  For i=1 : n all n flytraps                    //trapping phase
    At t=0,                              //first stimulation
    Evaluate Action Potential uₜ of flytrap i
    Charge Accumulation C of flytrap i  is determined by f(xᵢ)
    At t=rand()                          //second stimulation
    if t<=T then
        update uₜ, C of flytrap i
        evaluate the Object Status
    end if
  end for
  Rank the flytraps and find the current best     //sealing phase
  seal the best flytrap until another best flytrap arrives
end while
Post process results and visualization

---

The next parameter is the flytrap status, used to know the current status of the flytraps. The status of the flytrap will be either 0 or 1 or 2 (open or close or seal) is evaluated using the Eq. (6).

$$\delta(ft) = \begin{cases} 2, & S(C) = 1, \ \delta(ft) = 1 \\ 1, & 0 \leq t \leq T \\ 0, & \text{otherwise} \end{cases} \tag{6}$$

The flytrap will be initially in the opened the (0) state. If the trap closure is triggered by the prey (object) then the trap will be in closed (1) state. The time point of the first stimulation is taken as zero, here the t represents the second stimulation time. The T is the time threshold between two stimulations (say 30 s) [9, 10]. All of the closed traps will not be sealed. The flytrap with best fitness will be placed in sealed state. The sealed flytraps will be reopened when the fittest flytrap than it will be achieved.

# 5 Biclustering Using VFO

The VFO algorithm is novel optimization algorithm which is now applied for Biclustering Yeast data. Biclustering using VFO will retain the best trap until the best trap arrives. It will not stuck with the local optimum. The best bicluster will be kept idle until the new highly correlated biclusters are obtained. PSO [1], SA [1], hybrid PSO-SA [1] are some of the metaheuristic optimization methods which are also applied for gene biclustering process in this paper. The results obtained by applying VFO are compared with the results obtained by applying other algorithms.

## 5.1 Experimental Setup

The benchmark yeast dataset is taken for testing the Venus Flytrap algorithm. There are 2884 genes with 17 conditions are present in the yeast dataset. Initially K-Means algorithm [11] is used to cluster the yeast dataset with 30 row clusters and 3 column clusters. Then these row clusters and column clusters are combined to form 30 × 3 coclusters. The VFO, Particle Swarm Optimization (PSO) [12], Simulated Annealing (SA) and PSO-SA algorithms [1] are used further to find the optimal Biclusters based on the Average Correlation Value (ACV). The ACV measure is taken as the fitness measure in both the algorithms. These algorithms are implemented and executed in MATLAB. The Biclustering using Venus Flytrap Algorithm is compared with the PSO algorithm. The experimental results are given in the table. There are three type of coverages such as Gene Coverage, Condition Coverage, Matrix Coverage [13]. In this paper, the first two coverages are evaluated for the Biclusters of yeast dataset using the following Eqs. (7) and (8).

$$\text{Genes Coverage} = \frac{\text{Number of genes covered by the extracted biclusters}}{\text{Total number of genes in the matrix}} \qquad (7)$$

$$\text{Conditions Coverage} = \frac{\text{Number of conditions covered by the extracted biclusters}}{\text{Total number of genes in the matrix}} \qquad (8)$$

## 5.2 Results and Discussion

The Biclustering approach using Particle Swarm Optimization (PSO), Simulated Annealing (SA), PSO-SA, and VFO algorithms are applied to yeast dataset. The experimental results are listed in the following tables. The initial volume, final

**Table 1** Biclusters obtained using VFO

| Bicluster ID | Initial volume | Final volume | Final ACV |
|---|---|---|---|
| 1 | 834 | 1524 | 0.9699 |
| 2 | 984 | 1585 | 0.9432 |
| 3 | 702 | 644 | 0.9509 |
| 4 | 175 | 1125 | 0.9696 |
| 5 | 100 | 732 | 0.9616 |
| 6 | 1141 | 582 | 0.9513 |
| 7 | 602 | 642 | 0.9554 |
| 8 | 749 | 570 | 0.9537 |
| 9 | 348 | 518 | 0.9648 |
| 10 | 180 | 558 | 0.9543 |

**Table 2** Performance of biclustering using VFO, PSO, SA, PSO-SA algorithms

|  | VFO | PSO | SA | PSO-SA |
|---|---|---|---|---|
| No of Biclusters | 90 | 90 | 90 | 90 |
| Population size | 90 | 90 | 90 | 90 |
| Initial volume | 834 | 1622 | 1622 | 1622 |
| Initial ACV | 0.84 | 0.87 | 0.87 | 0.87 |
| Threshold | 0.92 | 0.92 | 0.92 | 0.92 |
| Gene coverage | 98.14 | 90.0 | 99.3 | 99.0 |
| Condition coverage | 100 | 97.4 | 87.8 | 88.3 |
| Final volume | 1524 | 1546 | 1325 | 2100 |
| Final ACV (gbest ACV) | 0.97 | 0.96 | 0.94 | 0.94 |

volume and final ACV of the best Biclusters of the best run out of 30 runs are listed in the Table 1.

The final ACV values show more biclusters are getting the Average Correlation Values greater than the ACV threshold value 0.92 [1]. The Table 2 shows the results obtained by applying the algorithms like VFA, PSO, SA and PSO-SA to yeast dataset. The yeast dataset has 2884 genes and 17 conditions. Initially 30 row clusters and 3 column clusters are taken. So the number of individuals in the population is 90. For PSO biclustering $v_{min} = -20$; $v_{max} = 20$; $w_{min} = 0.4$; $w_{max} = 0.9$; $c_1 = 2$; $c_2 = 2$ [1].

For SA the parameter values for SA Biclustering maximum temperature $T_{max} = 5$, the cool rate $\alpha = 0.9$, $t_{min} = 0.1$. The comparison between the Volume and ACV of Biclusters obtained using these four algorithms are represented using bar chart is shown in the Fig. 2. From the Fig. 2b it shows that VFO can discover the highly correlated bicluster than other algorithms. From the study conducted, it is evident that the modeling of the prey capturing behavior of the Venus Flytrap into Venus flytrap Optimization can extract highly correlated biclusters from the Yeast expression dataset.

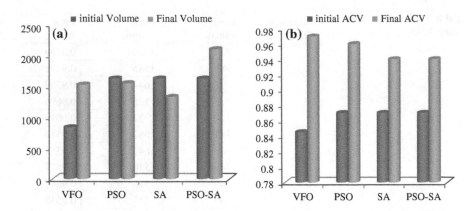

**Fig. 2** **a** Comparison of initial and final volume of best bicluster obtained using VFO, PSO, SA, PSO-SA. **b** Comparison of initial and final ACV of best bicluster obtained using VFO, PSO, SA, PSO-SA

# 6 Conclusion

In this paper, the Venus Flytrap Optimization algorithm has been designed based on the trap closure behavior of the plant Dionaea Muscipula. The implementation of VFO besides the parameter setup has also been discussed in this paper. The Venus flytrap optimization technique has been used to discover highly correlated optimal bicluster with minimal volume. The results are compared with results obtained by applying PSO, SA, PSO-SA algorithms. From the comparative analysis, it is shown that the VFO can discover more optimal bicluster. In further studies various measures like compactness, correctedness, significance, etc., can also be applied to evaluate the performance of VFO in predicting optimal bicluster.

# References

1. Thangavel, K., Bagyamani, J., Rathipriya, R.: Novel hybrid PSO-SA model for biclustering of expression data. Proc. Int. Conf. Commun. Technol. Syst. Design **30**, 1048–1055 (2011)
2. Tavazoie, S., Hughes, J.D., Campbell, M.J., Cho, R.J., Church, G.M.: Systematic determination of genetic network architecture. Nat. Genet. **22**:281–285
3. Darwin, C.: Insectivorous Plants. Murray, London (1875)
4. Volkov, A.G., Adesina, T., Jovanov, E.: Closing of Venus flytrap by electrical stimulation of motor cells. Plant Signal. Behav. **2**, 139–144 (2007)
5. Yang, Ruoting, Lenaghan, Scott C., Zhang, Mingjun, Xia, Lijin: A mathematical model on the closing and opening mechanism for Venus flytrap. Plant Signal. Behav. **5**(8), 968–978 (2010)
6. Volkov, A.G., Adesina, T., Jovanov, E.: Closing of Venus flytrap by electrical stimulation of motor cells. Plant Signal. Behav. **2**(3):139–145 (2007)
7. Thangavel, K., Rathipriya, R.: mining correlated bicluster from web usage data using discrete firefly algorithm based biclustering approach. Int. J. Math. Comput. Nat. Phys. Eng. **8**(4), 704–709 (2014)

8. Hodick, D., Sievers, A.: The action potential of Dionaeamuscipula Ellis. Planta **174**, 8–18 (1988)
9. Fagerberg, W.R., Howe, D.G.: A quantitative study of tissue dynamics in Venus's flytrap Dionaeamuscipula (Droseraceae) II. Trap reopening. Am. J. Bot. **83**:836–42 (1996)
10. Forterre, Y., Skotheim, J.M., Dumais, J., Mahadevan, L.: How the Venus flytrap snaps. Nature **433**:421–425 (2005)
11. Gowri, R., Rathipriya, R.: Extraction of Protein Sequence Motif Information using PSO K-Means. J. Network Inf. Secur. 8–13 (2014)
12. Balamurugan, R., Natarajan, A.M., Premalatha, K.: Comparative study on swarm intelligence techniques for biclustering of microarray gene expression data. Int. J. Comput. Control Quantum Inf. Eng. **8**(2), 323–329 (2014)
13. Saber, H.B., Elloumi, M.: DNA microarray data analysis: a new survey on biclustering. Int. J. Comput. Biol. **4**(1):21–37 (2015)

# Analytical Structure of a Fuzzy Logic Controller for Software Development Effort Estimation

S. Rama Sree and S.N.S.V.S.C. Ramesh

**Abstract** Most recently, attention has turned towards Machine learning techniques to predict software development cost as they are more apt when vague and inaccurate information is to be used. Based on the existing evidences, it is proved that a few of the problems associated with previous models are addressed by soft computing techniques. But, the need for accurate cost prediction in software project management is a challenge till today. In this paper, the analytical structure of a Takagi-Sugeno Fuzzy Logic Controller with two inputs and one output for software development effort estimation with a case study on NASA 93 dataset is discussed. The analytical study is also presented with two sample inputs. The Fuzzy models are developed using triangular and GBell membership functions. The results are compared using various assessment criteria. It has been observed that the fuzzy model with triangular membership function performed better than the other models.

**Keywords** Fuzzy logic controller · Analytical study · Fuzzy rules · Effort estimation · Criteria for assessment

## 1 Introduction

Fuzzy Logic [1] is used to solve any problem which is too difficult to be understood quantitatively. It is a simple and easy way of mapping an input space to an output space using certain rules. It has been effectively used for Software Effort Estimation. The literature shows, several authors like Iman Attarzadeh [2], Kirti

S. Rama Sree (✉)
Department of CSE, Aditya Engineering College, JNTUK, Kakinada, AP, India
e-mail: ramasree_p@rediffmail.com

S.N.S.V.S.C. Ramesh
Department of CSE, Sri Sai Aditya Institute of Science & Technology, JNTUK, Kakinada, AP, India
e-mail: ramesh_snsvsc@yahoo.co.in

© Springer India 2016
H.S. Behera and D.P. Mohapatra (eds.), *Computational Intelligence in Data Mining—Volume 1*, Advances in Intelligent Systems and Computing 410, DOI 10.1007/978-81-322-2734-2_22

Seth [3] etc. have proposed Fuzzy Logic Controller (FLC) as a tool for Software Cost estimation [4–6]. But, these authors have tuned the parameters of the FLC by trial and error method. In the present paper, an analytical structure of FLC is presented. Based on this analytical structure, the parameters are systematically tuned. The rulebase in FLC is of paramount importance. Identifying the proper rules is a challenging task.

## 2   Analytical Study of the Configuration of FLC with Two Inputs and One Output

The configuration of Two Input One Output Takagi Sugeno FLC is presented in Sect. 2.1. Section 2.2 presents the Effort estimation by a case study. Sugeno Type Fuzzy Inference System is considered as it is more compact, computationally efficient and works well for optimization techniques than Mamdani system.

### 2.1   Configuration of Two Input One Output FLC

Let the input variables to the FLC, M(n) and S(n) represent Mode and Size respectively [7–12] within the range of [−L, L]. The output of the FLC is Effort. Let EF(n) represent the output. Let n represent the current sample. Let the two input variables be fuzzified by the same input fuzzy sets. Assume there are $Q(Q \geq 1)$ triangular fuzzy sets for positive M(n) and S(n), Q fuzzy sets for negative M(n) and S(n) and one fuzzy set for nearly zero M(n) and S(n). Therefore there are $N = 2Q + 1$ fuzzy sets for M(n) and S(n).

A Fuzzy set for M(n) is denoted as $E_i$ and the corresponding Membership Function (MF) is designated as $\mu_i(M)$ where $-Q \leq i \leq Q$. Similarly the fuzzy sets for S(n) is denoted as $R_j$ and the corresponding MF is designated as $\mu_j(S)$ where $-Q \leq j \leq Q$. Figure 1 illustrates the input fuzzy sets over [−L, L]. The shape of fuzzy sets is identical and are uniformly distributed across [−L, L] with an overlap of 50 %. The base of the triangular MF is 2p. Using x(n) to represent either M(n) or S(n), the mathematical definition of $\mu_i(x)$ where $-Q \leq i \leq Q$ is

$$\mu_i(x) = \begin{cases} 0 & x \in [-\infty, (i-1)p] \\ \frac{x-[(i-1)p]}{p} & x \in [(i-1)p, ip] \\ \frac{(i+1)p-x}{p} & x \in [ip, (i+1)p] \\ 0 & x \in [(i+1)p, -\infty] \end{cases} \tag{1}$$

Assume $N^2$ Fuzzy rules be used to cover the entire $N * N$ possible combinations of the input fuzzy sets. The fuzzy rules are of the form

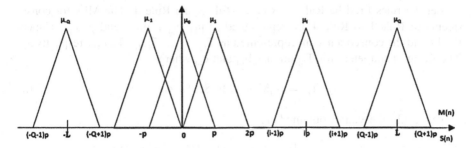

**Fig. 1** Illustrative definitions of triangular input fuzzy sets for M(n) and S(n)

$$\text{IF } M(n) \text{ is } E_i \text{ AND } S(n) \text{ is } R_j \text{ THEN } \Delta u(n) = a_{i,j}M(n) + b_{i,j}S(n) + c_{i,j}$$

where $\Delta u(n)$ is the contribution of this rule to the change of the FLC output, and $a_{i,j}$, $b_{i,j}$ and $c_{i,j}$ ($i, j = -Q, -Q + 1, ...., Q - 1, Q$) are design parameters. In the rules, AND operator is used and the resulting MF for the rule consequent is

$$\mu_{i,j}(\Delta u) = \prod(\mu_i(M), \mu_j(S)) \tag{2}$$

After defuzzification, the output of the FLC is given by

$$EF(n) = \Delta u(n) = \frac{\sum_{i=-Q}^{Q} \sum_{j=-Q}^{Q} \mu_{i,j}(\Delta u)[a_{i,j}M(n) + b_{i,j}S(n) + c_{i,j}]}{\sum_{i=-Q}^{Q} \sum_{j=-Q}^{Q} \mu_{i,j}(\Delta u)} \tag{3}$$

Let us now derive the structure of FLC with input variables x1, x2 representing M(n) and S(n) respectively. The input variables satisfies $ip \leq x1 \leq (i + 1)p$ and $jp \leq x2 \leq (j + 1)p$. After fuzzification, only the MFs for fuzzy sets $E_i$, $E_{i+1}$, $R_j$, $R_{j+1}$ are nonzero and the MFs for all the other fuzzy sets are zero. Therefore the fuzzy rules that are fired are only four.

Rule 1: IF $M(n)$ is $E_i$ AND $S(n)$ is $R_j$ THEN $\Delta u(n) = a_{i,j}M(n) + b_{i,j}S(n) + c_{i,j}$
Rule 2: IF $M(n)$ is $E_i$ AND $S(n)$ is $R_{j+1}$ THEN $\Delta u(n) = a_{i,j+1} M(n) + b_{i,j+1}$
$S(n) + c_{i,j+1}$
Rule 3: IF $M(n)$ is $E_{i+1}$ AND $S(n)$ is $R_j$ THEN $\Delta u(n) = a_{i+1,j} M(n) + b_{i+1,j}$
$S(n) + c_{i+1,j}$
Rule 4: IF $M(n)$ is $E_{i+1}$ AND $S(n)$ is $R_{j+1}$ THEN $\Delta u(n) = a_{i+1,j+1} M(n) + b_{i+1,j+1}$
$S(n) + c_{i+1,j+1}$

From (3), the obtained result is

$$\Delta u(n) = \frac{\sum_{k=0}^{1} \sum_{m=0}^{1} \mu_{i+k,j+m}(\Delta u)[a_{i+k,j+m}M(n) + b_{i+k,j+m}S(n) + c_{i+k,j+m}]}{\sum_{k=0}^{1} \sum_{m=0}^{1} \mu_{i+k,j+m}(\Delta u)}$$

$$\tag{4}$$

Let the rules fired be Rule 1, Rule 2, Rule 3 and Rule 4. The MFs for consequents of Rule 1 to Rule 4 are represented as $\mu_{i,j}$, $\mu_{i,j+1}$, $\mu_{i+1,j}$ and $\mu_{i+1,j+1}$ respectively and the consequents are represented as $T_{i,j}$, $T_{i,j+1}$, $T_{i+1,j}$, $T_{i+1,j+1}$ respectively. The design parameters for $T_{i,j}$ are $a_{i,j}$, $b_{i,j}$ and $c_{i,j}$ where

$$T_{i,j} = a_{i,j}M(n) + b_{i,j}S(n) + c_{i,j} \tag{5}$$

Using (2), the values obtained are

$$\mu_{i,j}(\Delta u) = \left[\left[\frac{(i+1)p - x1}{p}\right] * \left[\frac{(j+1)p - x2}{p}\right]\right] \tag{6}$$

$$\mu_{i,j+1}(\Delta u) = \left[\left[\frac{(i+1)p - x1}{p}\right] * \left[\frac{x2 - jp}{p}\right]\right] \tag{7}$$

$$\mu_{i+1,j}(\Delta u) = \left[\left[\frac{x1 - ip}{p}\right] * \left[\frac{(j+1)p - x2}{p}\right]\right] \tag{8}$$

$$\mu_{i+1,j+1}(\Delta u) = \left[\left[\frac{x1 - ip}{p}\right] * \left[\frac{x2 - jp}{p}\right]\right] \tag{9}$$

Using (4), the obtained result is

$$\Delta u(n) = \frac{\mu_{i,j} * T_{i,j} + \mu_{i,j+1} * T_{i,j+1} + \mu_{i+1,j} * T_{i+1,j} + \mu_{i+1,j+1} * T_{i+1,j+1}}{\mu_{i,j} + \mu_{i,j+1} + \mu_{i+1,j} + \mu_{i+1,j+1}} \tag{10}$$

The output of the FLC at n is given by

$$EF(n) = \frac{\left[\left[\frac{(i+1)p-x1}{p}\right] * \left[\frac{(j+1)p-x2}{p}\right]\right] * T_{i,j} + \left[\left[\frac{(i+1)p-x1}{p}\right] * \left[\frac{x2-jp}{p}\right]\right] * T_{i,j+1} + \left[\left[\frac{x1-ip}{p}\right] * \left[\frac{(j+1)p-x2}{p}\right]\right] * T_{i+1,j} + \left[\left[\frac{x1-ip}{p}\right] * \left[\frac{x2-jp}{p}\right]\right] * T_{i+1,j+1}}{\left[\left[\frac{(i+1)p-x1}{p}\right] * \left[\frac{(j+1)p-x2}{p}\right]\right] + \left[\left[\frac{(i+1)p-x1}{p}\right] * \left[\frac{x2-jp}{p}\right]\right] + \left[\left[\frac{x1-ip}{p}\right] * \left[\frac{(j+1)p-x2}{p}\right]\right] + \left[\frac{x1-ip}{p}\right] * \left[\frac{x2-jp}{p}\right]}$$

$$\tag{11}$$

## 2.2 Experimental Results of a Case Study with Two Inputs

For the purpose of effort estimation, NASA 93 dataset is taken. It includes the data of 93 projects taken from different NASA centers. The dataset includes 17 input parameters as in COCOMO. The inputs to FLC are Mode and Size, the output is estimated effort. The range of Mode is 1.05–1.2. The range of size is 0.9–980 KDSI. The range of Effort is 8–8211 person-months. Fuzzy Models are developed using Triangular Membership Function (TMF) and GBell Membership Function (GBellMF). The rulebase has 30 rules. The MFs for Mode are 3 and for that of size are 25.

**Table 1** Comparison of various fuzzy models with two inputs

| Model | VAF | MARE | VARE | Mean BRE | MMRE | Pred(30) |
|---|---|---|---|---|---|---|
| Fuzzy-TriMF | 96.53 | 28.53 | 10.51 | 0.61 | 54.81 | 66 |
| Fuzzy-GBellMF | 95.90 | 23.78 | 32.59 | 0.59 | 63.16 | 65 |
| COCOMO | 33.65 | 47.22 | 46.89 | 0.78 | 59.50 | 53 |

**Table 2** Estimated effort of fuzzy models with two inputs

| S. no. | Project ID | Mode | Size | Actual effort | Fuzzy-TriMF effort | Fuzzy-GBellMF effort |
|---|---|---|---|---|---|---|
| 2 | 39 | 1.12 | 8 | 42 | 42.14 | 43.83 |
| 3 | 18 | 1.12 | 31.5 | 60 | 67.46 | 80.99 |
| 4 | 1 | 1.12 | 25.9 | 117.6 | 118.6 | 113.77 |
| 5 | 32 | 1.12 | 35.5 | 192 | 192 | 172.59 |
| 6 | 28 | 1.12 | 48.5 | 239 | 245.3 | 268.04 |
| 7 | 47 | 1.12 | 190 | 420 | 448.12 | 483.07 |
| 8 | 41 | 1.12 | 177.9 | 1248 | 1183.38 | 1004.62 |
| 9 | 59 | 1.12 | 980 | 4560 | 4559.6 | 4581.44 |
| 10 | 90 | 1.2 | 233 | 8211 | 8210.84 | 7848.56 |

The criteria for selecting the best model is taken as Variance accounted For (VAF), Mean Absolute Relative Error(MARE), Variance Absolute Relative Error (VARE), Mean Balance Relative Error(Mean BRE), Mean Magnitude of Relative Error(MMRE) and Prediction(Pred(30)) [13]. A model with higher VAF, higher Pred (n), lower MARE, lower VARE, lower BRE and lower MMRE is preferred.

The comparison of various Fuzzy Models with two inputs and COCOMO using various criteria is depicted in Table 1. For the Fuzzy Model developed using TMF (Fuzzy-TriMF), VAF is 96.53 % and Pred(30) is 66 %. For the Fuzzy Model developed using GBellMF(Fuzzy-GBellMF), VAF is 95.9 % and Pred(30) is 65 %. The Effort Estimations of the Fuzzy Models for selective projects of NASA 93 dataset are shown in Table 2. From the results, it can be concluded that the Fuzzy Model developed using TMF is better than the Fuzzy Model developed using GBellMF or COCOMO [14]. Figure 2 shows the comparisons of actual effort with that of Fuzzy-TriMF.

## 2.3 Analytical Study of a Sample with Two Inputs

The FLC tuned with the NASA 93 dataset takes two inputs Mode and Size and yields effort as an output. Let the two inputs of FLC be x1 and x2 where x1 = 1.12 and x2 = 25.9. Consider the Fuzzy Model developed with TMF. From Table 2, for Project Id 1, the output for the present inputs is 118.6. The analytical study is

**Fig. 2** Estimated effort of fuzzy-TriMF versus actual effort

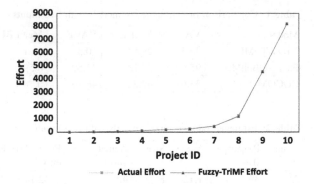

Actual Effort — Fuzzy-TriMF Effort

employed to get the output. The FLC has 30 rules in the rulebase. For the values of x1 and x2 where $1.05 \leq x1 \leq 1.2$ and $0 \leq x2 \leq 50$, the rules that are fired are only two i.e. Rule 7 and Rule 8. The design parameters of fuzzy set $E_2$ of x1 in Rule 7 and Rule 8 are 1.0499, 1.1244 and 1.199. The design parameters of fuzzy set $R_1$ for input x2 in Rule 7 are −39.895,0.9 and 41.69. The design parameters of fuzzy set $R_2$ for input x2 in Rule 8 are 0.899, 41.695 and 82.495. The mathematical definition of fuzzy set $E_2$ of x1 is represented as $\mu_2(1.12)$. The mathematical definition of fuzzy set $R_1$ of x2 is represented as $\mu_1(25.9)$. The mathematical definition of fuzzy set $R_2$ of x2 is represented as $\mu_2(25.9)$. Using (1), the membership values $\mu_2(1.12)$, $\mu_1(25.9)$, $\mu_2(25.9)$ are obtained.

$$\mu_2(1.12) = \frac{1.12 - 1.0499}{1.1244 - 1.0499} = 0.94094 \tag{12}$$

$$\mu_1(25.9) = \frac{41.69 - 25.9}{41.69 - 0.9} = 0.38710 \tag{13}$$

$$\mu_2(25.9) = \frac{25.9 - 0.899}{41.695 - 0.899} = 0.61282 \tag{14}$$

The MF of rule consequents of the fired rules Rule 7 and Rule 8 represented as $\mu_7(\Delta u)$ and $\mu_8(\Delta u)$ respectively are obtained using (2).

$$\mu_7(\Delta u) = \prod(\mu_2(1.12), \, \mu_1(25.9)) = 0.364 \tag{15}$$

$$\mu_8(\Delta u) = \prod(\mu_2(1.12), \, \mu_2(25.9)) = 0.577 \tag{16}$$

Let $T_7$ and $T_8$ represent the consequent of Rule 7 and Rule 8 respectively. The design parameters for $T_7$ are −380.309, 864.5895 and 339.5614. The design parameters for $T_8$ are −17,820, 864.8789 and 15910.7312. Using (5), the values of $T_7$ and $T_8$ are obtained

$$T_7 = -380.309 * 1.12 + 864.589 * 25.9 + (-339.561) = 21627.36 \qquad (17)$$

$$T_8 = -17820.019 * 1.12 + 864.879 * 25.9 + (-15910.731) = -13468.77 \quad (18)$$

Using (10) and substituting all the values, the output of the FLC obtained is

$$\Delta u(n) = \frac{\mu_7(\Delta u) * T_7 + \mu_8(\Delta u) * T_8}{\mu_7 + \mu_8} \qquad (19)$$

$$\Delta u(n) = \frac{0.364 * 21627.36 + 0.577 * -13468.77}{0.364 + 0.577} = 118.6018 \qquad (20)$$

Using (11), the final output of the FLC is calculated

$$\text{Effort EF(n)} = \Delta u(n) = 118.6018$$

Therefore, when the inputs Mode and Size are 1.12 and 25.9 respectively, the output Effort is 118.6. In this case study, the rules in the rulebase are 30, of which only 2 are fired. The execution time to check all the rules is increased.

# 3  Conclusion

In this paper, the application of Fuzzy Logic to Software Effort Estimation using two inputs is discussed. The detailed analytical structure of a two input FLC using triangular MFs for the Fuzzy sets is provided. The analytical study for sample inputs to the FLC is also presented. The same work can be mapped to the other type of MFs like Trapezoidal, Gauss, etc. The dataset used for this purpose is the NASA 93 dataset. Considering the various criteria for assessment of the models like VAF, MARE, VARE, Mean BRE, MMRE and Pred(30), it can be concluded that Fuzzy Models developed using triangular MFs provided better software development effort estimates.

# References

1. Babuska, R.: Fuzzy Modeling for Control. Kluwer Academic Publishers, Dordrecht (1999)
2. Attarzadeh, I., Ow, S.H.: Software development effort estimation based on a new fuzzy logic model. IJCTE 1(4) (2009)
3. Seth, K., Sharma, A., Seth, A.: Component selection efforts estimation—a fuzzy logic based approach. IJCSS 3(3), 210–215 (2009)
4. Saliu, M.O.: Adaptive fuzzy logic based framework for software development effort prediction. King Fahd University of Petroleum and Minerals (2003)
5. Prasad Reddy, P.V.G.D., Hari, CH.V.M.K., Jagadeesh, M.: Takagi-Sugeno fuzzy logic for software cost estimation using fuzzy operator. Int. J. Soft. Eng. 4(1) (2011)

6. Xu, Z., Khoshgoftaar, T.M.: Identification of fuzzy models of software cost estimation. Fuzzy Sets Sys. **145**, 141–163 (2004)
7. Alwadi, A., et al.: A practical two input two output Takagi-Sugeno fuzzy controller. Int. J. Fuzzy Syst. **5**(2), 123–130 (2003)
8. Ying, H.: An analytical study on structure, stability and design on general nonlinear Takagi-Sugeno fuzzy control systems. **34**(12), 1617–1623 (1998)
9. Ryder, J.: Fuzzy modeling of software effort prediction. In: Proceeding of IEEE Information Technology Conference, Syracuse, NY, pp: 53–56 (1998)
10. Sharma, V., Verma, H.K.: Optimized fuzzy logic based framework for effort estimation in software development. Int. J. Comput. Sci. Issues **7**(2), No 2, 30–38 (2010)
11. Zonglian, F., Xihui, L.: f-COCOMO: fuzzy constructive cost model in software engineering. In: Proceedings of IEEE International Conference on Fuzzy Systems, IEEE, pp. 331–337 (1992)
12. Ding, Y., Ying, H., Shao, S.: Typical Takagi-Sugeno PI and PD fuzzy controllers: analytical structures and stability analysis. Inf. Sci. **151**, 245–262 (2003)
13. Prasad Reddy, P.V.G.D., Sudha, K.R., Rama Sree, P.: Application of fuzzy logic approach to software effort estimation. Int. J. Adv. Comput. Sci. Appl. **2**(5), 87–92 (2011). ISSN: 2156-5570
14. Sandeep, K., Chopra, V.: Software development effort estimation using soft computing. Int. J. Mach. Learn. Comput. **2**(5) (2012)

# Malayalam Spell Checker Using N-Gram Method

P.H. Hema and C. Sunitha

**Abstract** Spell checker is a software tool which can detect incorrectly spelled words in a text document. Developing spell checker for a morphologically rich language like Malayalam is really tedious task. This paper mainly discusses about the construction of spell checker for Malayalam language. Since in Malayalam, many words can be derived from root word, it will be impossible to include all the words in a lexicon. So a hybrid method of different techniques can improve the performance of a spell checker. The method explained is an n-gram based approach and which will be inexpensive to construct without deep linguistic knowledge. Along with n gram, a Minimum edit distance algorithm is also added to detect the errors due to addition, deletion, or interchange of letters in a word. This will improve the efficiency of the spell checker. This approach will be useful when less linguistic resources are available for Malayalam language. And also the performance analysis is done with an existing method of spell checking in Malayalam language.

**Keywords** Dictionary lookups · N gram · Non word error · Real word error

## 1 Introduction

Spell checker is a tool used by the machines to process natural language. A spell checker can detect and correct misspelled words in a text document. A misspelled word can be an incorrectly spelled word which exists in an existing corpus or the word can be in shortened form. Spell checker can perform as an individual application or it can be a part of other applications [1]. A spell checker mainly includes

P.H. Hema · C. Sunitha (✉)
Department of Computer Science & Engineering, Vidya Academy
of Science & Technology, Thalakkottukara P. O, Thrissur, Kerala, India
e-mail: sunitha@vidyaacademy.ac.in

P.H. Hema
e-mail: hemaphari@gmail.com

© Springer India 2016

H.S. Behera and D.P. Mohapatra (eds.), *Computational Intelligence
in Data Mining—Volume 1*, Advances in Intelligent Systems
and Computing 410, DOI 10.1007/978-81-322-2734-2_23

two components—Error detection and Suggestion prediction modules. Error detection module detects the errors in a text document and Suggestion prediction module provides possible candidate corrections for the misspelled word. The most important requirement for any spell checker is a dictionary of different possible words of that language which will act as a corpus. How much we can improve the size of a dictionary that much we can improve the performance of a spell checker otherwise some error word may not detected by the spell checker. Mainly we can have two types of spelling mistakes in a text document. Error due to invalid words in a language is called non word error and in Real word errors the word will be correct but it will be incorrect in that context [2]. The task of spelling correction can be classified into three classes based on the errors in the text:

(1) Non-word error detection: The non word errors can be detected by using Dictionary-lookups, N-gram analysis, and Word-boundary analysis methods.
(2) Isolated word error correction: It corrects misspellings without considering the context.
(3) Context-dependent error detection and correction: It is the Real word error detection process. It is a context dependent error detection method.

Today we have many spell checkers for different languages including Malayalam. Because of the morphological richness, it is really tedious task to include all the Malayalam words into a dictionary. So other than a pure dictionary based approach we have a Rule cum dictionary based approach for detecting the misspelled words in Malayalam text document [3]. In which a morphological analyzer is incorporated along with dictionary to detect misspelled words. But sometimes the linguistic resources (post tags, Morphological analyzer, tagged corpora and lexicon), for developing such a spell checker might be unavailable. And also the person who deals with it must have deep linguistic knowledge. To overcome these problems we can go for a statistical approach for developing spell checker.

## 2 Related Work

Developing spell checker for Indian language is something different from those for Roman scripts. Due to the morphological richness of Indian languages like Malayalam, Oriya, Assamese, Tamil, Kannada and Telugu, generating spell checker for these languages are really complicated task [3]. In those languages case endings, post tags and suffixes are added with the pronouns, nouns, verbs or adverbs. Also one or more suffixes can combine with the base word. In some Indian languages like Hindi, case ending will not agglutinate with the base form. And in the case of Indian languages the combinations like verb-verb, noun-noun, verb-noun etc. are not permitted. For Dravidian language like Malayalam many words can be created from the root word. So it will be difficult to incorporate all the words in the dictionary. Purely dictionary based approach will be difficult for Malayalam language. Now we have a Dictionary cum Rule based method for spell checking in Malayalam language. It incorporates a morphological analyzer and a morphological generator to deal with flexible agglutination found in the language where suffixes can be combined with root forms of words.

Santhosh et al. developed the spell checker based on "Rule cum Dictionary based approach" [4]. This spell checker mainly includes two important modules, a language module and an engine module. It identifies part of speech for all words, classifies each word based on its behavior and tense change, classifies nouns based on their behavior, list all suffixes, and list all post-positions or auxiliary words and models rules for formation of junctions (Sanhdi). The engine module first checks the validation for the input word by comparing word with the dictionary. The words which are not there in the dictionary will be detected and it will be passed to the morphological analyzer. The analyzer will detect the error words from that. After that morphological generator will accept the output from the analyzer and it will provide suggestion for the error word. For the suggestion generation a comparison must be done with input word and words in the list from left to right and most similar 10 words have to be selected. The point from which no similar words found is noted as start point. Same comparison must be done from right to left to find the end point. Then we want to find out the intersection of these words. If no similar word is found then the start point value will be decreased by the value one, end point value is increased by one and the process is repeated. The output from the engine module is then passed to the Morphological generator to get the proper inflected forms of words and these are shown as the candidate corrections.

# 3 Proposed Method

The method for error detection and correction is described below (Fig. 1). Word unigram, syllable bigram, and syllable trigram for the words in the corpus must be pre compiled and stored [5].

## 3.1 Pre-processing Module

Tokenization of input Malayalam Unicode text is the major task of this module. In which it will tokenize the text document into a list of unique words. After tokenization the word list will be moved to the permutation generation module.

## 3.2 Permutation Generation Module

Our spell checker depends on the phonetic similarity between the letters. Sometimes the phonetic similarity between the letters may cause the spelling error

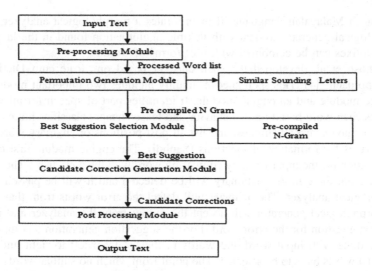

**Fig. 1** Overall architecture of the spell checker

in text document. The similar sounding letters can be detected by using the soundex
algorithm. A previous work has been done on the similar sounding letters in
Malayalam language. Through that work twenty groups of similar sounding letters
were detected. We have the list as { അ , ആ , ഃ },{ ഇ, ഈ, ി ,ീ }, { ഉ , ഊ ,
ു , ൂ },{എ) , ഏ,ഐ), ൊ, ോ, ൊ }, , { ഒ ,ഓ , ഔ ,ൊ ,ോ ,ൌ},
{ക,ഖ, ഗ, ഘ } , {ങ, ഞ } ,{ച , ഛ ,ജ ,ഝ} , {ഞ}, {ട, ഠ,ഡ ,ഢ} , { ത , ഥ
, ദ , ധ } , { പ , ഫ , ബ , ഭ } , {ന} ,{മ} , {യ} , { ര ,റ ,ഋ } , { ല , ള , ഴ }
, {ഹ }, {ശ , ഷ , സ } . After tokenization the unique Malayalam word list form
form the pre processing module will be accepted by the permutation generation
module. Then by replacing each letter in the word with the similar sounding letter in
the same letter group, the module will generate permutation for each word. In the
generated permutation list there will be correct word, the given input word and
incorrect word. For Example if the input word is written as ഭാശ, then the permutations will be generated as ഭാസ, ഭാഷ, ബാശ etc.

## 3.3 Best Suggestion Selection Module

It is the error detection and best suggestion generation module. The pre compiled N
grams are used in this module. A dictionary of unique words along with pre
compiled word unigram, syllable bigram and syllable trigram is used in this section

for efficient performance. In this module the words generated from the permutation generation module is ranked by using the pre-compiled word unigram frequencies. Pre-compiled word unigram frequency can be taken from the dictionary. By using this frequency best suggestion can be selected from the words. The highest frequency word can be selected as best suggestion. If any of the permuted is not there in the dictionary, then syllable by gram and syllable trigram frequencies can be opted for finding best suggestion. If the word is not there in the dictionary then the word unigram frequency will be zero. Then we will go for further steps. If the word generated through permutation contains more than three syllables it will be splitted into overlapping three syllable chunk sequences. Then total trigram frequency for each word can be obtained by adding the precompiled trigram frequencies. If the trigram frequencies for the words are zero then it will divide the word into overlapping sequence of two syllable chunks and use the bigram frequency form the dictionary to get the score of the word. Like this if the permuted word contain two syllables the bigram frequency can be used to identify the score of the word. After that the generated words will be sorted according to the value of the n gram frequency. Then the highest frequency word can be selected as best suggestion.

*Example 1* Best suggestion for a misspelled word using word unigram frequency

Input: ഭാസു
Permutation list: ഭാസു, ഭാഷ, ബാശ

> Step 1: Obtain the unigram frequencies from the dictionary.
> ഭാസു:0, ഭാഷ:1 , ബാശ : 0
> Step 2: choose highest frequency word as best suggestion.
> Best suggestion: ഭാഷ

*Example 2* Best suggestion for a word using syllable trigram frequency

Input : മലയാഴം

> Permutation list: മലയാഴം, മലയാളം, മലയാലം, മളയാഴം,
> മഴയാഴം, മലയഅ്ഴം, മലയഅ്ഴം, മലയഅ്ളം, മലയഅളം
>
> Step 1: divide the word into overlapping sequence of three syllable chunks
> മലയാളം = മലയ + ലയാ + യാള + ാളം
> Step 2: derive trigram frequency for above three syllable chunks from the dictionary.
> മലയ: 2  ലയാ: 2  യാള: 4  ാളം: 4
> Step 3: Add trigram frequencies to get the total score of the word.
> മലയാളം= = 2 + 2 + 4 + 4 = 12
> Similarly trigram frequency for all permuted word will be calculated like this.

Step 4:  Output best suggestion as the word with highest frequency.
         Best suggestion: മലയാളം

Similarly Best suggestion can be detected by using bigram frequency. To find the syllable bigram frequency, divide the word into overlapping sequence of two syllable chunks and will find out the bigram frequency for that word. He word with highest bigram frequency will be selected as best suggestion.

## 3.4  Candidate Correction Generation Module

This module helps to detect the errors due to adding extra letters into a word, loss of letters from a word or interchange of letters in a word. By adding this method along with the best suggestion generation we can improve the performance of the Malayalam spell checker. Minimum Edit distance Algorithm can be used to detect these types of errors [4]. Minimum Edit Distance mainly provides the number of transformations needed to reach the correct word from the source word. This method just wants to know about the list of possible destination words from the dictionary and the cost of transformations needed in the case of Substitution, insertion, Deletion errors. For that we just need a source string and a collection of acceptable final strings from the dictionary. Then from the above collection the algorithm will select candidate corrections which will be the acceptable destination strings. The set of candidate corrections must be large and accurate for covering possible solutions and also to avoid unwanted solutions. For that we want to set a threshold value, which will be the integer square root of input word. Then select words from the dictionary which having length with in the threshold value. Then from the list we want to find out the minimum edit distance of the first word, and set it as candidate correction. Then set the threshold as the value calculated in the above step. Then modify the list with words having length within the revised threshold. By repeating the steps we will get a list of possible candidate corrections.

## 3.5  Post Processing Module

In this session it will list the set of suggestions from the above modules, it will include a best suggestion word and a list of candidate corrections.

# 4 Results and Discussions

The system is designed in such a way that: When we give a Malayalam misspelled word as input to the spell checker, it will provide correct word along with some possible suggestions. For the N gram frequency detection we have created a dictionary of Malayalam words. About 80,000 words were used to create the dictionary. And for those words in the dictionary n gram frequencies were calculated. Without using any kind of linguistic resources, such as post tags, morphological analyzer, linguistic rules we can simply detect and correct the error word. A minimum edit distance algorithm also incorporated for improving efficiency of the spell checker. Through that we will get possible suggestions for a misspelled word. Figure 2 shows the output for existing system and Fig. 3 shows our proposed system output.

When we give a correct word as input to the spell checker and if the word is there in the dictionary then it will detect it as correct word. And if the word entered is correct but not there in the dictionary, then spell checker will display the word as best suggestion. Figure 4 shows the output for the word which is there in the dictionary. Figure 5 shows the other.

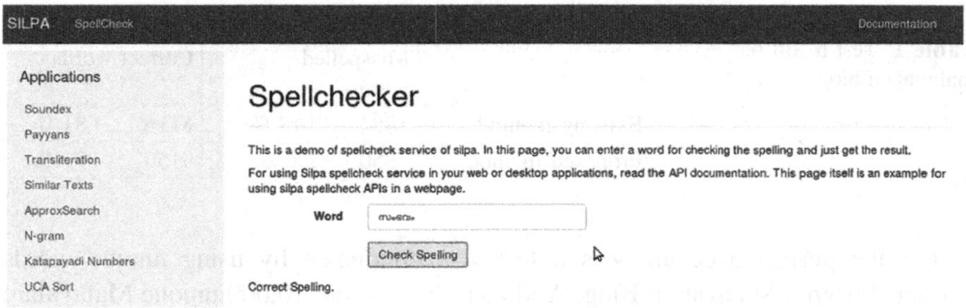

**Fig. 2** Existing system

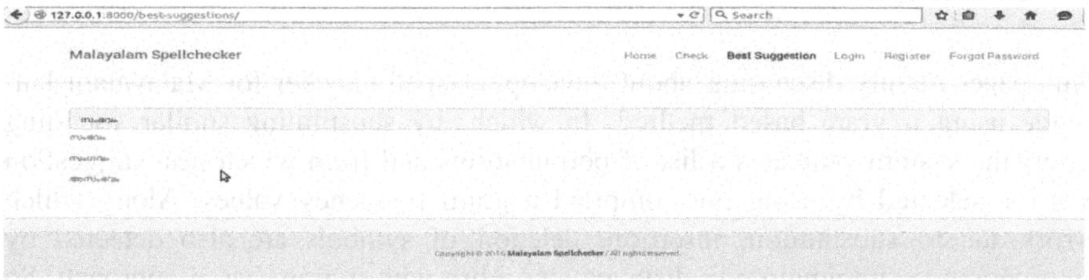

**Fig. 3** Proposed system

**Fig. 4** Output for words in the dictionary

**Fig. 5** Output for words which is not in dictionary

| **Table 1** Test result on malayalam blog | | Misspelled words | | Correct words | |
|---|---|---|---|---|---|
| | Existing method | 1884 | 18 % | 8116 | 81 % |
| | Proposed method | 850 | 8 % | 9150 | 91 % |

For the performance analysis a test was conducted by using unique words extracted from a Malayalam Blog. And there were about 10,000 unique Malayalam words in the list. The numbers of correct and incorrect words were detected by using two methods. And the result is presented in Table 1.

# 5   Conclusion

This paper mainly discussing about developing spell checker for Malayalam language using n gram based method. In which, by substituting similar sounding letters the system generates a list of permutations and from which best suggestion word is selected by using pre compiled n gram frequency values. Along which errors due to substitution, insertion, deletion of symbols are also detected by incorporating a minimum edit distance algorithm with n gram based approach. So the system will provide a list of words which includes a best suggestion and candidate corrections for an error word. Here the minimum edit distance algorithm is incorporated to improve the performance of the spell checker. This spell checker will be useful when the linguistic resources are unavailable for Malayalam language. And also it is really easy to handle by any people who have no deep linguistic knowledge. The efficiency of the spell checker can be improved by adding more words to the dictionary which is used by the edit distance algorithm and also by including morphological analyzer to support inflections.

# References

1. Kaur, R.: Spell checker for Gurmukhi script. In: Computer Science and Engineering Department. Thapar University, June 2010
2. Gupta, N., Mathur, P.: Spell checking techniques in NLP: a survey. In: Int. J. Adv. Res. Comput. Sci. Soft. Eng. 2(12), 217–22 (2012)
3. Santhosh, T., Varghese, K.G., Sulochana, R., Kumar, R.: Malayalam spell checker. In: Proceedings of the International Conference on Universal Knowledge and Language—2002, Goa, India (2002)
4. Jayalatharachchl, E., Asan, A., Wasa, A., Ruvan Eerasmg, E.: Data-driven spell checking: the synergy of two algorithms for spelling error detection and correction. In: The International Conference on Advances in ICT for Emerging Regions—iCTer, pp. 007–013. (2012)
5. Wasala, A., Weerasinghe, R., Pushpananda, R., Liyanage, C., Jayalatharachchi, E.: A data-driven approach to checking and correcting spelling errors in sinhala. Int. J. Adv. ICT Emerg. Reg. 03(01), (2010)

# Sample Classification Based on Gene Subset Selection

**Sunanda Das and Asit Kumar Das**

**Abstract** Microarray datasets contain genetic information of patients analysis of which can reveal new findings about the cause and subsequent treatment of any disease. With an objective to extract biologically relevant information from the datasets, many techniques are used in gene analysis. In the paper, the concepts like functional dependency and closure of an attribute of database technology are applied to find the most important gene subset and based on which the samples of the gene datasets are classified as normal and disease samples. The gene dependency is defined as the number of genes dependent on a particular gene using gene similarity measurement on collected samples. The closure of a gene is computed using gene dependency set which helps to know how many genes are logically implied by it. Finally, the minimum number of genes whose closure logically implies all the genes in the dataset is selected for sample classification.

**Keywords** Gene selection · Gene dependency · Closure of a gene · Sample classification

## 1 Introduction

Now-a-days, gene analysis like differentiating normal tissues from cancer tissues [1] and predicting protein fold from its sequence [2] are very important task in bioinformatics. A critical issue in analysis is feature selection which is now a big

S. Das (✉)
Department of Computer Science and Engineering, Neotia Institute of Technology,
Management and Science, Diamond Harbour, South 24-Pargana 743368, India
e-mail: das.sunanda2012@gmail.com

A.K. Das
Department of Computer Science and Technology, Indian Institute of Engineering
Science and Technology, Shibpur, Howrah 711103, India
e-mail: akdas@cs.iiests.ac.in

© Springer India 2016                                                                                    227
H.S. Behera and D.P. Mohapatra (eds.), *Computational Intelligence
in Data Mining—Volume 1*, Advances in Intelligent Systems
and Computing 410, DOI 10.1007/978-81-322-2734-2_24

attention to the researchers in the recent years for gaining meaningful knowledge from huge volume of microarray data. In data mining, the knowledge discovery [3] is the process of finding useful patterns in raw data. Today in the field of the medical domain, medical data mining creates great potential for exploring the hidden patterns. The existing techniques [4, 5] of medical data analysis are researching with (i) Treatment of incomplete knowledge (ii) Management of inconsistent information and (iii) Manipulation of various levels of representation of data. But doing this with huge volume of features (here, the genes) is a big issue. This difficulty can be handled by feature selection method which identifies and removes the irrelevant and redundant features in the data to a great extent. Now-a-days, many more intelligent methods like neural networks, decision trees, fuzzy sets and expert systems are applied in the medical fields [6]. But still it cannot manage inconsistent information. So, selection of feature subset [7–9] properly is a vital task for classification problems. Consequently, many learning algorithms can be used to obtain a set of rules in IF-THEN form, from a decision table. We have used Microarray data set for our study, which has been widely used in cancer classification experiments.

The remaining part of this paper is organized as follows: Sect. 2 describes the gene sub set selection using gene dependency and closure of the gene findings. The experimental results and the effectiveness of the method are demonstrated in Sect. 3 and finally conclusion is made in Sect. 4.

## 2   Gene Subset Selection

Let $MDS = (U, C, D)$ is the labeled microarray gene expression dataset; where $U$ is the set of samples, $C = \{g_1, g_2, \ldots, g_n\}$ is the condition attribute set contained all the genes of the dataset, $D = \{d_1, d_2\}$ is the set of decision attributes representing whether the samples are normal or cancerous. For any particular cancer disease, as all genes are not important, a relevance analysis of genes to select only the important genes is necessary.

### 2.1   Data Preprocessing

In many cases, algorithms require discrete data as it represents the data in more generalized form and as a result gives better result than those using continuous values for gene expression data. Microarray data discretization is one of the most important task for preprocessing. Several number of discretization methods like Equal Width Discretization (EWD), Equal Frequency Discretization (EFD), Kmeans Discretization are available for preprocessing of gene expression data. Here chimerge discretization [10] is applied on some publicly available microarray data which are available in http://datam.i2r.a-star.edu.sg/datasets/krbd/.

## 2.2 Similarity Based Gene Dependency

Let $g_i = \{s_{i_1} s_{i_2} \ldots s_{i_k}\}$ and $g_j = \{s_{j_1} s_{j_2} \ldots s_{j_k}\}$ are two sets of discrete sample values corresponding to the universe of samples U. Similarity factor $S_{ij}$ between two genes $g_i$ and $g_j$ are computed $\{g_i \rightarrow g_{i_1} g_{i_2} \ldots g_{i_k}\}$. $\{g_i \rightarrow g_{i_1} g_{i_2} \ldots g_{i_k}\}$ using Jaccard's coefficient described in Eq. (1).

$$S_{ij} = \frac{\{s_{i_1} s_{i_2} \ldots s_{i_k}\} \cap \{s_{j_1} s_{j_2} \ldots s_{j_k}\}}{\{s_{i_1} s_{i_2} \ldots s_{i_k}\} \cup \{s_{j_1} s_{j_2} \ldots s_{j_k}\}} \tag{1}$$

Obviously, $0 \leq S_{ij} \leq 1$ and greater its value implies higher the correlation among $g_i$ and $g_j$. $S_{ij} = 0$ implies that $g_i$ and $g_j$ have no common value and thus the two genes are considered as the most dissimilar genes and $S_{ij} = 1$ implies that both are one of the most similar genes in the set C.

Using Eq. (1), a n x n two dimensional matrix S is generated, where n is the total number of genes. For ith row of the matrix, gene dependency $g_i \rightarrow g_{j_1} g_{j_2} \ldots g_{j_k}$ is obtained such that $S_{ij_t} > \delta_i$ where $t = 1, 2, \ldots, k$. For any gene dependency $g_i \rightarrow g_{j_1} g_{j_2} \ldots g_{j_k}$ for gene $g_i$, antecedent $g_i$ is a single gene and consequent $g_{j_1} g_{j_2} \ldots g_{j_k}$ is a collection of genes. $\delta_i$ is the average similarity of gene $g_i$ with all other genes and genes with similarity less than $\delta_i$ are considered as independent genes to $g_i$. So the gene dependency $g_i \rightarrow g_{j_1} g_{j_2} \ldots g_{j_k}$ implies that all the genes in the right side are similar to $g_i$ and so they are considered as dependent to $g_i$. Thus, n gene dependency one for each gene is obtained and a gene dependency set *GD* of n elements is formed.

## 2.3 Closure of a Gene

The closure of a gene $g_i$ with respect to a set of gene dependency is the set of genes which logically implied by $g_i$. So, when a set of genes are implied by a gene $g_i$, we can select only the gene $g_i$ as the representative of this set of genes and remove other genes as unnecessary genes. So, our goal is to select minimum number of genes whose closure with respect to the gene dependency set GD contains all genes in the data set. This minimum gene set is considered as the most informative genes for disease identification. This gene subset is selected using the concept of closure of a gene(s). Let $g$ be a gene(s), *GD* is the gene dependency set, $g^+$ is the closure of $g$ with respect to *GD* which are logically determined by gene(s) $g$. In gene subset selection, we set the importance of genes based on their average similarity value computed in Sect. 2.2, greater the value assumes more important the gene is. The most important gene is named as gene g and its closure is computed using *GD* (described in algorithm below). If the closure set contains all the genes in set C then $g$ is the final gene subset (here, a single gene). Otherwise, from the genes not included in the closure set, the most important gene is selected as g and closure of it is computed using *GD*. If the union of two closure sets is the whole gene set C then the process is terminated; otherwise, selecting another gene the same process is repeated and finally closure of the selected genes contains all genes in C. The gene subset selected by the algorithm whose closure is the whole gene set is considered as the most important genes in the dataset. The overall gene subset selection algorithm is described below:

**Algorithm: Gene_Subset_Selection(*GD*, C)**

```
1.  C₁ = C //copy of genes
2.  G = φ //selected gene subset, initially empty
3.  Repeat
4.     Set g = gene in C₁ with the most average similarity
5.     G = G ∪ {g} /*selected gene*/
6.     g⁺ := g; //closure of g computed using step 7 to 11
7.     Repeat
8.        For each gene dependency β → γ in GD do
9.           If β ⊆ g⁺
10.             then g⁺ := g⁺ ∪ γ;
11.     While (changes to g⁺);
12.     If (g⁺ = = C) //algorithm termination condition
13.        then break;
14.     C₁ = C₁ - {g⁺}
15.  Until(g⁺ = = C); //continue to find the closure
16.  Return(G);
```

## 3    Result and Discussions

In the experiment, various kinds of microarray data like leukemia, Colon, prostate cancer and lung cancer are used which are publicly available as training and test dataset, summarized below in Table 1 where each dataset contains two types of samples, one group is normal and other is cancerous. The proposed method

**Table 1** Summary of gene expression (training/testing) dataset

| Dataset | No. of genes | Class name | No. of training samples (class 1/class 2) | No. of test samples (class 1/class 2) |
|---|---|---|---|---|
| Leukemia | 7129 | ALL/AML | 38 (27/11) | 34 (20/14) |
| Colon cancer | 2000 | Negative/positive | 62 (40/22) | 22 (14/8) |
| Prostate cancer | 12,600 | Tumor/normal | 102 (52/50) | 34 (25/9) |
| Lung cancer | 12,533 | MPM/ADCA | 32 (16/16) | 149 (15/134) |

**Table 2** Selected gene subset by the proposed method

| Dataset | Selected genes |
|---|---|
| Leukemia | g874, g894, g1082, g1615, g1685, g2015, g2233, g879 |
| Colon cancer | g143, g249, g765, g1042, g1153, g1423 |
| Prostate cancer | g6185, g8965, g10494, g6462, g7557 |
| Lung cancer | g2549, g7765, g7249, g3508, g9863 |

computes the gene subset based on the concept of functional dependency and closure of an attribute of database technology. The method selects genes for each dataset as the most important gene subset which is listed in Table 2 and then Tables 3 and 4 described accuracy of the Rule based classifier J48 and the probability based Naïve bayes classifier on the reduced datasets together with some other statistical measurements.

Also some rule based classifiers are run based on the selected genes for each dataset and the rules and accuracy of the classifiers are listed in Tables 5, 6, 7 and 8. The 10-fold cross validation is used for accuracy measurement.

**Table 3** Some statistical measurements obtained by J48 classifier

| Dataset | No. of genes in reduct | Correctly classified samples (total samples) | Classification accuracy (%) | Kappa statistics | Root mean squared error | Mean absolute error |
|---|---|---|---|---|---|---|
| Leukemia | 8 | 32 (38) | 84.21 | 0.6162 | 0.3647 | 0.1658 |
| Colon cancer | 6 | 48 (62) | 77.41 | 0.5167 | 0.425 | 0.2436 |
| Prostate cancer | 5 | 90 (102) | 88.23 | 0.7652 | 0.1768 | 0.1368 |
| Lung cancer | 5 | 31 (32) | 96.87 | 0.9375 | 0.3168 | 0.0313 |

**Table 4** Some statistical measurements obtained by Naive Bayes classifier

| Dataset | No. of genes in reduct | Correctly classified samples (total samples) | Classification accuracy (%) | Kappa statistics | Root mean squared error | Mean absolute error |
|---|---|---|---|---|---|---|
| Leukemia | 8 | 34 (38) | 91.89 | 0.773 | 0.2651 | 0.0796 |
| Colon cancer | 6 | 54 (62) | 87.09 | 0.7123 | 0.3431 | 0.1286 |
| Prostate cancer | 5 | 92 (102) | 92.15 | 0.843 | 0.2476 | 0.0785 |
| Lung cancer | 5 | 31 (32) | 96.87 | 0.9375 | 0.122 | 0.0216 |

**Table 5** Rule set using selected genes and its performance for leukemia dataset

| Rule based classifier | No. of rules | Rule set | Accuracy (%) |
|---|---|---|---|
| J48 | 3 | IF(g 2233 ≦ 80 and g1685 ≦ 1628): AML | 84.21 |
|  |  | IF(g2233 ≦ 80 and g1685 > 1628): ALL |  |
|  |  | IF(g2233 > 80): ALL |  |
| PART | 3 | IF(g2233 > 80): ALL | 84.21 |
|  |  | IF(g1685 ≦ 1628): AML |  |
|  |  | IF(g 2233 ≦ 80 and g1685 > 1628): ALL |  |
| Ridor | 2 | IF(g2233 < 80.5): ALL | 81.57 |
|  |  | IF(g2233 > 80.5): AML |  |
| JRIP | 2 | IF(g2233 ≦ 80): AML | 84.21 |
|  |  | IF(g2233 > 80): ALL |  |

**Table 6** Rule set using selected genes and its performance for colon cancer dataset

| Rule based classifier | No. of rules | Rule set | Accuracy (%) |
|---|---|---|---|
| J48 | 5 | IF(g765 ≦ 749, g1042 ≦ 237 and g143 ≦ 336): negative | 77.41 |
|  |  | IF(g765 ≦ 749, g1042 ≦ 237, g143 > 336 and g1153 > 693: negetive) |  |
|  |  | IF(g765 ≦ 749, g1042 ≦ 237, g143 > 336 and g1153 ≦ 693: positive) |  |
|  |  | IF(g765 ≦ 749, g1042 > 237): negative |  |
|  |  | IF(g765 > 749): positive |  |
| PART | 4 | IF(g765 ≦ 749 and g1042 > 237): negative | 79.03 |
|  |  | IF(g143 > 336, g765 ≦ 749 and g1153 ≦ 693: positive |  |
|  |  | IF(g765 > 749): positive |  |
|  |  | Except all: negative |  |
| Ridor | 3 | IF(g249 ≦ 977) and (g1042 > 242): negative | 75.80 |
|  |  | IF(g249 ≦ 707) and (g143 ≦ 659.5): negative |  |
|  |  | Except all: positive |  |
| JRIP | 3 | IF(g249 > = 1820): positive | 83.87 |
|  |  | IF(g1042 ≦ 112): positive |  |
|  |  | Except all: negative |  |

The accuracy of four classifiers on given four datasets are plotted in Figs. 1, 2, 3 and 4. From Fig. 4, it is observed that accuracy of all four classifiers are very high for Lung Cancer data. So the proposed method is best fitted with Lung cancer data.

**Table 7** Rule set using selected genes and its performance for prostate cancer dataset

| Rule based classifier | No. of rules | Rule set | Accuracy (%) |
|---|---|---|---|
| J48 | 4 | IF(g10494 <= 29, g8965 <= 210 and g6462 <= 28): Tumor | 88.23 |
| | | IF(g10494 <= 29, g8965 <= 210 and g6462 > 28): Normal | |
| | | IF(g10494 <= 29 and g8965 > 210): Tumor | |
| | | g10494 > 29: Normal | |
| PART | 4 | IF(g10494 > 29): Normal | 88.23 |
| | | IF(g8965 > 210): Tumor | |
| | | IF(g6462 <= 28): Tumor | |
| | | Except all: Normal | |
| Ridor | 4 | IF((g6185 <= 74) and (g8965 <= 166.5): Normal | 90.19 |
| | | IF((g10494 > 29.5) and (g8965 > 173): Normal | |
| | | IF((g8965 <= 209) and (g10494 > 12) and (g10494 <= 22.5): Normal | |
| | | Except all: Tumor | |
| JRIP | 4 | IF(g6185 <= 69): Normal | 84.31 |
| | | IF(g7557 > = 97 and g10494 > = 55): Normal | |
| | | IF(g8965 <= 145): Normal | |
| | | Except all: Tumor | |

**Table 8** Rule set using selected genes and its performance for lung cancer dataset

| Rule based classifier | No. of rules | Rule set | Accuracy (%) |
|---|---|---|---|
| J48 | 2 | IF(g2549 <= 54.9): ADCA | 96.87 |
| | | IF(g2549 > 54.9): MPM | |
| PART | 2 | IF(g2549 <= 54.9): ADCA | 96.87 |
| | | Except all: MPM | |
| Ridor | 2 | IF(g2549 > 73.45): MPM | 96.87 |
| | | Except all: ADCA | |
| JRIP | 2 | IF(g2549 <= 54.9): ADCA | 96.87 |
| | | Except all: MPM | |

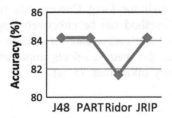

**Fig. 1** Classifier's accuracy for leukemia data

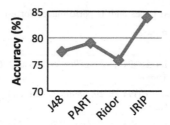

**Fig. 2** Classifier's accuracy for colon data

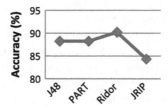

**Fig. 3** Classifier's accuracy for prostate data

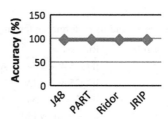

**Fig. 4** Classifier's accuracy for lung data

The method is also compared with two existing methods namely the wrapper method and the work in paper [7]. The accuracies of various classifiers obtained by the different methods are listed in Table 9. The result shows that accuracy of all seven classifiers is very high for Lung Cancer data reduced by the proposed method. So the proposed method can be effectively applied for important gene selection before analyzing Lung cancer data for disease detection. Also for other datasets we can claim that the method selects important genes as the classifier provides acceptable accuracy (more than 75 %).

**Table 9** Comparison of classification results in (%) for four microarray data

| Dataset | Methods (No. of selected genes) | Classifiers | | | | | | | | |
|---|---|---|---|---|---|---|---|---|---|---|
| | | KSTAR | NB | J48 | PART | MLP | Bagging | MCC | Average accuracy |
| Leukemia | Paper [7] (5) | 96 | 97 | 99 | 98 | 97 | 94 | 97 | 97 |
| | Wrapper (162) | 75 | 81 | 75 | 75 | 89 | 75 | 92 | 80 |
| | Proposed (8) | 84 | 92 | 84 | 84 | 100 | 84 | 95 | 89 |
| Colon cancer | Paper [7] (6) | 92 | 91 | 87 | 91 | 93 | 91 | 90 | 92 |
| | Wrapper (1) | 53 | 52 | 61 | 61 | 65 | 55 | 63 | 59 |
| | Proposed (6) | 86 | 87 | 77 | 79 | 89 | 87 | 87 | 85 |
| Prostate cancer | Paper [7] (6) | 94 | 95 | 94 | 96 | 93 | 95 | 97 | 95 |
| | Wrapper (1) | 46 | 52 | 51 | 51 | 51 | 44 | 55 | 50 |
| | Proposed (5) | 88 | 92 | 88 | 88 | 91 | 91 | 90 | 90 |
| Lung cancer | Paper [7] (4) | 94 | 96 | 94 | 85 | 93 | 92 | 93 | 92 |
| | Wrapper (37) | 88 | 78 | 75 | 75 | 78 | 84 | 69 | 78 |
| | Proposed (5) | 97 | 97 | 97 | 97 | 100 | 97 | 100 | 98 |

# 4 Conclusion

The method provides gene subset with very few genes and so the rules generated by the classifiers are easy to understand and more preferable to biologists and clinicians. The results have proven the efficacy of the method through their application to several microarray datasets. The gene subset selection by the proposed method not only gives accurate classification of cancer samples, but also identifies biologically important genes for the models. The proposed method generates gene sub set using closure property based on the dependency of genes. The generated classification rules on the basis of significant genes by some traditional classifiers namely J48, PART, Ridor and JRIP and their accuracies for various microarray dataset shows the goodness of the proposed method of sample classification for cancer detection.

# References

1. Alon, U., Barkai, N., Notterman, D.A.: Broad patterns of gene expression revealed by clustering analysis of tumor and normal colon tissues probed by oligonucleotide arrays. PNAS **96**, 6745–6750 (1999)
2. Ding, C., Dubchak, I.: Multi-class protein fold recognition using support vector machines and neural networks. Bioinformatics **17**, 349–358 (2001)
3. Tsky-Shapiro, G., Smyth, P., Uthurusamy, R.: From Data Mining to Knowledge Discovery: An Overview in Advances in Knowledge Discovery and Data Mining, pp. 1–36. (1996)
4. Lavrajc, N., Keravnou, E., Zupan, B.: Intelligent Data Analysis in Medicine and Pharmacology. Kluwer Academic Publishers (1997)
5. Wolf, S., Oliver, H., Herbert, S., Michael, M.: Intelligent data mining for medical quality management. In: Proceedings of the Fifth International Workshop on Intelligent Data Analysis in Medicine and Pharmacology, Berlin, Germany (2000)
6. Ye, C.Z., Yang, J., Geng, D.Y., Zhou, Y., Chen, N.Y.: Fuzzy rules to predict degree of malignancy in brain glioma. Med. Biol. Comput. Eng. **40**(2), 145–152 (2002)
7. Das, S., Das, A.K.: An approach towards most cancerous gene Selection from microarray data. ICCIDM **3**, 641–648 (2014)
8. Das, A.K., Pati, S.K.: Gene subset selection for cancer classification using statistical and rough set approach, pp. 294–302. Evol. Memetic Comput., Swarm (2012)
9. Das, A.K., Pati, S.K., Chakrabarty, S.: Reduct generation of microarray dataset using rough set and graph theory for unsupervised learning. In: Proceedings of the Second International Conference on Computational Science, Engineering and Information Technology, pp. 555–561. (2012)
10. Kerber, R., ChiMerge.: Discretization of numeric attributes. In: Proceedings of AAAI-92. Ninth International Conference on Artificial Intelligence, pp. 123–128. AAAI-Press, (1992)

# A Multi Criteria Document Clustering Approach Using Genetic Algorithm

D. Mustafi, G. Sahoo and A. Mustafi

**Abstract** In this work we present a multi criteria based clustering algorithm and demonstrate its usefulness in clustering documents. The algorithm proposes various metrices to judge the veracity of the clusters formed and then finds a near optimal solution that ensures good fitness scores for the all metrices. In view of the complexity of optimizing multiple clustering goals using classical optimization techniques, the paper proposes the use of an evolutionary strategy in the form of Genetic algorithm to quickly find a near optimal cluster set that satisfies all the cluster goodness criteria. The use of Genetic algorithm also inherently allows us to overcome the problem of converging to locally optimal solutions and find a global optima. The results obtained using the proposed algorithm have been compared with the outputs from standard classical algorithms and the performances have been compared.

**Keywords** Nearest neighbor · Crossover · Genetic algorithm · Chromosomes

## 1 Introduction

The unprecedented proliferation of digital documents into all aspects of our lives provides a great software challenge. Many large document repositories often require meaningful segregation of the documents into sets based on similarity metrices to facilitate retrieval while not sacrificing speed. While some amount of manual interference is acceptable, automated document clustering algorithms play a

D. Mustafi (✉) · G. Sahoo · A. Mustafi
Department of CSE, Birla Institute of Technology, Mesra 835215, India
e-mail: debjani.mustafi@bitmesra.ac.in

G. Sahoo
e-mail: gsahoo@bitmesra.ac.in

A. Mustafi
e-mail: abhijit@bitmesra.ac.in

© Springer India 2016                                                                  237
H.S. Behera and D.P. Mohapatra (eds.), *Computational Intelligence
in Data Mining—Volume 1*, Advances in Intelligent Systems
and Computing 410, DOI 10.1007/978-81-322-2734-2_25

significant role in lubricating this process. Many different clustering paradigms have been proposed in literature, but the partition based K-Means [8] method and its variants have remained the principal clustering algorithms used in the vast majority of cases. With its excellent speed and adaptability the K-Means performs extremely well with proper tuning. However, as has been documented elsewhere [6] a badly tuned K-means algorithm is prone to local minima convergence and incorrect cluster recognition. Also, in the case of document clustering it is often required to use multiple similarity parameters to ensure that the clusters are well separated. This is a challenge as documents tend to share a lot of common words even when their topics are completely different. An N-gram approach may be a solution to this but comes at the cost of huge computational load. It is believed that evolutionary algorithms can provide a viable alternative while clustering large document corpuses because of their ability to work with multiple objectives simultaneously and their exploration capabilities which negate local convergence. In this paper we propose a multi criteria GA based algorithm [6] for clustering documents, which performs excellently in comparison to the standard K-means algorithm and does not require exceptionally high end hardware resources.

## 2 Mathematical Model

The clustering of text data documents requires the documents to be represented in the form of vectors traditionally known as the Vectors space model [2]. The model takes into consideration the frequency of a term in a document and assigns a suitable weight for the same which is referred to as term frequency (TF). The assessment of more relevant terms is also very important for correct prediction of results and so along with the term frequency the sparsity of terms among the documents is also considered. This is commonly known as the Inverse Document Frequency (IDF) factor. Thus, the significance of a term in a document can be estimated by TF-IDF value [1]. Cosine similarity [5] is a commonly used similarity measure which measures the similarity of two documents in the vector space. It is defined as $\cos\left(\vec{d_1}, \vec{d_2}\right) = \vec{d_1} \cdot \vec{d_2} / \left\|\vec{d_1}\right\| \cdot \left\|\vec{d_2}\right\|$ where $\vec{d_1}, \vec{d_2}$ are two document vectors, and "$\cdot$" denotes the dot product and $\|\cdot\|$ denotes the norm of a vector. Documents found to be in close proximity of each other are then grouped to be in the same cluster, where the number of clusters to be formed is usually a user driven input. Any crisp clustering algorithm is considered to be valid if and only if it satisfies the following conditions:

1. A document vector must be assigned to only a single cluster.
2. A cluster must not be empty.
3. All documents must be assigned a cluster.

# 3  Proposed Method

## 3.1  Basic Principles and Flow Chart of GA

GA is a popular evolutionary search optimization technique where parameters are encoded in the form of strings known as chromosomes. The collection of these chromosomes is referred to as population [6]. Each chromosomes is associated with a fitness value indicative of its goodness in the search space with regards to a fitness function. The chromosomes [4] with best fitness value are selected as the population in the next generation following the Darwin's theory of survival for the fittest. Crossover [3] and mutation operations are performed to generate off springs which improves the overall health of the chromosomes over generations. The basic flow chart [9] of GA is shown in the Fig. 1.

**Fig. 1** Flowchart of simple GA

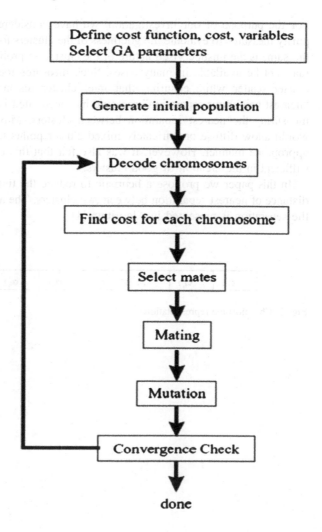

## 3.2 Chromosome Representation for Clustering

Every document $d_i$ in the data set D has been represented in the form of TF-IDF vector which is then reduced to "$p$" Singular Value Decomposed (SVD) components [7]. Our chromosomes represent a collection of "k" centroids, each having "$p$" numbers of components, as shown in Fig. 2. Each chromosome thus represents an initial estimate of "k" centroids, against which the clustering is performed. In each iteration, the GA simply finds better and better centroids till the termination condition is reached.

## 3.3 The Choice of the Fitness Function

For the purpose of our investigation, we have considered purely internal cluster purity measures to evaluate the goodness of the clusters formed. Our motivation for the same is the unsupervised nature of the clustering problem, where apriori inputs may not be available in many cases. Such measures focus on reducing the intra cluster scatter while ensuring that inter cluster distances are maximized. The "nearest neighbour separation" measure has been cited in literature as a means of measuring the nearest separation between clusters. Maximizing such a measure would allow diffuse or intricately mixed cluster points to be separated in a more appropriate manner. However, it was also felt that this one metrics would not be sufficient in the creation of good clusters.

In this paper we propose a heuristic to reduce the time required in finding the distance of nearest separation between two clusters. The algorithm used for finding the separation is presented in Algorithm [1].

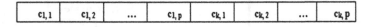

**Fig. 2** Chromosome representation

**Algorithm 1.** Algorithm to find the nearest separation between two clusters

---
FIND NEAREST SEPARATION (C1, C2, T) // C1 and C2 are two clusters T
is a scalar threshold between 0 and 1
---
1.  D1 = {distance from centroid of C1 to all points in C1}
2.  D2 = {distance from centroid of C2 to all points in C2}
3.  PT1 = {all points in C1 with a distance greater than T*max(D1) from centroid of C1}
4.  PT2 = {all points in C2 with a distance greater than T*max(D2) from centroid of C1}
5.  NNS = find minimum distance between any pair of points (i, j) where i $\in$ PT1 and j$\in$ PT2
6.  return NNS

Based on this, we propose the use of four different metrices to measure the cluster validity of the clusters returned by the algorithm. These measures are presented in equations [1–4]

$$obj1 = \frac{1}{k} \sum_{i=1}^{k} \{\cos(c_i, c_j)\} \tag{1}$$

$$obj2 = \frac{1}{k} \sum_{i=1}^{k} \left[ \frac{1}{n} \sum_{d_j \in C_i} \cos(d_j, c_i) \right] \tag{2}$$

$$obj3 = \min \left[ \cos \left\{ \sum_{d_i > T*\max(D_i)} \cos(d_i, c_i), \sum_{d_j > T*\max(D_j)} \cos(d_j, c_j) \right\} \right] \tag{3}$$

$$obj4 = k^{\wedge}[|\text{No. of unique clusters in assignments}|] \tag{4}$$

The final fitness function has been obtained by incorporating the objective functions defined in equations [1–4]. Thus the fitness function for the ith chromosome is given as

$$F_i = \alpha \frac{obj2}{obj1 * obj4} + \beta \frac{1}{obj3} \tag{5}$$

where $\alpha$ and $\beta$ are two scalar quantities. It is to be observed that while the first component, controls the inter cluster distances and intra cluster scatter, the second component controls the nearest neighbour separation. By choosing different values of the scalar parameters $\alpha$ and $\beta$ the user can control the amount of diffusivity between the various clusters.

# 4  Implementation and Results

To demonstrate the performance of the proposed method two different test beds were used. The first test bed contained a synthetically generated dataset having high diffusivity, while the second test bed contained a collection of documents from the BBC and BBC Sports corpus. The details about the datasets used for the experiments is shown in Table 1. The simulation of the synthetic datasets and the text processing used in the implementation haven been carried out in Python, while the actual clustering algorithms were implemented using Matlab. The results obtained were compared with a standard implementation of K-Means with the maximum number of iterations set to 300 and number of iterations being set to 10 (to prevent local convergence).

## 4.1  Results on the Synthetic Datasets

In the case of the synthetic dataset, the clusters can be seen to be quite diffuse (refer to Figs. 3, 4, 5 and 6). However, the GA based algorithm performs quite well in comparison to the K-Means algorithm and produces well defined clusters, even in this case. The parameters used by the GA algorithm in both the cases is shown in Table 2. To speed up the running of the GA, all functions were vectorized in MATLAB allowing for parallel computations on multiple chromosomes.

**Table 1**  Synthetic dataset 1

| Dataset attribute | Value |
| --- | --- |
| No. of samples | 600 |
| No. of features | 3 |
| Data type of features | Floating point |
| No. of clusters | 4 |
| Standard dev. of clusters | 0.5 |

**Fig. 3**  Two dim. view of synthetic dataset 1

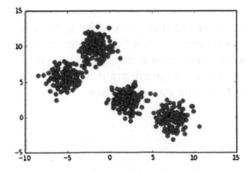

**Fig. 4** Result of K-means clustering on synthetic dataset 1

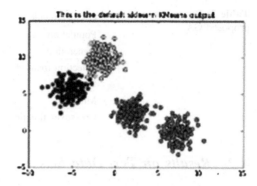

**Fig. 5** Clustering of synthetic dataset using proposed algorithm

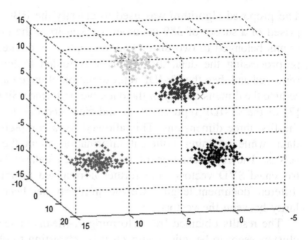

**Fig. 6** Convergence of the proposed algorithm on synthetic dataset 1

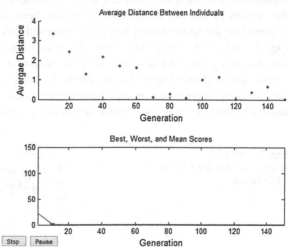

**Table 2** Parameters used by proposed GA

| GA parameter | Value |
|---|---|
| Population size | 10 * no. of features |
| Generations | 250 |
| Selection method | Tournament selection |
| Crossover | Two point crossover |
| Mutation | Adaptive mutation |
| Crossover function | 0.8 |

## 4.2   Results on Text Data Sets

The proposed algorithm was used to classify the BBC Sports dataset which comprised of a subset of 389 documents taken from the original corpus. These documents belonged to two distinct sports topics i.e. cricket and football. As a part of preprocessing the data, the corpus was folded to lower case, stop words were removed and then the TF-IDF representation of the dataset was obtained. To further reduce the dimensionality of the data, we performed Singular Value Decomposition [7] of the TF-IDF representation and truncated the decomposition to contain only the first three dimensions. This allowed for a comprehensible visualization of the data, while retaining sufficient information about the document contents themselves. The representation of this truncated dataset is presented in Fig. 7. The truncated SVD vectors were further normalized to unit length to compensate for different document lengths. Cosine similarity has been used to measure the similarity between the vectors.

The results obtained by performing K-Means is seen in Fig. 8, and while the clusters seem to be quite compact, it is interesting to observe that documents near the borders of the two clusters seem to have been wrongly clustered. The results obtained on the same dataset using the proposed GA based algorithm is shown in Fig. 9, and is observed that while the most of the results replicate the results obtained using the K-Means algorithm, the proposed algorithm seems to have done a better job in classifying the border cases. This can be attributed to the fact that not only does the algorithm consider inter cluster distance and intra cluster scatter, it also takes into account the nearest neighbour separation. The convergence of the

**Fig. 7** Two dim. representation using first two SVD vectors

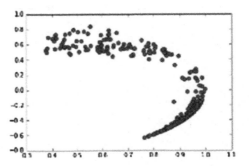

**Fig. 8** Result of K-means clustering for two clusters

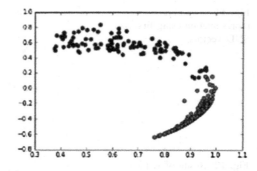

**Fig. 9** Clustering using the proposed algorithm

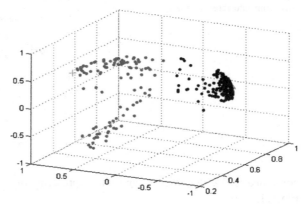

**Fig. 10** Convergence of the proposed algorithm on the BBC sports dataset

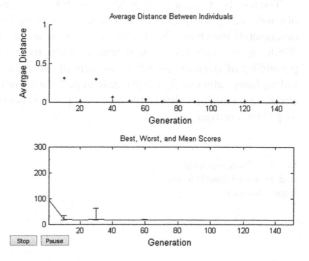

algorithm is seen in Fig. 10 and it can be seen that again the algorithm converges in a small number of iterations.

To further test the algorithm the experiment was repeated on the BBC dataset, containing 3 different topics i.e. Sports, Politics and Technology. The corpus

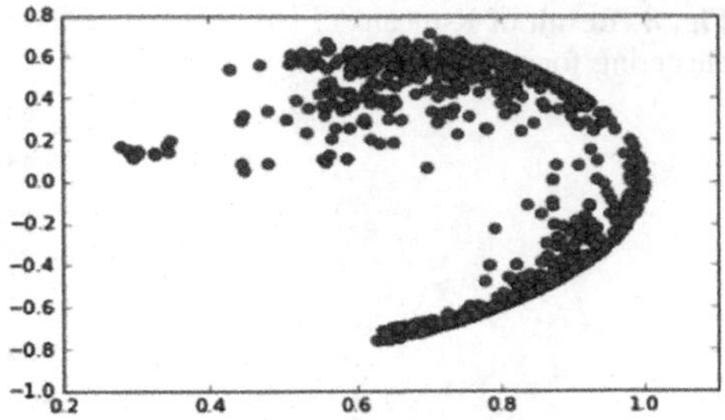

**Fig. 11** Two dim.
representation using first two
SVD vectors

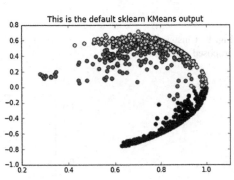

**Fig. 12** Result of K-means
clustering into three partition

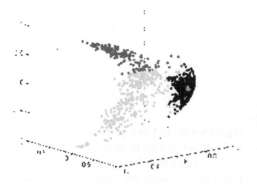

contained 1012 documents, and the representation of the dataset is shown in
Fig. 11.

The results obtained using K-Means with K = 3 is seen in Fig. 12, and the results
obtained using the proposed algorithm is seen in Figs. 13 and 14. The results
demonstrate the change in the clusters obtained while varying the values of α and β.
While using large values of α tend to form more compact clusters, there is a
possibility of overlap around the borders of the clusters. On the other hand, providing lager values of β, it is possible to pull the clusters away from each other. By
choosing a suitable value of the tuple <α, β>, it is possible to fine tune the clusters
as per requirement.

**Fig. 13** Clustering results
with proposed algorithm for
large value of α

**Fig. 14** Clustering results with proposed algorithm for large value of β

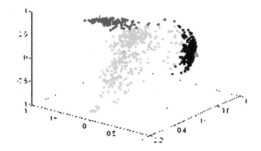

## 5 Conclusion

In this article, we propose a solution to the document clustering problem using genetic algorithms. In order to demonstrate the effectiveness of the GA based document clustering algorithm, several synthetic and real-life data sets with large number of dimensions and the number of clusters have been considered. This paper proposes a modified fitness function using the nearest neighbor criteria to improve the quality of the clusters formed. The speed of convergence of our proposed algorithm is comparable to the classical K-Means algorithm. The scalar parameter α and β can be adjusted by the user to control the quality of the cluster. Further work is in progress to improve the algorithm, particularly in the automatic setting of α and β.

## References

1. Akter, R., Chung, Y.: An evolutionary approach for document clustering. IERI Procedia **4**, 370–375 (2013)
2. Kalogeratos, A., Likas, A.: Document clustering using synthetic cluster prototypes. Data Knowl. Eng. **70**(3), 284–306 (2011)
3. Matthews, S.G., Gongora, M.A., Hopgood, A.A., Ahmadi, S.: Web usage mining with evolutionaryextraction of temporal fuzzy association rules. Knowl.-Based Syst. **54**, 66–72 (2013)
4. Mukhopadhyay, A., Maulik, U., Bandyopadhyay, S., Coello Coello, C.: A survey of multiobjective evolutionary algorithms for data mining: part i. IEEE Trans. Evol. Comput. **18**(1), 4–19 (2014)
5. Nasir, J.A., Varlamis, I., Karim, A., Tsatsaronis, G.: Semantic smoothing for text clustering. Knowl.-Based Syst. **54**, 216–229 (2013)
6. Premalatha, K., Natarajan, A.M.: Genetic algorithm for document clustering with simultaneous and ranked mutation. Modern Appl. Sci. **3**(2), (2009)
7. Rana, C., Jain, S.K.: An evolutionary clustering algorithm based on temporal features for dynamic recommender systems. Swarm Evol. Comput. **14**, 21–30 (2014)
8. Singh, V.K., Tiwari, N., Garg, S.: Document clustering using k-means, heuristic k-means and fuzzy c-means. In: Computational Intelligence and Communication Networks (CICN), 2011 International Conference on, pp. 297–301. IEEE (2011)
9. Song, W., Qiao, Y., Park, S.C., Qian, X.: A hybrid evolutionary computation approach with its application for optimizing text document clustering. Expert Syst. Appl. **42**(5), 2517–2524 (2015)

# Positive and Negative Association Rule Mining Using Correlation Threshold and Dual Confidence Approach

Animesh Paul

**Abstract** Association Rule Generation has reformed into an important area in the research of data mining. Association rule mining is a significant method to discover hidden relationships and correlations among items in a set of transactions. It consists of finding frequent itemsets from which strong association rules of the form A => B are generated. These rules are used in classification, cluster analysis and other data mining tasks. This paper presents an extensive approach to the traditional Apriori algorithm for generating positive and negative rules. However, the general approaches based on the traditional support–confidence framework may cause to generate a large number of contradictory association rules. In order to solve such problems, a correlation coefficient is determined and augmented to the mining algorithm for generating association rules. This algorithm is known as the Positive and Negative Association Rules generating (PNAR) algorithm. An improved PNAR algorithm is proposed in this paper. The experimental result shows that the algorithm proposed in this paper can reduce the degree of redundant and contradictory rules, and generate rules which are interesting on the basis of a correlation measure and dual confidence approach.

**Keywords** Data mining · Itemset · Frequent itemset · Infrequent itemset · Apriori · Positive and negative association rules · Minimum support · Different minimum support · Minimum confidence · Dual confidence · Correlation coefficient · Correlation threshold

A. Paul (✉)
Department of Computer Science and Engineering, National Institute of Technology,
Mizoram, Chaltlang, Aizawl 796012, Mizoram, India
e-mail: ani.nitmz@gmail.com; animeshpaul44@gmail.com

© Springer India 2016                                                                                      249
H.S. Behera and D.P. Mohapatra (eds.), *Computational Intelligence
in Data Mining—Volume 1*, Advances in Intelligent Systems
and Computing 410, DOI 10.1007/978-81-322-2734-2_26

# 1 Introduction

Association Rule Mining is a discipline of Data Mining which is aimed at discovering the frequent patterns among itemsets in a transactional database. It extracts interesting associations, frequent patterns and correlations among sets of items in the data repositories and transactional databases in particular [1]. Association rules provide an efficient and effective way to discover and represent data dependencies as well as correlations between attributes in a database [1]. Studies in the past have suggested that negative association rules have a high degree of accuracy at handling massive unstructured data [2]. However, these techniques also suffer drawbacks [3].

To eliminate these limitations, in this paper, we put forward an improved algorithm for generating positive as well as negative association rules because the general techniques are often based on simple measures and standards. In this paper, we determine the positive and negative association rules from a given dataset and prune contradictory positive and negative association rules.

# 2 Prior Work

Rakesh Agrawal initially specified the formal statement of Association rule mining in 1993 [2]. The first algorithm in this discipline was suggested by R. Agrawal and R. Srikant in 1994 for generating frequent itemsets for Boolean associations [4]. Apriori applies an iterative method known as a level wise search, where n itemsets are used to discover (n + 1)-itemsets using the unique apriori property which states that "All nonempty subsets of a frequent itemset is also frequent [4] as well as all nonempty supersets of an infrequent itemset is also infrequent [4]".

The primary purpose of Association rule mining is to discover association rules that fulfill a certain predefined support and confidence threshold. The problem is subdivided into two phases. The first phase generates itemsets whose support counts exceed a given predefined threshold [4]. Such an itemset is termed a frequent itemset [3, 4]. The next phase generates association rules from frequent itemsets where the rule bears a certain minimal confidence threshold. Such a rule is termed a strong rule [4]. The first phase problem can be two separated into candidate large itemsets $(C_k)$ generation process and frequent itemsets generation $(L_k)$ process [3]. Those itemsets that are expected to be large or frequent are known candidate itemsets $(C_k)$.

Other methods which followed Apriori algorithm [4] include correlation analysis [5], FP-Growth Tree analysis [6, 7], AIS [4], DHP [8], Partition [9] were adopted to improve performance in the later years.

# 3 Preliminary Terms and Definitions

## 3.1 Concepts and Definitions

Let $A = \{A1, A2,..., Am\}$ be a set of m different attributes of a transactional database, T be the transaction that comprises a set of items so that $T \subseteq A$. Let D be a transactional database with different transactions T's. An association rule is an implication of the form of $X \Rightarrow Y$, where $X, Y \subset A$ are sets of items, termed itemsets, and $X \cap Y = \emptyset$ [5]. An itemset of length p is called a p-length itemset. A transaction T contains itemset X if $X \subseteq T$. X is termed *antecedent*. Y is termed *consequent*.

Support (s) indicates the ratio of the number of transactions that contain both *antecedent* and *consequent* terms to the overall transactions in the database. Absolute support of X is the total number of transactions T (such that $X \subseteq T$) in the entire transactional database. Relative support is the ratio of the absolute support to the total number of transactions in the transactional database [2]. The Rule $X \Rightarrow Y$ bears a *support* (denoted by *supp*) s in a database if s% of database transactions out of all transactions contains both X and Y i.e.

$$supp(X => Y) = (supp(XUY))/\text{Number of database records} = P(XUY) [2].$$
(1)

Confidence(c) indicates the ratio of the number of transactions that contain both *antecedent* and *consequent* terms to the overall transactions that contain only the *antecedent* term [2]. The Rule $X \Rightarrow Y$ has a *confidence* (denoted by *confd*) in database if c% of transactions out of all transactions that contain X also contain Y i.e.

$$confd(X => Y) = (supp(XUY))/supp(X) = P(Y|X) [2].$$
(2)

## 3.2 Apriori Rule

The *support-confidence framework* [2, 4] for association rule mining can be categorized into the following two tasks [2, 5]:

1. Generate frequent itemsets: Itemsets that meet the *support threshold (minsup/msu)* value are generated.
2. Rules that bear a *confidence threshold (minconf/mc)* are generated using this technique: For every frequent itemset F and any $G \subset F$, let $H = F - G$. If the rule $H \Rightarrow G$ exceeds the *confidence* threshold, then it is a qualified rule.

The rules rendered are called *interesting positive rules*.

A *frequent itemset* (denoted as FL) [7] is an itemset that fulfills the *msu*. Similarly, an *infrequent itemset* (denoted as IL) is an itemset that does not fulfill the user-specified *msu* [7].

An iterative approach is used to first generate the set of frequent-1 itemsets $L_1$, then the set of frequent-2 itemsets $L_2$, and so on until for a value of r when the set $L_r$ becomes NULL. The algorithm is then terminated. The set of candidates $C_k$ is generated by performing a $(k-2)$-join operation on the frequent-$(k-1)$ itemsets $L_{k-1}$ during the $k$-th iteration of the procedure [5]. The itemsets in this set $C_k$ are potential *candidates* for frequent itemsets, and the final set of obtained frequent itemsets $L_k$ is a subset of $C_k$. Each element of $C_k$ is to be validated against the transactional database to see if it indeed belongs to $L_k$ [5].

The validation of the candidate itemset $C_k$ against the transaction database appears to be a constriction for the algorithm. In order to improve the efficiency of the algorithm, the apriori property [2, 4] is introduced which states that all subsets of a frequent itemset are also frequent, and all supersets of an infrequent itemset are also infrequent.

### 3.3 Negative Association Rules

The absence of the itemset A means negation of the itemset and is indicated by $\neg$A, [10]. A rule of the form A => B is a positive association rule whereas rules of other forms (A => $\neg$B, $\neg$A => B, $\neg$A => $\neg$B) are negative association rules [5, 11].

The support and confidence of the negative association rules are expressed in terms of those of the positive association rules [10]. The support is determined by the following formulae [5, 10]:

$$supp(\neg A) = 1 - supp(A) \tag{3}$$

$$supp(A \cup \neg B) = supp(A) - supp(A \cup B) \tag{4}$$

$$supp(\neg A \cup B) = supp(B) - supp(A \cup B) \tag{5}$$

$$supp(\neg A \cup \neg B) = 1 - supp(A) - supp(B) + supp(A \cup B) \tag{6}$$

The confidence is derived from the following expressions [5, 10]:

$$confd(A => \neg B) = (supp(A) - supp(A \cup B))/supp(A) = 1 - confd(A => B). \tag{7}$$

$$confd(\neg A= > B) = (supp(B) - supp(AUB))/(1 - supp(A)) \qquad (8)$$

$$confd(\neg A= > \neg B) = (1 - supp(A) - supp(B) + supp(AUB))/(1 - supp(A))$$
$$= 1 - confd(\neg A= > \neg B)$$

$$(9)$$

The *negative association rules* generation process seeks to determine rules of the three types with their *support* and *confidence* values meeting the user-specified *support* and *confidence* thresholds respectively. These turn out to be *interesting rules.*

The number of negative association rules in the entire search space for all non-frequent itemsets in any database is exponentially large, which is also the primary reason that the study of negative association rules is more challenging than their positive counterparts [10]. While computing the support of all non-frequent itemsets is not only surrealistic but also impractical, it must be taken care that the negative rules mined must not be contradictory to each other as well as not redundant. Any negative association rule mining process may be divided into three sub-problems [5] (Ref to Sect. 5):

# 4 Parameters Involved

This paper presents an improved *PNAR* algorithm over the original *PNAR* algorithm, considering [5] as the base paper of proposal. It uses a correlation coefficient measure and dual confidence approach to achieve this [12].

## 4.1 Correlation Coefficient

While mining positive and negative association rules simultaneously, the mining rules appear to be contradictory quite often. For example, the rules A => ¬B and ¬A => B are contradictory. In order to eliminate such contradictions, we identify the nature of association by the correlation coefficient value (denoted as *corrl* (A, B)) [5, 11]:

$$corrl(A, B) = supp(AUB) - supp(A){*}supp(B) \qquad (10)$$

The value of correlation coefficient may exist in the following three situations:

1. If *corrl(A,B)* > 0, then *A* and *B* exhibits positive correlation.
2. If *corrl(A,B)* = 0, then *A* and *B* are independent.
3. If *corrl(A,B)* < 0, then *A* and *B* exhibits negative correlation.

## 4.2 Dual Confidence and Correlation Threshold

Let us consider two items $A$ and $B$, with low support counts. $supp(AUB)$ is still smaller. The $confd(A => B)$ may be large or small, but $confd(\neg A => \neg B)$ is very large [12]. If all types of rules use the same confidence value, a critical circumstance might arise. If confidence is very less, then a large number of rules will be generated. Also, the presence of a large number of $(\neg A => \neg B)$ type rules is impractical and unnecessary. If confidence is greater, many important positive association rules may be erroneously eliminated. Therefore, the best solution here is to use different confidence thresholds, and thus, we set two confidence thresholds, *Pminconfd* and *Nminconfd* [12]. *Pminconfd* represents the confidence threshold of the rules of the type $A => B$ and $\neg A => \neg B$ and *Nminconfd* represents the confidence threshold of the rules of type $A => \neg B$ and $\neg A => B$ [12]. Also, to prune the search space of negative rules, we use another threshold called different minimum support (*dms*) only for generating negative rules [5, 12].

From the analysis of Sect. 3.3 it can be proved that

$$Pminconfd + Nminconfd = 1(\text{Unity Constraint})[16] \qquad (11)$$

The paper [13] proposed that if a rule meets

$$supp(AUB) - supp(A)*supp(B) \to 0 \qquad (12)$$

then, the rule is not interesting i.e. only if the correlation coefficient value of the rule surpasses a certain threshold, the rule will be termed an interesting one [12]. Therefore, we set the correlation threshold value (*mincorr*) to a certain value ($\to 0$, typically 0.1), so that only those association rules which satisfy

$$supp(\text{AUB}) - supp(\text{A})*supp(\text{B}) > = mincorr \qquad (13)$$

will be termed interesting rules [12, 14]. For the negative association rules,

$$(supp(AUB) - supp(A)*supp(B)) < 0 \qquad (14)$$

is negative, therefore, the absolute value has to be determined, i.e., if $A => B$ is an interesting rule, iff

$$|supp(\text{AUB}) - supp(\text{A})*supp(\text{B})| > = mincorr \qquad (15)$$

## 5 Algorithm Design

Positive and negative association rule generation process can be categorized into three sub problems, similarly to how positive rules are generated [5, 11, 13].

(1) Generate sets FL(frequent itemsets) and IL(infrequent itemsets) for every iteration.
(2) Generate rules of the type A => B and ¬A => ¬B in FL.
(3) Generate rules of the type A => ¬B and ¬A => B in IL.

Consider *DB* to be a transactional database, and *msu* (minimum support), *dms* (different minimum support) and *P_mc* (confidence threshold generating positive rules), *N_mc* (confidence threshold generating negative rules), *mincorr* (correlation threshold) are specified by the user. The proposed algorithm for generating positive and negative association rules using a correlation coefficient measure and dual confidence approach is an improvement to the original PNAR algorithm proposed in [5] as well as [12].

```
Algorithm: Positive and Negative Association Rules
*Improved_Algorithm
Input: DB, msu, dms, P_mc, N_mc, mincorr
Output: ARL (PARL U NARL)
/*Set of all Association Rules*/
/*Generate FL: set of frequent itemsets and IL: set of
infrequent itemsets*/

(1) PARL ← Ø; NARL ← Ø; /* Positive and Negative Asso-
ciation Rules*/
(2) FL ← Ø; IL ← Ø;
(3) Scan DB and find L₁ and NL₁ if it exists
(4) FL ← FL U L₁; IL ← IL U NL₁;
(5) for (r=2; L_{r-1} != Ø; r++){
(6) C_r = L_{r-1} |X| L_{r-1} /* The Join Step */
(7) For each i € C_r if any subset of i is not present in
L_{r-1} then
(8) C_r = C_r - {i} /* Apriori Rule */
(9) For each i € C_r {
(10)   s = supp(i)
(11)   if (s >= msu) then {
(12)   L_r ← L_r U {i}
(13)   FL ← FL U L_r } /* end if */
(14)   else {
(15)   NL_r ← NL_r U {i}
(16)   IL ← IL U NL_r }/*end else */
```

```
/*Generate Positive Association Rules in FL*/

(17)   For every frequent itemset i in FL
(18)   For subsets A,B of i such that AUB=i and A∩B=∅ {
(19)   corrl(A,B) = supp(AUB) - supp(A)*supp(B)
(20)   if |corrl(A,B)| >= mincorr {/*only then interesting
[13],[16]*/
(21)   if  corrl(A,B)  >  0  then  {/*Extract  (A=>B)  and
(¬A=>¬B)*/
(22)   if confd(A=>B) >= P_mc then
(23)   PARL ← PARL U {A=>B}; /* Positive Rules[16] */
(24)   if supp(¬A=>¬B) >=dms and confd(¬A=>¬B) >=P_mc then
(25)   NARL ← NARL U {¬A=>¬B}; /* Positive Rules [16]*/
(26)   } /* end if of (21) */
(27)   } /* end if of (20) */
(28)   }/* end For each of (18) */

/*Generate Negative Association Rules in IL*/

(29)   For every infrequent itemset i in IL
(30)   For subsets A,B of i such that AUB=i and A∩B=∅ {
(31)   corrl(A,B) = supp(AUB) - supp(A)*supp(B)
(32)   if |corrl(A,B)| >= mincorr {/*only then interesting
[13],[16]*/
(33)   if  corrl(A,B)  <  0  then  {/*Extract  (¬A=>B)  and
(A=>¬B)*/
(34)   if supp(A=>¬B) >= dms and confd(A=>¬B) >= N_mc then
(35)   NARL ← NARL U {A=>¬B}; /* Negative Rules [16]*/
(36)   if supp(¬A=>B) >= dms and confd(¬A=>B) >= N_mc then
(37)   NARL ← NARL U {¬A=>B}; /* Negative Rules [16]*/
(38)   } /*end if of (33)*/
(39)   }/*end if of (32)*/
(40)   }/*end For each of (30)*/
(41)   }/*end For each of (9)*/
(42)   }/*end For of (5)*/
(43)   ARL = PARL U NARL /*Set of Association Rules */
(44)   Return(ARL);
```

This Algorithm extracts positive association rules in FL, as well as negative association rules in IL [5]. While generating negative association rules, a different support threshold is brought in place to improve the utility of the frequent negative itemsets. With a correlation threshold and dual confidence approach, the algorithm generates all interesting as well as valid association rules quickly.

## 6 Experimental Results

Our experiments were performed on the following thin dataset to study the nature and behaviour of the algorithm [7] (Table 1).

In the following tables all frequent k—itemset are denoted by $L_k$. Given that $msu = 0.2$, all the positive and negative frequent itemsets are hereby generated in Tables (3, 4 and 5) by our algorithm. Table 2 lists the frequent-1 itemsets when $msu = 0.2$ (and so on until Table 4).

Table 5 lists a comparison between Apriori [4], SRM [15], *PNAR* [5] and our proposed algorithm. Table 4 helps us identify that the improved *PNAR* can reduce the number of the frequent-n itemsets. Our algorithm thus, helps in pruning the search space effectively and improves the efficiency as well as amends past proposed mining methods.

## 7 Analysis of Results

From our experimental computations, we observe that the maximum correlation coefficient for positive rules (of type A => B and ¬A => ¬B) is 0.25 and the maximum negative correlation coefficient for rules (of type ¬A => B and A => ¬B)

**Table 1** Transactional database

| TID | Items |
|-----|-------|
| 1.  | G, A, C, F, I, D |
| 2   | B, H, J, C |
| 3   | G, I, E, C, D |
| 4   | H, J, B, F, A |
| 5   | I, A, D, C |
| 6   | E, J, I, G |
| 7   | G, H, F, B, D |
| 8   | J, F, C, B |
| 9   | I, E, C, B |
| 10  | J, F, D, H, A |
| 11  | J, E, B, G, A |
| 12  | I, F, D, H, B |
| 13  | F, G, E, B, C, I |
| 14  | D, J, A, H, B |
| 15  | G, A, C, E, F, I |
| 16  | H, F, I, C, B, D |
| 17  | A, F, J, C, H, E |
| 18  | E, G, C, D, A, B |
| 19  | I, J, C, H, A, B |
| 20  | E, C, I, A, G, D, J |

**Table 2** Frequent-1 itemsets ($L_1$) when $msu = 0.2$

| $L_1$ | *supp* |
|-------|--------|
| A | 0.60 |
| B | 0.60 |
| C | 0.60 |
| D | 0.50 |
| E | 0.45 |
| F | 0.50 |
| G | 0.45 |
| H | 0.45 |
| A | 0.55 |
| J | 0.50 |

**Table 3** Frequent-2 itemsets ($L_2$) when $msu = 0.2$

| $L_2$ | *supp* |
|-------|--------|
| AC | 0.70 |
| AI | 0.60 |
| CI | 0.80 |
| HJ | 0.60 |
| ¬AC | 0.40 |
| C¬D | 0.60 |
| B¬C | 0.60 |
| A¬B | 0.60 |
| A¬F | 0.70 |
| A¬C | 0.50 |

**Table 4** Frequent-3 itemsets ($L_3$) when $msu = 0.2$

| $L_3$ | *supp* |
|-------|--------|
| ACI | 0.25 |
| AHJ | 0.25 |
| CEG | 0.25 |
| CDG | 0.20 |
| AB¬G | 1.00 |
| ¬BCD | 1.00 |
| ¬ACF | 0.95 |
| ¬AEG | 0.95 |
| ¬AGI | 0.95 |
| A¬BH | 0.90 |

is −0.44. Percentage of rules having correlation coefficient greater than 0.20 is 30.05 % and the percentage of rules having correlation coefficient less than −0.25 is 35.75 %. Our objective is to eliminate highly negative correlated rules constituting in this 35.75 % as well as to increase the former figure beyond 30.05 %. From our

**Table 5** Comparison of all algorithms with frequent positive($L_k$) itemsets

| Algorithm | minsup | dms | mc/P_mc | dmc/N_mc | mincorr | $L_1$ | $L_2$ | $L_3$ |
|---|---|---|---|---|---|---|---|---|
| Apriori | 0.20 | | mc = 0.70 | | | 10 | 30 | 15 |
| SRM | 0.20 | | mc = 0.70 | | | 20 | 42 | 26 |
| PNAR | 0.20 | 0.50 | mc = 0.70 | dmc = 0.15 | | 10 | 40 | 22 |
| Improvedpnar | 0.20 | 0.50 | P_mc = 0.70 | N_mc = 0.30 | 0.10 | 10 | 35 | 18 |

experimental analysis, we identify the rules which have a high negative correlation coefficient and can therefore discard them from our study. Also, we can try to identify those rules which have a high positive correlation coefficient since they are strong positive rules. The range of correlation function is from −1 to +1. Any rule which has a correlation coefficient close to zero ($corr \rightarrow 0$) will be redundant because the *antecedent* and *consequent* terms will be independent in such a situation.

## 8 Conclusion and Future Work

We have made an improvement to the traditional *PNAR* algorithm [5] for effectively generating positive as well as negative association rules in transactional databases. We have adopted the dual confidence and correlation threshold approach where P_mc and N_mc values sum to give unity. We have developed pruning strategies for pruning the search space and the correlation threshold to monitor the interestingness and validity of the rules mined. Our objective is to find rules which have a high positive correlation coefficient value i.e. which have high positive correlation and eliminate rules which are highly negative correlation.

In future, this algorithm may be performed for various large datasets as well as other factors, techniques and parameters can be taken into consideration namely Hash-based technique [6], F-P growth tree analysis [6], chi square correlation analysis [6], Bayesian approach [6] and Kulczynski measures [6] which will help generate more interesting association rules [16]. Recent works have been made in these disciplines.

## References

1. Junwei, L.U.O., Bo, Z.: Research on mining positive and negative association rules. In: International Conference on Computer and Communication Technologies in Agricultural Engineering, pp. 302–304. (2010)
2. Agrawal, R., Imielinski, T., Swami, A.: Mining association rules between sets of items in massive databases. In: ACM SIGMOID International Conference on Management of Data, ACM, Washington D.C., pp. 207–216. (1993)

3. Brin, S., Motwani, R., Silverstein, C.: Beyond market generalizing association rules to correlations. In: Processing of the ACM SIGMOID Conference, pp. 265–276. ACM Press (1997)
4. Agrawal, R., Srikant, R.: Fast algorithms for mining association rules. In: 20th International Conference on Very Large Databases (VLDB), pp. 487–499. Santiago, Chile (1994)
5. Zhu, H., Xu, Z.: An effective algorithm for mining positive and negative association rules. In: International Conference on Computer Science and Software Engineering, pp. 455–458. (2008)
6. Han, J., Kamber, M., Pei, J.: Mining frequent patterns, associations and correlations: basic concepts and methods., advanced pattern mining. In: DATA MINING Concepts and Techniques, 3rd edn. Waltham, MA 02451, USA: Elsevier, pp. 243–317. (2011)
7. Chen, M., Han, J., Yu, P.: Data mining: an overview from a database perspective. IEEE Trans. Knowl. Data Eng. **8**(6), 866–883 (1996)
8. Park, J.S., Chen, M.S., Yu, P.S.: An effective hash-based algorithm for mining association rules. In: ACM SIGMOID International Conference on Management of Data, pp. 175–186. San Jose, California (1995)
9. Savasere, A., Omiecinski, E., Navathe, S.: An Efficient algorithm for mining association rules in large databases. In: International Conference on Very Large Databases (VLDB), pp. 1–24. Zurich, Switzerland (1995)
10. Dong, X., Wang, S., Song, H., Lu, Y.: Study on negative association rules. Trans. Beijing Inst. Technol. **24**(11), 978–981 (2004)
11. Wu, X., Zhang, C., Zhang, S.: Efficient mining of both positive and negative association rules. ACM Trans. Inf. Syst. **22**(3), 381–405 (2004)
12. Piao, X., Wang, Z., Liu, G.: Research on mining positive and negative association rules based on dual confidence. In: Fifth International Conference on Internet Computing for Science and Engineering, pp. 102–105. (2010)
13. Geng, L., Hamilton, H.J.: Interestingness measures for data mining: a survey. ACM Comput. Surv. 1–32 (2006)
14. Jalali-Heravi, M., Zaiane, O.R.: A study on interestingness measures for associative classifiers. In: Proceedings of the ACM Symposium on Applied Computing, pp. 1039–1046. (2010)
15. Teng, W., Hsieh, M., Chen, M.: On the mining of substitution rules for statistically dependent items. In: International Conference on Data Mining, pp. 442–449. (2002)
16. Kavitha, J.K., Manjula, D.: Recent study of exploring association rules from dynamic datasets. Int. J. Adv. Res. Comput. Sci. Manag. Stud. **2**(11), 236–243 (2014)

# Extractive Text Summarization Using Lexical Association and Graph Based Text Analysis

R.V.V. Murali Krishna and Ch. Satyananda Reddy

**Abstract** Keyword extraction is an important phase in automatic text summarization process because it directly affects the relevance of the system generated summary. There are many procedures for extracting keywords, but all of these aim to find the words that directly represent the topic of the document. Identifying lexical association between terms is one of the existing techniques proposed for determining the topic of the document. In this paper, we have made use of lexical association and graph based ranking techniques for retrieving keywords from a source text and subsequently to assign them a relative weight. The individual weights of the extracted keywords are used to rank the sentences in the source text. Our summarization system is tested with DUC 2002 dataset and is found to be effective when compared to the existing context based summarization systems.

**Keywords** Extractive text summarization · Lexical association · PageRank

## 1 Introduction

Automatic Text summarization has received a great deal of attention because of the large volumes of information available on the World Wide Web. Text summarization is the process of condensing a given text without losing the gist of information so that readers can quickly get some understanding of that text. Hovy et al. [1] defined the summary as a text which is based on one or more texts; it has the

R.V.V. Murali Krishna (✉)
IT Department, G.V.P College of Engineering (A), Visakhapatnam, India
e-mail: rvvmuralikrishna@gmail.com

Ch. Satyananda Reddy
CS & SE Department, Andhra University, Visakhapatnam, India
e-mail: satyanandau@yahoo.com

© Springer India 2016
H.S. Behera and D.P. Mohapatra (eds.), *Computational Intelligence in Data Mining—Volume 1*, Advances in Intelligent Systems and Computing 410, DOI 10.1007/978-81-322-2734-2_27

most important information of the main texts and its content is less than half of the main texts. Due to the increased use of online information access using smart phones, the text summarization techniques are receiving a great deal of attention. Text summarization techniques can be classified as Extractive and Abstractive based on the summary generated [2]. When a summary contains the sentences of the source document in their original form it is called as an Extractive summary. When sentences in the source document are modified to form a summary then that summary is called as abstractive summary.

Summarization techniques focus on the extraction of keywords from the source document. Keywords have an impact on the relevance of system generated summary. Studies have shown that context based keyword extraction improves the quality of system generated summaries [3–5]. Goyal et al. [6] have proposed "a context based word indexing model for document summarization". In this paper we adopt their text summarization process with few variations. In short, our text summarization process can be described as a three step process towards achievement of a summary. First step is focused on the extraction of topical words (words that describe the topic of the document) in the document. In the Second step topical words are assigned a relative weight. Important sentences are identified in the third step by utilizing the weights of the topical words.

## 2 Related Work

Several techniques have been proposed for Automatic text summarization, which are either directly or indirectly dependent on the keywords in the source text. The quality of an automatic summarization system mainly depends on the strength of the keyword extraction algorithm. Keyword extraction algorithms makes use of statistical analysis of the given text and linguistic knowledge to identify the keywords in the source text. Statistical analysis involves calculation of term frequency [7], inverse document frequency [8], word co-occurrence [9, 10] and other measures. Linguistic knowledge provides information of language related features like Parts of Speech (POS) of terms, synonyms and grammar. Goyal et al. [6] proposed that statistical analysis of the text document could lead to identification of the context of the document. They have defined the concept of topical words to describe the context of a document and described a procedure to identify topical words in the text document. This procedure is based on the assumption that the ratio of topical words is higher in a summary of a document than in original document. Also it is considered that the topical words are more related to each other than non topical words. The topical words extracted from a given text document are assigned a context sensitive weight using a graph based text ranking algorithm. The context sensitive weights along with neighborhood knowledge [11] of the sentences are used in ranking the sentences.

Mihalcea et al. [12] proposed a graph representation for a text document in which words form the vertices and the relationship between words is represented by edges between the vertices. Distance between the words in a given sentence is used to in relating the words. Two words are related if and only if they appear within a distance of N words in the given text. The significance of each vertex in the graph is calculated by TexRank [12] algorithm obtained by slightly modifying the existing PageRank algorithm [13]. TextRank algorithm considers edge weights while ranking the vertices. Aggarwal et al. [14] proposed the concept of distance graphs to graphically represent a given text and made few observations on the use of each of these representations. Distance graphs can be classified based on the chosen distance between a given pair of words to be represented as vertices. In a distance graph of order k, two vertices are related to each other only if their corresponding words do not have more than k words in between them. Also a distance graph preserves the order of the words in the sentences as it is a directed graph Sub-headings.

## 3 Proposed Keyword Extraction Process

The proposed keyword extraction process consists of three stages. At each stage a group of keywords is generated which is subsequently used in the later stages. Also at each stage the source text document (document to be summarized) is modified to suit the text processing involved at that level. The source text document is represented as a bag of words during text processing at stage-1. During text processing at stage-2 the source text document is represented as a collection of sentences containing only the keywords generated in stage-1. Source Text document is represented as a graph in stage-3.

### 3.1 Stages in Keyword Extraction Process

#### 3.1.1 Stage-1

In stage-1, the source text document is treated as a bag of words. Let us name it as WordsBag. Stop words are removed from this WordsBag and the remaining words are tagged with their parts of speech using a POS tagger [15, 16]. Words with parts of speech other than verbs, nouns, adverbs and adjectives are removed from the WordsBag. Let us call these words in the WordsBag as stage-1-keywords.

#### 3.1.2 Stage-2

In stage-2, the source text document is modified such that each sentence in the document retains only the words belonging to stage-1 keywords while preserving their original sequence in that sentence. Using this Modified text document, Lexical association is calculated for each unique pair of stage-1 keywords. If there are N stage-1 keywords the total number of unique pairs will be not more than (N * N − N)/2.

## Calculation of Lexical Association

Lexical association defines the relation between a pair of words in a given text. Statistical relationship is one form of lexical association between words. This relationship is calculated by considering the positions of the words in the given text. Goyal et al. [6] have derived a formula for lexical association using term frequency, term co-occurrence and inverse document frequency. Term co-occurrence is calculated based on Co-occurrence of words in various documents. Lexical association indicates the overall relationship at corpus level and is expressed in terms of some probability. In our experiments we have used a more precise and direct co-occurrence measure for calculating lexical association. A sentence level co-occurrence measure is defined as it would give a document level (not at corpus level) lexical association between two words. We have defined a co-occurrence measure for calculating lexical association between two words in the document in a scale of Zero to One.

$$\text{lexicalassociation}(w1, w2)\frac{\text{CO-OCCURRENCES}(W1, W2)}{\text{ISF}(W1) + \text{ISF}(W2) - \text{CO-OCCURRENCES}(W1, W2)}$$

$$(1)$$

where CO-OCCURRENCES$(W1, W2)$ gives the number of sentences in which words $w1$ and $w2$ appear together, and ISF $(w1)$ gives the number of sentences in the source text document in which the word $w1$ appears and ISF $(w2)$ gives the number of sentences in which the word $w2$ appears.

## Calculation of Average Lexical Association Among the Words of a Document

Let $(X, Y)$ represent ith pair of words among the word pairs formed with stage-2 keywords and LexicalAssociation$_i$(X,Y) be the lexical association between the words X and Y. Let us consider there are k unique pairs of stage-2 keywords in document D.

$$\text{AverageLexicalAssociation}(D) = \frac{\sum_1^k \text{LexicalAssociation}_i(X, Y)}{k} \qquad (2)$$

where AverageLexicalAssociation (D) represents the average of lexical associations calculated on the pairs of stage-2 keywords for document D.

## Obtaining Stage-2 Keywords from Stage-1 Keywords

Goyal et al. [6] defined a topical word as the keyword that is closely associated with other keywords in the text. This ideology has been implemented in our work to obtain stage-2 keywords from stage-1 keywords. The stage-1 keywords which maintain strong lexical association with maximum number of other stage-1 keywords are chosen as stage-2 keywords. Two keywords are said to have strong lexical association if the lexical association between them is greater than the AverageLexicalAssociation value for that document.

### 3.1.3   Stage-3

In stage-3, the source text document is modified such that each sentence in the document retains only the words belonging to stage-2 keywords without changing their sequence in the text. Let us call this document as stage-3 document. Using this stage-3 document, a graph is constructed.

**Representation of the Text Document as a Graph**

Consider a document D with the following three sentences

**Sentence 1:**   Rama is the king of Ayodhya
**Sentence 2:**   Ravana fought with Rama
**Sentence 3:**   Rama killed Ravana

Assume the words Rama, king, Ayodhya, Ravana, fought, killed form the stage-2 keywords for the document D. After removing words in each sentence of the document which are not present in stage-2 keywords, we obtain Stage-3 document.

**Stage-3 document**

**Sentence 1:**   Rama is the king of Ayodhya
**Sentence 2:**   Ravana fought Rama
**Sentence 3:**   Rama killed Ravana

**Graph Construction**

Let G(V,E) be the graph representation of document D where set V represents vertices and set E represents edges between the vertices. All the stage-2 keywords of document D form the set V. In other words they are chosen as the vertices of the graph G. Edges are drawn between the keywords based on the order in which they appear in the sentences of the source text document. Let us consider sentence 1 of stage-3 document. Label positions of words in that sentence.

| Pos:0 | Pos:1 | Pos:2 |
|-------|-------|-------|
| Rama | king | Ayodhya |

It is assumed that the word at position i in a sentence points to the word at position i + 1 (succeeding word). In sentence 1, the word *Rama* is pointing to the word *king* and the word *king* is pointing to the word *Ayodhya*. If $R_i$ denotes a group

of relations for ith sentence and the symbol "→" denote the relation, then we obtain
the following relations

$$\text{Rama} \rightarrow \text{king, king} \rightarrow \text{Ayodhya} \qquad (R_1)$$
$$\text{Ravana} \rightarrow \text{fought, fought} \rightarrow \text{Rama} \qquad (R_2)$$
$$\text{Rama} \rightarrow \text{killed, killed} \rightarrow \text{Ravana} \qquad (R_3)$$

Figure 1 presents a graphical representation of the document with the relations
obtained from $R_1$, $R_2$, and $R_3$.

**Assigning Relative Weight to the Vertices**

To calculate the importance of a vertex we have used the PageRank algorithm
proposed by Brin and Page [13].

$$s(v_i) = (1 - d) + d* \sum_{j \in in(v_i)} \frac{1}{|\text{out}(v_j)|} s(v_j) \tag{3}$$

where $In(v_i)$ is the set of nodes pointing the node $v_i$, $Out(v_j)$ is the set of nodes
which are pointed by the node $v_i$, $S(v_j)$ is the weightage of the node $v_i$ and d is
defined as damping factor and was assigned a value of 0.85.

Goyal et al. [6] have considered different values for damping factor (d) in the
evaluation of their context summarization system. They observed that when
damping factor value is initialized to 0.85, their Graph based ranking algorithm has
yielded better results. So in our experiments we have considered 0.85 as the
damping factor. Each of the stage-3 keywords is assigned a weight using the
PageRank algorithm. Keywords having weightage more than the average weight of
all words are selected and considered as stage-3 keywords. We have considered
stage-3 keywords as topical words.

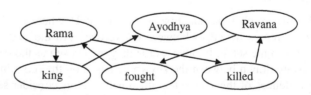

**Fig. 1** Stage-3 representation of document D

## 3.2 Sentence Selection Based on Topical Words

If $w_0$ to $w_n$ are the topical words in a sentence then weight of the sentence is calculated as

$$\text{sentence weight} = \sum_{i=0}^{m} w_i. \tag{4}$$

The sentences in the document are ranked according to their weights in decreasing order. Based on the user's choice the top N sentences can be retrieved as an extractive summary of the document.

## 4 Evaluation of the Proposed Technique

The proposed summarization system is evaluated with DUC 2002 data set [17] provided by National Institute of Standards and Technology (NIST). It contains text documents along with their summaries written by professionals. The text documents in DUC 2002 are single newswire/newspaper articles. We have experimented with 533 text documents wherein each document is associated with 10–30 human written summaries. ROUGE-1 and ROUGE-2 are two widely accepted measures for comparing system summaries with human summaries [18]. Three values Precision, Recall and F-Measure are computed in ROUGE evaluation.

Precision = $\frac{A}{B}$ where A = Total no of common Unigrams/Bigrams in System summary and human summary and B = Total no of Unigrams/Bigrams in the system summary.

Recall = $\frac{A}{C}$ where A = Total no of common Unigrams/Bigrams in System summary and human summary and C = Total no of Unigrams/Bigrams in the human summary.

$$\text{F-Measure} = 2 * \frac{\text{Precission}}{*} \text{RecallPrecission} + \text{Recall} \tag{5}$$

The formula for calculation of Recall value using ROUGE $-$ N [18] is

$$\text{ROUGE} - \text{N} = \frac{\sum_{S \in \{\text{RefSum}\}} \sum_{n\text{-gram} \in S} \text{Count}_{\text{match}}(n - \text{gram})}{\sum_{S \in \{\text{RefSum}\}} \sum_{n\text{-gram} \in S} \text{Count}(n - \text{gram})} \tag{6}$$

where N indicates the length of the n-gram, Count(n-gram) gives number of n-grams in the reference summaries and Count$_{\text{match}}$(n-gram) is the total number of n-grams, co-occurring in the candidate summary and the reference summary.

## 5   Results and Discussion

To find the relevance of proposed lexical association measure, average Lexical association in DUC 2002 source text documents and in their corresponding summaries is computed. We have used the simple average formula to calculate the average lexical association on a set of text documents. Let DocSet represent the set of N text documents.

$$\text{AverageLexicalAssociation}(\text{DocSet}) = \frac{\sum_1^N \text{AverageLexicalAssociation}(D_i)}{N} \quad (7)$$

where $\text{AverageLexicalAssociation}(D_i)$ is the average of lexical associations of all unique pairs of the stage-2 keywords for the ith Text document.

We have compared average lexical association in each text document with that of its corresponding summaries. To achieve this we have reorganized the documents in DUC2002 dataset, such that each source text document and its corresponding human summaries are placed in one folder. We are able to create 533 folders. For a fair comparison between source document and summaries, we have calculated the average lexical association in all the summaries by considering individual average lexical associations of the summaries. Figure 2 shows the comparison of average lexical associations in manual summaries and source text documents with respect to the proposed lexical association measure. We observed that keywords are more closely associated in the summaries than in the source documents. Also the average lexical association in the summaries is slightly increasing as the size of the source document is increasing. From Table 1 we can

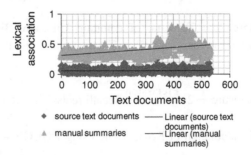

**Fig. 2** AverageLexicalAssociation obtained for each source text document and its corresponding human written summaries

**Table 1** AverageLexicalAssociation computed over 533 DUC source text documents and their corresponding 8310 human written summaries

|  | Source text documents | Manual summaries |
|---|---|---|
| AverageLexicalAssociation | 0.05779 | 0.40207 |

observe that average lexical association in manual summaries is nearly 8 times more than the average lexical association in the source text documents.

**Calculation of Sentence Overlapping in the Human Summaries**

Let S denote the set of sentences representing the union of all the human written summary sentences written for a given text document. Let us define $FQ(S_i)$ for ith sentence in set S such that $FQ(S_i)$ gives the number of human written summaries containing $S_i$. In an ideal case $FQ(S_i)$ is equal to the total no. of human written summaries and it can be said that $S_i$ is unanimously accepted as an important sentence. Since sentences that are present in all the human summaries are very few, we have experimented with three categories of summary sentences. Category-I represents the summary sentences that are present in at least 75 % summaries of the total human summaries. Similarly category-II represents the summary sentences present in at least 50 % summaries and category-III represents the summary sentences present in at least 25 % summaries of the total human summaries. We observed that 422 Source Text documents of DUC2002 dataset contained category-III sentences. 358 source text documents contained Category-II sentences. 129 source text documents contained Category-I sentences. From Fig. 3 we can observe that FQ $(S_i)$ of source document sentences gradually decreased with the increase in their position (Line Number) in the document. So we can conclude that, professionals deputed for DUC2002 summarization task have given higher priority to the first 10 sentences of the source document.

For the sentences in each of the 533 system generated summaries we have recorded the corresponding line numbers in the source text documents. We have observed that the first few lines of a source text document appeared in most of the system generated summaries. From Fig. 4 we can observe that more than 350

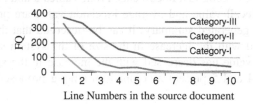

Fig. 3 Line numbers of category-I, II and III sentences in their source text documents

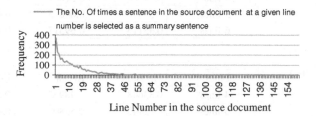

Fig. 4 Line numbers of summary sentences in the source text document

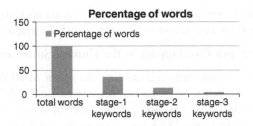

**Fig. 5** Percentage of stage-1 keywords, stage-2 keywords and stage-3 keywords with respect to total words in source text document

system generated summaries included the first line of the source text document. Second line in the source text document is included by more than 240 System summaries. From Figs. 3 and 4 we can conclude that the proposed summarization process has simulated the behavior of human professionals.

We have recorded the number of stage-1 keywords, stage-2 keywords and stage-3 keywords obtained for each of the 533 source text documents. From Fig. 5 we can observe that 35 % of the words in the text document form the group of Stage-1 keywords. Stage-2 keywords make 12 % of total words and the number of Stage-3 keywords fall below 4 % of the total words in the text document. The percentage of Stage-3 keywords indicates the effectiveness of the proposed keyword extraction process in eliminating unimportant words. Tables 2 and 3 present the average Rouge scores of the proposed system when evaluated with DUC 2002 dataset.

Since most of the human written summaries are consisting of 4–6 sentences, we have generated three sets of summaries of lengths 4 sentences, 5 sentences and 6 sentences on the DUC 2002 text documents. From Tables 2 and 3 we can conclude that the average Recall value and average Precision value are proportional to the size of the system generated summary. It can be observed that the average Recall value has increased with the increase in the size of the system summary. But the average Precision value has decreased with the increase in the size of the system summary. Tables 4 and 5 present a comparison among proposed system, context

**Table 2** ROUGE-1 evaluation

| Lines | Precision | Recall | F-measure |
| --- | --- | --- | --- |
| 4 lines | 0.39796 | 0.57681 | 0.46715 |
| 5 lines | 0.37496 | 0.64675 | 0.47113 |
| 6 lines | 0.35115 | 0.69517 | 0.46334 |

**Table 3** ROUGE-2 evaluation

| Lines | Precision | Recall | F-measure |
| --- | --- | --- | --- |
| 4 lines | 0.33648 | 0.48965 | 0.39775 |
| 5 lines | 0.32143 | 0.55802 | 0.40480 |
| 6 lines | 0.30403 | 0.60640 | 0.40125 |

**Table 4** ROUGE scores obtained for a 100 words system generated summary

| Quality measure | Proposed system | UniformLink+bern+neB |
|---|---|---|
| ROUGE-1 | 0.48645 | 0.46432 |
| ROUGE-2 | 0.39927 | 0.20701 |

**Table 5** ROUGE-1 scores obtained for a 100 words system generated summary

| Summarization system | ROUGE-1 score | | |
|---|---|---|---|
| | Basic | Stemmed | Stemmed and no stopwords |
| Proposed System | 0.48645 | 0.56274 | 0.45925 |
| TextRank | 0.4708 | 0.4904 | 0.4229 |

based summarization system (UniformLink+bern+neB [6]) and graph based text summarization system (TextRank [12]). We can observe that the recall values for the proposed system are better than the recall values of the existing systems.

# 6 Conclusions

In this paper a keyword extraction methodology is proposed to find keywords based on the context. This methodology is based on the calculation of lexical association between the words in the text. Lexical association and Parts of speech of the words are used to obtain keywords from the text. These keywords are assigned a weight based on graph based ranking algorithm and are further reduced to a small set of keywords. The final set of keywords was utilized to find important sentences in the given text. It is observed that the proposed system exhibited better ROUGE scores compared to the existing context based summarization system. When we observed human written summaries in the DUC 2002 data set we found that most of the summaries contained first few lines of the source documents. So we are experimenting on utilizing the first few lines of the source document to extract the keywords instead of the whole document. This can reduce the time required for keyword extraction process and may improve the precision of keyword extraction algorithm.

# References

1. Hovy, E.H., Lin, C.Y.: Automated text summarization in SUMMARIST, pp. 81–94. MIT Press (1999)
2. Gholamrezazadeh, S., Salehi, M.A., Gholamzadeh, B.: A comprehensive survey on text summarization systems. In: 2nd International Conference on Computer Science and Its Applications, pp. 1, 6, 10–12. (2009)

3. Kamble, P., Dharmadhikari, S.C.: Context based topical document summarization. Data Mining Knowl. Eng. **6**, 146–150 (2014)
4. Pawar, D.D., Bewoor, M.S., Patil, S.H.: Context based indexing in text summarization using lexical association. Int. J. Eng. Res. Technol. **2**(12), (2013)
5. Ferreira, R., Freitas, F., de Souza Cabral, L., Lins, R.D., Lima, R., França, G., Simske, S.J., Favaro, L.: A context based text summarization system. In: Document Analysis Systems (DAS), 11th IAPR International Workshop, pp. 66–70. (2014)
6. Goyal, P., Behera, L., McGinnity, T.M.: A context-based word indexing model for document summarization. IEEE Trans. Knowl. Data Eng. **25**(8), 1693–1705 (2013)
7. Matsuo, Y., Ishizuka, M.: Keyword extraction from a single document using word co-occurrence statistical information. In: FLAIRS Conference, AAAI Press, pp. 392–396. (2003)
8. Lott, B.: Survey of Keyword Extraction Techniques. UNM Education (2012)
9. Wartena, C., Brussee, R., Slakhorst, W.: Keyword extraction using word co-occurrence. In: Workshop on Database and Expert Systems Applications (DEXA), IEEE, pp. 54–58. (2010)
10. Rajaraman, A., Ullman, J.D.: Data Mining. Mining of Massive Datasets, pp. 1–17. (2011)
11. Wan, X., Xiao, J.: Exploiting neighborhood knowledge for single document summarization and keyphrase extraction. ACM Trans. Inf. Syst. **28**, 8:1–8:34 (2010)
12. Mihalcea, R., Tarau, P.: Textrank: bringing order into texts. In Lin, D., Wu, D. (eds.), Proceedings of EMNLP, pp. 404 (2004)
13. Brin, S., Page, L.: The anatomy of a large-scale hypertextual Web search engine. Comput. Netw. ISDN Syst. **30**, 1–7 (1998)
14. Aggarwal, C.C., Zhao, P.: Towards graphical models for text processing. Knowl. Inf. Syst. **36**(1), 1–21 (2013)
15. Toutanova, K., Klein, D., Manning, C., Singer, Y.: Feature-rich part-of-speech tagging with a cyclic dependency network. In: Proceedings of HLTNAACL, pp. 252–259. (2003)
16. Toutanova, K., Manning, C.D.: Enriching the knowledge sources used in a maximum entropy part-of-speech tagger. In: Proceedings of the Joint SIGDAT Conference on Empirical Methods in Natural Language Processing and Very Large Corpora (EMNLP/VLC-2000), pp. 63–70. (2000)
17. Over, P., Liggett, W.: Introduction to DUC: an intrinsic evaluation of generic news text summarization systems. In: Proceedings of DUC workshop Text Summarization. (2002)
18. Lin, C.Y., Hovy, E.H.: Automatic evaluation of summaries using N-gram co-occurrence statistics. In: Proceedings of 2003 Language Technology Conference (HLT-NAACL), pp. 71–78. (2003)

# Design of Linear Phase Band Stop Filter Using Fusion Based DEPSO Algorithm

Judhisthir Dash, Rajkishore Swain and Bivas Dam

**Abstract** This manuscript prepares a modern hybrid technique by combining the usual particle swarm optimization (PSO) technique with differential evolution (DE) technique, named as fusion based differential evolution particle swarm optimization (DEPSO) in order to enhance the global search abilities, while solving impulse response of linear phase digital band stop filter. The linear phase band stop filter is mostly considered as a nonlinear multimodal problem. The DEPSO is a heuristic based global search method modelled on considering the advantages of both of the methods to enhance the superiority of result and convergence speed. The performance of the intended DEPSO based approach has been contrasted with few renowned optimization techniques such as PSO, comprehensive learning PSO (CLPSO), craziness based PSO (CRPSO) and DE. The proposed DEPSO based result confirms the supremacy of solving design problems of FIR filters.

**Keywords** Particle swarm optimization (PSO) · Differential evolution particle swarm optimization (DEPSO) · FIR filters

## 1 Introduction

The main purpose of digital filter is to apply for signal separation or signal restoration. A digital filter is merely modelled as a discrete time signal convolution. Digital filters are the fundamental structures in most of the digital communication

J. Dash (✉)
Department of ECE, Silicon Institute of Technology, Bhubaneswar, India
e-mail: jd.08@aol.in

R. Swain
Department of EEE, Silicon Institute of Technology, Bhubaneswar, India
e-mail: r_kswain@yahoo.co.in

B. Dam
Department of IEE, Jadavpur University, Kolkata, India
e-mail: bd@iee.jusl.ac.in

© Springer India 2016
H.S. Behera and D.P. Mohapatra (eds.), *Computational Intelligence in Data Mining—Volume 1*, Advances in Intelligent Systems and Computing 410, DOI 10.1007/978-81-322-2734-2_28

systems. Basically, digital filter is finite impulse response type or infinite impulse response type [1, 2]. If the filter weights are properly symmetrical about the middle coefficient, then the finite impulse response (FIR) filter can be assured to have precise linear phase characteristic at all frequencies of the pass band [3, 4]. The filter design procedure can be modelled as a non linear complex optimization dilemma in which the cost function is considered as an error function required to be minimized. The usual gradient based classical methods are not appropriate for linear phase filter optimization due to premature convergence leading to suboptimal solutions [3, 5].

Therefore, modern heuristic search techniques have been applied to design the best possible filters with the tuning of the parameters along with higher stop band attenuation as reported in the literatures [5, 6]. In order to reduce the difficulties of premature convergence and to enhance the searching ability of the particles, the authors, in this manuscript have modified the conventional particle swarm optimization (PSO) by the mutation operator of differential evolution (DE) to improve its global search procedure, named as the fusion based differential evolution particle swarm optimization (DEPSO) [7, 8]. Based on the proposed DEPSO approach, this document presents a high-quality, efficient set of results to illustrate its superiority.

## 2 Problem Formulations

Transfer function of a linear phase band stop digital filter is described as

$$H(z) = \sum_{n=0}^{N} h(n)z^{-n} \tag{1}$$

In Eq. (1), N represents the order of the filter having (N + 1) number of coefficients in the impulse response, h(n). This manuscript describes optimal design of an even order digital band stop (BS) linear phase filter with symmetric h(n). Since the coefficients of h(n) are symmetrical about the middle point, the dimension is reduced to 50 %, i.e., only (N/2 + 1) number of coefficients are required to be searched to recover the total (N + 1) number of coefficients [4, 6]. For the optimal filter design, the required filter parameters are: pass band $(\omega_p)$ and stop band $(\omega_s)$ edge frequencies; pass band $(\delta_p)$ and stop band $(\delta_s)$ ripples; stop band attenuation and transition width [9]. These parameters are mainly controlled by the filter coefficients. Let us assume that the filter coefficients of particle vector are $\{h_0, h_1...$ $h_N\}$. The particle vectors are scattered in an L-dimensional search space, with L = (N/2 + 1) for the BS digital filter of order N. Then the frequency response of the linear phase filter will be considered as described in Eq. (2).

$$H(\omega_k) = \sum_{n=0}^{N} h(n)e^{-j\omega_k n} \tag{2}$$

With $\omega_k = (2\pi k/N)$ and $H(\omega_k)$ as the discrete Fourier transform vector known as the BS digital filter frequency response [10]. The frequency $[0, \pi]$ response is sampled with M sampling points as shown in Eqs. (3) and (4).

$$H_a(\omega) = [H_a(\omega_1), H_a(\omega_2), \ldots, H_a(\omega_M)]^T \tag{3}$$

$$H_I(\omega) = [H_I(\omega_1), H_I(\omega_2), \ldots, H_I(\omega_M)]^T \tag{4}$$

where $H_a(\omega)$ represents actual magnitude response of the designed filter and $H_I(\omega)$ is the ideal response of the designed BS filter as stated below:

$$H_I(\omega) = \begin{cases} 0 \text{ for } \omega_{ls} \leq \omega \leq \omega_{us} \\ 1 \text{ otherwise} \end{cases} \tag{5}$$

With $\omega_{ls}$ and $\omega_{us}$ as the lower and the upper stop band cut-off frequencies respectively in the ideal linear phase BS filter as described in Eq. (5). For this work, the authors have applied the same advantageous error fitness function to accomplish high stop band attenuation with successful reduction in pass band and stop band ripples [6, 7]. The error fitness function, C applied for this filter design is described in Eq. (6) with ‖ or, abs indicating the absolute value allowing small amount of ripples $\delta_p$ and $\delta_s$ in the pass band and stop band respectively due to Gibb's phenomena.

$$C = \sum abs[abs(|H_a(\omega)| - 1) - \delta_p] + \sum abs(|H_a(\omega)| - \delta_s) \tag{6}$$

## 3 Fusion Based DEPSO Algorithm

### 3.1 Particle Swarm Optimization (PSO)

PSO is the most popular population based search technique mostly motivated by the inspection of usual habits of bird flocking. It is initialized by a population of random solution vectors, known as particles. A swarm of particles move in an L dimensional search space for solution. It searches the optimal value through several iteration cycles. Through every iteration cycle, the velocities of the particles are adjusted as the sum of their inertia velocities, along with local best values (components) and global best values. The position of each particle is also modified by adding its updated velocity to the present position [3, 7]. The velocities and position vectors are updated using the Eqs. (7) and (8) respectively. After checking the limits of the filter coefficients, the updated fitness of the particle vectors are computed based on the least amount fitness errors, considering the smallest as the global best (gbest). At the end of the iteration cycles, the final gbest is considered as the solution of this technique.

$$V_i^{(k+1)} = w * V_i^k + C_1 * r_1 * (pbest_i^k - P_i^k) \\ + C_2 * r_2 * (gbest_i^k - P_i^k) \tag{7}$$

$$P_i^{(k+1)} = P_i^k + V_i^{(k+1)} \tag{8}$$

## 3.2 Differential Evolution (DE)

DE is also one of the most popular, efficient population based random search algorithms, widely used for optimization [11, 12]. In this algorithm, solution is achieved by these three main operations such as mutation, crossover, and selection. First each particle is mutated with the weighted difference between any other two random population particles to form a new particle, called the kid. The new kid recombines with the parent under certain criterion, like, crossover rate, CR [12]. Then the fitness of both, the new kid and the parent is evaluated and compared. The kid is chosen to the next iteration if it has an improved fitness, C than the parent.

## 3.3 Fusion Based Differential Evolution Particle Swarm Optimization (DEPSO)

It is an efficient error function minimization technique developed by fusing the unique mutation operator of the DE algorithm with the conventional PSO technique, which is applied to mutate the global best particle for providing fast convergence and improved performance in the search space. The hybridization of PSO by DE results in an improved technique of solution known as the DEPSO algorithm [7, 8]. In DEPSO, new kid is formed through the mutation of the gbest (parent) particle during each iteration cycle, considering a Gaussian distribution. Here, four population particles are arbitrarily preferred during mutation as described in Eq. (9). The weighted (F) difference between these particles is added to mutate the parent $(P_{gl})$ to produce a baby $(T_{il})$ conditionally, according to Eq. (10) [12]. Where $\delta_{il}$ is the weighted difference for each l-th dimension with $r_3$, a random number between [0, 1]. Subsequently the error fitness of the kid is calculated and compared to replace the current iteration parent (existing gbest) in case it has an improved fitness, otherwise the existing gbest continues to the next iteration.

$$\delta_{il} = F \frac{(P_{1l} - P_{2l}) + (P_{3l} - P_{4l})}{2} \tag{9}$$

If $r_3 < CR$ or $j = j_{rand}$ then,

$$T_{il} = P_{gl} + \delta_{il} \tag{10}$$

## 3.4 Application of DEPSO to Linear Phase BS Filter Design

Step 1: Initialize $n_p$ number of particle vectors in the L-dimensional searching space with lower and upper limits of filter coefficients, $h_{min} = -1$, $h_{max} = 1$; upper limit of iteration cycles, and select suitable values of $C_1, C_2$.

Step 2: Compute the initial pbest (personal best) solution vectors and the gbest (group best) solution vector.

Step 3: Update the velocities and position vectors in accordance with Eqs. (7) and (8) respectively, and check the limits of the filter coefficients.

Step 4: compute the updated fitness of the particle vectors based on minimum fitness value; Update the pbest vectors and gbest vectors.

Step 5: Mutate gbest vector if $r_3 < CR$ or $j = j_{rand}$, using Eqs. (9) and (10) and update gbest based on kid's fitness value.

Step 6: Continue iteration from step 3–6 till the upper limit of the iteration cycle is reached. Final, gbest vector is the L numbers of optimal filter coefficients, from which whole (N + 1) optimal coefficients are formed by copying and concatenating these for getting the optimum frequency response (2) [3, 6].

# 4 Simulation Results

The proposed fusion based DEPSO method has been implemented to design the linear phase band stop digital filter of order 36. The program was run in MATLAB 7.5 version/core (TM) 2 duo processor/3.00 GHz/2 GB RAM. In comparison with other existing methods, the DEPSO is found to be more efficient owing to its search space capability in addition to better convergence characteristic. In this simulation, 128 numbers of samples are considered [6]. The filter parameters considered for the above design problem are: Lower pass band (normalized) cut-off frequency, $\omega_{lp} = 0.3$; Lower stop band cut-off frequency, $\omega_{ls} = 0.35$; Upper stop band cut-off frequency, $\omega_{us} = 0.7$; Upper pass band cut-off frequency, $\omega_{up} = 0.75$; Pass band ripple, $\delta_p = 0.03$; Stop band ripple, $\delta_s = 0.003$. The following parameters are

**Table 1** Optimal coefficients of linear phase BS digital filters of order 36

| h(n) | PSO [6] | CLPSO [6] | CRPSO [6] | DE | DEPSO |
|------|---------|-----------|-----------|-----|-------|
| h(1) = h(37) | −0.0011 | −0.0046 | −0.0050 | 0.0015 | −0.0003 |
| h(2) = h(36) | −0.0262 | −0.0311 | −0.0250 | −0.0127 | −0.0123 |
| h(3) = h(35) | 0.0057 | 0.0047 | 0.0015 | 0.0056 | 0.0028 |
| h(4) = h(34) | −0.0025 | −0.0081 | 0.0018 | 0.0035 | 0.0020 |
| h(5) = h(33) | 0.0104 | 0.0061 | 0.0074 | 0.0106 | 0.0050 |
| h(6) = h(32) | 0.0333 | 0.0278 | 0.0312 | 0.0268 | 0.0252 |
| h(7) = h(31) | −0.0202 | −0.0257 | −0.0193 | −0.0198 | −0.0195 |
| h(8) = h(30) | −0.0247 | −0.0268 | −0.0272 | −0.0196 | −0.0205 |
| h(9) = h(29) | −0.0004 | −0.0020 | −0.0034 | −0.0021 | 0.0010 |
| h(10) = h(28) | −0.0267 | −0.0259 | −0.0232 | −0.0216 | −0.0223 |
| h(11) = h(27) | 0.0568 | 0.0533 | 0.0484 | 0.0519 | 0.0522 |
| h(12) = h(26) | 0.0413 | 0.0342 | 0.0466 | 0.0426 | 0.0412 |
| h(13) = h(25) | −0.0528 | −0.0549 | −0.0479 | −0.0509 | −0.0519 |
| h(14) = h(24) | −0.0047 | −0.0013 | −0.0000 | 0.0001 | −0.0001 |
| h(15) = h(23) | −0.0861 | −0.0897 | −0.0803 | −0.0851 | −0.0852 |
| h(16) = h(22) | −0.0527 | −0.0516 | −0.0502 | −0.0476 | −0.0464 |
| h(17) = h(21) | 0.2993 | 0.2934 | 0.2979 | 0.2969 | 0.2966 |
| h(18) = h(20) | 0.0328 | 0.0271 | 0.0328 | 0.0285 | 0.0294 |
| h(19) | 0.5992 | 0.5939 | 0.5993 | 0.5998 | 0.5994 |

judiciously chosen for obtaining better optimal solution in different techniques. Population size = 100; No. of iteration = 400; Limits of coefficients = −1, 1; $v_{min}$, $v_{max}$ = 0.01, 0.9; $w_{min}$, $w_{max}$ = 0.4, 0.9; CR = 0.5; F = 0.4.

The finest optimum coefficients of the designed BS filter have been searched by DEPSO algorithm and are listed along with other techniques in Table 1 [9]. Tables 2 and 3 review the comparative findings of various performance parameters with DEPSO algorithm. Figures 1 and 2 compare the optimal magnitude response of the linear phase BS filters with DEPSO technique. The proposed DEPSO based design acquires the highest stop band attenuation of 57.21 dB; Minimum pass band

**Table 2** Findings of the performance parameters of BS digital filters

| Algorithm | Stop band attenuation (dB) | | | Variance | Normalized | |
|-----------|------|-----|------|----------|-----------------------|-------|
|           | Max  | Min | Mean |          | Transition width BW |       |
| PSO [6]   | 26.0 | 71.2 | 36.91 | 45.33 | 0.08 | 0.427 |
| CLPSO [6] | 27.0 | 74.5 | 39.73 | 41.01 | 0.081 | 0.428 |
| CRPSO [6] | 29.4 | 59.6 | 45.28 | 35.19 | 0.09 | 0.426 |
| DE        | 30.5 | 75.8 | 53.95 | 25.16 | 0.082 | 0.424 |
| **DEPSO** | **31.4** | **80.0** | **57.21** | **21.48** | **0.087** | **0.421** |

**Table 3** Comparative findings of pass band and stop band ripple

| Algorithm | Pass band ripple (absolute) | | | Stop band ripple (absolute) | | |
|---|---|---|---|---|---|---|
| | Max | Min | Mean | Max | Min | Mean |
| PSO [6] | 0.088 | 0.02 | 0.076 | 0.056 | 0.003 | 0.034 |
| CLPSO [6] | 0.16 | 0.03 | 0.129 | 0.043 | 0.001 | 0.024 |
| CRPSO [6] | 0.072 | 0.02 | 0.068 | 0.034 | 0.008 | 0.031 |
| DE | 0.034 | 0.016 | 0.04 | 0.03 | 0.0024 | 0.0186 |
| **DEPSO** | **0.038** | **0.01** | **0.031** | **0.027** | **0.001** | **0.013** |

**Fig. 1** Magnitude (dB) response of the BS filters of order 36

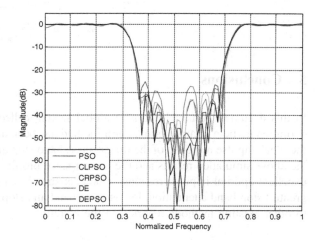

**Fig. 2** Magnitude (absolute) response of the BS filters of order 36

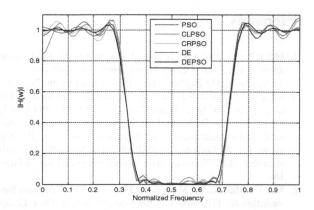

ripple (absolute) = 0.031; Minimum stop band ripple (absolute) = 0.013. The convergence characteristic of the fitness errors are shown in Fig. 3. The DEPSO converges to smallest error fitness in comparison to DE, CRPSO, CLPSO and PSO. From the Fig. 3, it is observed that the DEPSO has acquired the lowest error fitness value of 3.52 in 215 iteration cycles in the above proposed design.

**Fig. 3** Convergence graph of linear phase BS filters of order 36

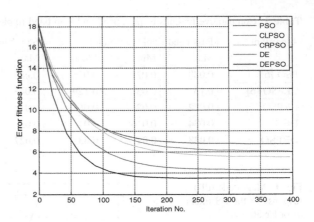

# 5  Conclusions

In this document, a novel heuristic global search, called fusion based differential evolution particle swarm optimization method has been demonstrated for designing the ideal impulse response of the linear phase band stop digital filter. It is clear from the above simulation results and the convergence profiles that the DEPSO provides excellent optimal impulse response characteristics in terms of convergence speed, greatest stop band attenuation, lowest pass band and stop band ripples as compared to other existing methods reported in the literatures.

# References

1. McClellan, J., Parks, T.: A unified approach to the design of optimum FIR linear-phase digital filters. IEEE Trans. Circuit Theory **20**(6), 697–701 (1973)
2. Schlichtharle, D.: Digital Filters, Basics and Design, 2nd edn. Springer (2011)
3. Saha, S.K., Ghosal, S.P., Kar, R., Mandal, D.: Cat Swarm Optimization algorithm for optimal linear phase FIR filter design. ISA Trans. **52**, 781–794 (2013)
4. Saha, S.K., Kar, R., Mandal, D., Ghoshal, S.P.: Seeker optimisation algorithm: application to the design of linear phase finite impulse response filter. IET Signal Proc. **6**(8), 763–771 (2012)
5. Wade, G., A. Roberts, A., Williams, G.: Multiplier-less FIR filter design using a genetic algorithm, IEEE Transaction on Vision, Image and Signal Processing, vol.141(3), (1994) 175–180
6. Kar, R., Mandal, D., Mandal, S., Ghoshal, S.P.: Craziness based particle swarm optimization algorithm for FIR band stop filter design. Swarm Evol. Comput. **7**, 58–64 (2012)
7. Mondal, S., Mallick, P., Mandal D., Kar, R., Ghoshal, S.P.: Optimal FIR band pass filter design using novel particle swarm optimization algorithm. In: IEEE Symposium on Humanities, Science and Engineering Research (SHUSER), pp. 141–146. (2012)
8. Chattopadhyay, S., Sanyal, S.K., Chandra, A.: Optimization of control parameter of differential evolution algorithm for efficient design of FIR filters. In: 13th IEEE Conference on Computer and Information Technology, pp. 267–272. (2010)

9. Mandal, S., Ghoshal, S.P., Kar, R., Mandal, D., Shiva S.C.: Non-recursive FIR band pass filter optimization by improved particle swarm optimization. In: Proceedings of the InConINDIA, AISC 132, pp. 405–412. Springer, Berlin Heidelberg New York (2012)
10. Mukherjee, S., Kar, R., Mandal, D., Mondal, S., Ghoshal, S.P.: Linear phase low pass FIR filter design using improved particle swarm optimization. In: IEEE Student Conference on Research and Development (SCOReD), pp. 358–363. (2011)
11. Sharma S., Arya, L.D., Katiyal, S.: Design of linear-phase digital FIR filter using differential evolution optimization with ripple constraint. In: IEEE Conference on Computing for Sustainable Global Development (INDIACom), pp. 474–480. (2014)
12. Vaisakh, K., Praveena, P., Rao, S.R.M.: DEPSO and bacterial foraging optimization based dynamic economic dispatch with non-smooth fuel cost functions. In: 2009 World Congress on Nature & Biologically Inspired Computing (NaBIC), pp. 152–157 (2009)

# A Survey on Face Detection and Person Re-identification

M.K. Vidhyalakshmi and E. Poovammal

**Abstract** Today surveillance systems are used widely for security purposes to monitor people in public places. A fully automated system is capable of analyzing the information in the image or video through face detection, face tracking and recognition. The face detection is a technique to identify all the face in the image or video. Automated facial recognition system identifies or verifies a person from an image or a video by comparing features from the image and the face database. When surveillance system is used to monitor human for locating or tracking or analyzing the activities, the challenge of identification of a person is really a hard task. In this paper we survey the techniques involved in face detection and person re-identification.

**Keywords** Surveillance · Face detection · Person re-identification

## 1 Introduction

Video surveillance is increasingly used in many public and private places. Normally CCTV will be used to monitor a particular place. A large volume of visual data is obtained from the video camera continuously. At present humans are employed to monitor the data from the cameras. This method is not successful many times as humans have limited abilities. Continuous monitoring of the data from the cameras is very tedious and error prone for a human. Automated surveillance system is capable of tracking the face, identify people without human

M.K. Vidhyalakshmi (✉)
Department of Electronics and Communication Engineering,
Tagore Engineering College, Chennai 600127, India
e-mail: mkvlakshmi@yahoo.co.in

E. Poovammal
Department of Computer Science Engineering, SRM University,
Chennai 603203, India

© Springer India 2016                                                          283
H.S. Behera and D.P. Mohapatra (eds.), *Computational Intelligence in Data Mining—Volume 1*, Advances in Intelligent Systems and Computing 410, DOI 10.1007/978-81-322-2734-2_29

intervention. In this paper a survey of the two important techniques namely the face detection and person re-identification has been done. In Sect. 2 we discuss about importance of face detection and methods of face detection. The popularly used Viola Jones face detection algorithm is also discussed in Sect. 2.1. In Sect. 2.2 improvements in face detection algorithm has been surveyed. In Sect. 3 we discussed about some of the recent person re-identification techniques.

## 2 Face Detection

With the technology advancement, biometric recognition finds its place in access control, person identification, surveillance etc. Among the biometrics Face detection is very important as it is the primary step in facial recognition, facial expression, and gender recognition systems. The face detection algorithm finds the location of the face in the image or video frame. The challenges that are associated with the face detection are mentioned in [1]. Detector performances vary with the pose of the person. Some detectors are trained only to detect frontal view. There are few detectors which are trained for profile, 45° rotations. Detectors fail in real time when the expressions of the person changes when compared to image in the database. Certain components like beards, glasses, mustaches etc. may affect the face detection. Faces can be occluded by either objects or other faces.

Yang et al. [2] categorized the methods of face detection into four. Knowledge-based methods are based on certain set of rules that describe face. Feature invariant methods find certain features which do not change when pose illumination or viewing point changes. Template matching methods use several templates describing the face are stored as a whole or features. Detection is performed by computing correlation between input image and stored templates. Appearance-based methods uses set of training facial images are used to train a model. The learned models are used for detection.

### 2.1 Viola Jones Face Detector

Among the different face detection methods appearance based method is widely popular nowadays due to advancements in storage and computational efficiency. Viola and Jones [3] face detection algorithm is made a great influence in computer vision field in 2000s. They made three contributions: The integral image, Ada boost classifier and the attentional cascade structure. Viola and Jones used the integral image which was used originally to find mipmaps in computer graphics. Haar-like features are computed using integral image for fast computation. The integral image is constructed using the Eq. 1:

**Fig. 1** Features used by
Viola and Jones

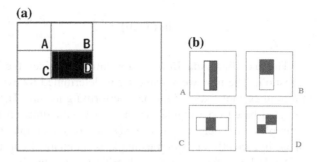

$$S(x, y) = I(x', y') + S(x - 1, y) + S(x, y - 1) - S(x - 1, y - 1) \qquad (1)$$

where S(x; y) is the integral image at pixel location (x; y) and I (x'; y') is the original
image. The integral image computes the sum of any rectangular area efficiently, as
shown in Fig. 1a. The sum of pixels in region *ABCD* can be calculated using Eq. 2:

$$\Sigma I(x, y) = S(D) + S(A) - S(B) - S(C) \qquad (2)$$

Using only 4 reference values sum of the values in original image can be found
out.

Viola and Jones used three kinds of features as shown in Fig. 1a, b represents
two rectangular feature which finds the difference between the sum of the pixels in
two rectangular regions which have same size and shape, C represents three rectangular feature which calculates the sum in two outside rectangles deducted from
the sum within the center rectangle and D represents four-rectangle feature which
calculates the difference among diagonal pairs of rectangles. Using integral image
the computation of two rectangle features involves six array references, the three
rectangle features involves eight array references, and the four rectangle features
involves nine array references. Viola-Jones found that a detector with a base resolution of 24 * 24 pixels gives a total of approximately 160,000 different features.
These features are computationally efficient.

Boosting is a method in which set of classification functions are pooled based on
a weighted majority vote. Here weights are associated based on functions. Larger
weights are specified for good classification function, while poor functions are
specified with lesser weights. Ada Boost is a forceful mechanism for choosing a set
of good classification functions which still have significant variety. Among all
160,000 different features, some features give almost steadily high values when on
top portion of a face. In order to find these features Viola-Jones use a modified
version of the Ada Boost algorithm developed by Freund and Schapire in 1996 [4].
A weak classifier ($h(x, f, p, \theta)$) consists of a feature ($f$), a threshold ($\theta$) and a polarity
($p$) is described by the Eq. 3.

$$h(x,f,p,\theta)=\begin{cases} 1 & if\ pf(x)<p\theta \\ 0 & otherwise \end{cases} \qquad (3)$$

The modified Ada Boost algorithm determines the best feature, polarity and threshold. The new weak classifier is determined by evaluating each feature on all training examples to find the best performing feature. The best performing feature is selected based on weighted error, which is a function of weights in the training examples. The weight of a correctly classified example is decreased and the weight of an in-correctly classified example is maintained constant.

Viola-Jones face detection algorithm scans the image many times in different sizes. The evaluated sub windows containing non faces will be discarded using cascaded classifiers. Such cascaded classifiers contain different stages, each stage being a strong classifier. The stages fixes whether given stage is face or non-face. If it is a non-face it is discarded if the sub window is a face, it is passed on to the next stage of the cascade structure as shown in the Fig. 2.

## 2.2 Advancements in Feature Set and Boosting Algorithms

The Haar-like features used in Viola-Jones algorithm [3] has good computational efficiency and perform well for frontal face detection. Many researchers modified the feature set introduced by [3] for multi view face detection. In [5] S. Li et al. proposed 3 types of rectangular features with flexible sizes and distances. Such

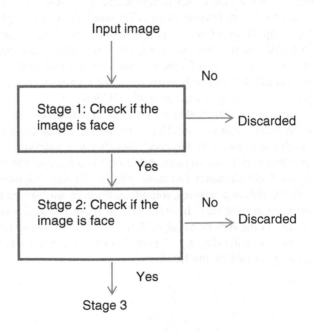

**Fig. 2** Stages of cascaded classifiers

features are non-symmetrical which is better for non-symmetrical characteristics of non-frontal faces. In another work by Viola and Jones [6] diagonal filters has been proposed for profile and non-upright faces. They used separate detectors for different views. Along with the 3 types of original set of filters that were used in [3] diagonal filters are used to emphasis diagonal structures in the image. They consist of four overlapping rectangles in which, the difference is found between the summation of the pixels in the dark gray shaded region and the summation in the light gray shaded region. Using an integral image, the diagonal filter can be calculated by only observing at the 16 corner pixels. Huang et al. [7] suggested a sparse feature set which strengthens the features' discrimination ability with less computational cost. Each sparse feature has a granule. The granules are computed efficiently using either image pyramids or through integer image. The single sparse feature consists of maximum of 8 granules. Here heuristic search scheme has been used to add granules one by one to sparse feature. The multi scaled search is applied to reduce computation. Another limitation of Haar features is that it lacks in robustness to detection under varied illumination. In [8] modified census transform was adopted to generate features that are insensitive to illumination variation. It compares intensity means of the neighborhood pixels. The pixels' local structure is represented with index number. The weak classifiers are created by examining the distributions of the index numbers. In [9] local binary patterns are introduced that are robust to illumination variation. In [10] Yan et al. proposed locally assembled binary feature, which gave good performance on standard face detection data sets. Local binary pattern and modified transform are sensitive to noise. In order to overcome the noise semi local structure patterns (SLSP) was used in [11]. These features are robust to both illumination variations and noise.

In [12], the authors proposed linear projection features which are so powerful acts as face templates. These features improve the convergence speed of the classifier. The computational load of linear features are higher than Haar-like features. Baluja et al. [13] used pixel pairs as features to increase the speed of computation at the cost of reduced discrimination power. Pang et al. [14] proposes distributed object detection framework (DOD) which uses HOG based features for real time face detection that has greater speed. They used cell based HOG (CHOG) in which the features in one cell are not shared with overlapping blocks. This method is used to reduce the dimension of traditional block based HOG. The advantage of this method is that the dimension of the vectors of CHOG-DOD is small. In [15] the author proposes half template to reduce the burden of computation of face search. By using this time complexity is reduced and speed of detection is increased. Recently in [16] pyramid like face detection scheme has been proposed which has 3 stages namely coarse, shift and refine. This reduces computation overhead.

In [17] the author proposes a method to acquire training images for facial attribute detection obtained from photos of community. The community contributed photos has tags, comments, location etc. On searching with the keywords results in many false positive due to an uncontrolled annotation quality. The detector is degraded when learned with noisy data. This paper deals with a method to measure the quality of training images. The authors use a feature selection mechanism which measures the discriminant capability of each feature by discriminability voting. This voting is done based on pseudo positive (negatives) retrieved by textual relevance. The features are then optimized by discriminability and mutual similarity. The selected features are then used to evaluate the quality in the aspect of visual. The features are then added with the context information.

Researches also made several changes in the original boosting algorithm. Detailed discussion has been done in [18]. Huang et al. [19] proposed a novel method called Vector boosting for multi view face detector. It uses a coarse to fine approach which divides entire face space into smaller subspaces. To achieve high speed and accuracy WFS (Width-First-Search) tree structure has been used. It is the extension of the real Ada Boost [20]. The vector boosting algorithm trains the predictor at the branch nodes of the tree which have multi component output vectors. The advantage of this method is that it covers face space of ±45° rotation in plane (RIP) and ±90° rotation off plane (ROP). In another recent paper [21] the author proposes a multi view face detection method which requires only single reference image that contains only 2 manually indicated reference points. The advantage of this method is that it requires only minimal manual intervention for training. This method uses probabilistic classifier.

Zhang et al. [22] proposes a method for multi spectral face detection which can be done by combining large scale visible face images and few multispectral face images. This novel boosting method is called regularized transfer boosting (R-Trboost). This deals the data from different domain with different weights particularly more attention is given to the target domain. Overfitting problem is prevented by manifold regularization to achieve label/score smoothness among target data. The features and Boosting algorithms has been listed in the Table 1.

# 3 Person Re-identification

Person re-identification is the task of matching people across surveillance cameras at various time and place. It has a critical importance in surveillance system. The person re-identification task is challenging because the images of person captured by one camera May vary significantly from other or the images of the same person May be viewed differently by the same camera at a later time due to illumination condition, pose changes and scale. In [23] the issues of person re identification is divided into inter-camera and intra-camera. inter-camera issues include illumination changes, disjoint fields of view, different entry poses in different cameras, similarity of clothing; rapid changes in person's clothing; and blurred images. Intra-camera issues include background illumination changes; low resolution of CCTVs and occlusion in frames. Ali Saghafi [23] grouped the person re-identification methods into two groups. The first one is signature based methods which extract color, shape, texture etc. These methods are called appearance based methods. The second method is based on gait. Gait is a biometric that is based on how people move the limbs. Vezzani et al. [24] have given an outstanding comprehensives survey of

**Table 1** Features and boosting algorithms used for face detection

| S. no | Feature/boosting algorithm | Advantage | Type of detection |
|---|---|---|---|
| 1 | Haar features [3] | Good computational efficiency | Frontal face detection |
| 2 | Rectangular features [5] | Multi view face detection | Non frontal face detection |
| 3 | Diagonal features [6] | Separate detectors for different views | Profile and non-upright face detection |
| 4 | Sparse features [7] | Strengthens the discrimination ability and less computational cost | Multi view face detection |
| 5 | Pyramid like cascade structure | Increases computational overhead | Frontal face detection |
| 6 | Modified census transform based features [8] | Insensitive to illumination variation | Frontal face detection |
| 7 | Local binary patterns [9] | Robust to illumination variation | Frontal face detection |
| 8 | Linear projection features [12] | Acts as face template and speed of classifier is increased | Frontal face detection |
| 9 | Pixel pair features [13] | Increased speed of computation | Frontal face detection |
| 10 | Half template features [15] | Computation speed increases | Frontal face detection |
| 11 | Vector boosting [19] | High speed and accuracy | Multi view face detection |
| 12 | Regularized transfer boosting (R-Trboost) [22] | Prevents overfitting problem | Multispectral face image detection |

research advances in Re-identification over the past 15 years. In this paper some the recent developments in the field of person re identification is provided.

According to recent survey person re-identification techniques can be grouped into methods based on feature representation and methods based on model learning. The first one takes into consideration of features that discriminates different persons but invariant for the same person. The most common features extracted are color, shape, texture, and gradient, and resolution. The papers [25–30] are based on the feature based person re-identification. These feature based methods requires complex computations levy stringent conditions on feature extraction. The model based methods use relationship between different cameras. Support vector machines and distance metric learning methods belong to this type. Some of the recent works has been reviewed here. In [31] person re-identification is done by Robust Canonical Correlation analysis (ROCCA). Robust canonical correlation is a multi-view base data analysis technique which is used for several tasks like face recognition. Le An et al. used this method along with shrinkage estimation and smoothing technique to address the challenge of high dimension feature trained with small number of samples in person re-identification task. ROCCA estimates the data co-variance matrices for learning feature transformations. This learned transformation matrices projects the data from multiple cameras into coherent subspace. The data of a same person will have maximum correlation in the subspace. Liu [32] proposed relaxed margin components analysis (RMCA) for person re-identification. This paper addresses the issue of generating more flexible margin. The proposed method defines the margin by considering information from nearest point of both intra and interclasses. They also proposed Kernalised Relaxation margin component analysis (KRMCA) for further increasing the accuracy. In [33] a novel person re-identification method which uses dissimilarities in a low dimensional space is proposed. This method is based on localized error between the images of a same person and the images of 2 different persons viewed in 2 dis-joint cameras. These dissimilarities projected in a linear subspace with local Eigen similarities which is learned using Principal component analysis. This method uses supervised classification framework to discriminate the positive and negative points in subspace. In [34] Datum adaptive local metric learning method for person re-identification method was proposed. This method learns one local feature projection for each feature sample and maps the samples in a common discriminate space. The author uses local Co-ordinate coding for avoiding computational cost. Anchor points are generated from clusters in the feature space and projection bases are learned for anchor points. A feature is denoted as weighted linear combination of these anchor points and the local feature projection of this sample is denoted by weighted linear combination of projection bases.

Ma et al. [35] proposed multi task maximally collapsing metric learning model to overcome the challenges like illumination variation, view angle variation and background clutter. It uses multiple Mahalanobis distance metrics for each camera pair which are related to each other. In [36] a adaptive ranking support vector machine based person re-identification method is proposed which cope ups the problem of non-availability of label information of target images by using only the matched and unmatched image pairs from source cameras. This method finds positive mean of target images by using negative and un-labelled data at target domain and labelled data from source domain.

# 4 Conclusions

In this paper we have presented a survey on the recent papers on face detection and person-re identification. We begin the paper by discussing the need for automated surveillance. We discussed about the problems namely pose variation, illumination variation, occlusions that are associated with face detection. The popularly used Face detector has been described briefly. We then surveyed the papers in which original Viola-Jones algorithm had been modified to detect multi view face detection, to reduce the computational complexity, to increase the speed of detection and to do detection in various illumination. We then discussed about the methods used for person re-identification and current state of art in the field of person re-identification. We consider that research in the field of face detection, person re-identification will increase the efficiency of smart video surveillance system which provides more secured environment.

# References

1. Yang, M.-H., Kriegman, D.J., Ahuja, N.: Detecting faces in images: a survey. IEEE Trans. PAMI **24**(1), 34–58 (2002)
2. Freund, Y., Schapire, R.E.: A Decision-Theoretic Generalization of On-line Learning and an Application to Boosting in Computational Learning Theory, pp. 23–37. Eurocolt 95, Springer
3. Viola, P., Jones, M.: Rapid object detection using a boosted cascade of simple features. In: Proceedings of CVPR (2001)
4. Tao, L., Zheng, Y., Du, Y.: A failure-aware explicit shape regression model for facial landmark detection in video. IEEE Signal Process. Lett. **21**(2), 244 (2014)
5. Li, S., Zhu, L., Zhang, Z., Blake, A., Zhang, H., Shum, H.: Statistical learning of multi-view face detection. In: Proceedings of ECCV (2002)
6. Jones. M., Viola, P.: Fast multi-view face detection. In: Technical report, Mitsubishi Electric Research Laboratories, TR2003-96 (2003)
7. Huang, C., Ai, H., Li, Y., Lao, S.: Learning sparse features in granular space for multi-view face detection. In: International Conference on Automatic Face and Gesture Recognition (2006)
8. Froba, B., Ernst, A.: Face detection with the modified census transform. In: IEEE International Conference on Automatic Face and Gesture Recognition (2004)
9. Ojala, T., Pietikäinen, M., Mäenpää, T.: Multiresolution gray-scale and rotation invariant texture classification with local binary patterns. IEEE Trans. PAMI **24**, 971–987 (2002)
10. Yan, S., Shan, S., Chen, X., Gao, W.: Locally assembled binary (LAB) feature with feature-centric cascade for fast and accurate face detection. In: Proceedings of CVPR (2008)
11. Jeong, K., Choi, Jaesik, Jang, G.-J.: Semi-local structure patterns for robust face detection. IEEE Signal Process. Lett. **22**(9), 1400–1403 (2015)
12. Wang, P., Ji, Q.: Learning discriminant features for multiview face and eye detection. In: Proceedings of CVPR (2005)
13. Baluja, S., Sahami, M., Rowley, H.A.: Efficient face orientation discrimination. In: Proceedings of ICIP (2004)
14. Pang, Y., Zhang, K., Yuan, Y., Wang, K.: Distributed object detection with linear SVMs. IEEE Trans. Cybern. **44**(11), (2014)
15. Chen, W., Sun, T., Yang, X., Wang, L.: Face detection based on half face-template. In: The Ninth International Conference on Electronic Measurement & Instruments ICEMI'2009
16. Wang, Qiang, Wu, Jing, Long, Chengnian, Li, Bo: P-FAD: real-time face detection scheme on embedded smart cameras. IEEE J. Emerging Sel. Top. Circuits Syst. **3**(2), 210–222 (2013)
17. Chen, Y.-Y., Hsu, W.H., Mark Liao, H.-Y.: Automatic training image acquisition and effective feature selection from community-contributed photos for facial attribute detection. IEEE Trans. Multimedia **15**(6), 1388 (2013)
18. Zhang, C., Zhang, Z.: A survey of recent advances in face detection. In: Technical Report, MSR-TR-2010-66, June 2010
19. Huang, C., Ai, H., Li, Y., Lao, S.: Vector boosting for rotation invariant multi-view face detection. In: Proceedings of the Tenth IEEE International Conference on Computer Vision (ICCV'05) 1550-5499/05
20. Huang, C., Ai, H., Li, Y., Lao, S.: Vector boosting for rotation invariant multi-view face detection. In: Proceedings of ICCV (2005)

21. Hassan Anvar, S.M., Yau, W.-Y., Teoh, E.K.: Multiview face detection and registration requiring minimal manual intervention. IEEE Trans. Pattern Anal. Mach. Intell. **35**(10), 2484–2497 (2013)
22. Zhang, Z., Yi, D., Lei, Z., Li, S.Z.: Regularized transfer boosting for face detection across spectrum. IEEE Signal Process. Lett. **19**(3), (2012)
23. Saghafi, M.A., Hussain, A., Zaman, H.B., Md. Saad, M.H.: Review of person re-identification techniques. In: IET Computer Vision (2013)
24. Vezzani, R., Baltieri, D., Cucchiara, R.: People re-identification in surveillance and forensics: a survey. University of Modena and Reggio Emilia. ACM Comput. Surv. **1**(1), Article 1, (2013)
25. Kviatkovsky, I., Adam, A., Rivlin, E.: Color invariants for person reidentification. IEEE Trans. Patt. Anal. Mach. Intell. **35**(7), 1622–1634 (2013)
26. Kuo, C.-H., Khamis, S., Shet, V.: Person re-identification using semantic color names and rank boost. In: Proceedings of WACV, pp. 281–287 (2013)
27. Yang, Y., Yang, J., Yan, J., Liao, S., Yi, D., Li, S.: Salient color names for person re-identification. In: Proceedings ECCV, pp. 536–551 (2014)
28. Liu, C., Gong, S., Loy, C.C.: On-the-fly feature importance mining for person re-identification. Pattern Recogn. **47**(4), 1602–1615 (2014)
29. Zhao, R., Ouyang, W., Wang, X.: Learning mid-level filters for person re-identification. In: Proceedings of CVPR, pp. 144–151 (2014)
30. Farenzena, M., Bazzani, L., Perina, A., Murino, V., Cristani, M.: Person re-identification by symmetry-driven accumulation of local features. In: Proceedings of CVPR Workshop, pp. 2360–2367 (2010)
31. An, L., Yang, S., Bhanu, B.: Person re-identification by robust canonical correlation analysis. IEEE Signal Process. Lett. **22**(8), (2015)
32. Liu, H., Qi, M., Jiang, J.: Kernelized relaxed margin components analysis for person re-identification. IEEE Signal Process. Lett. **22**(7), (2015)
33. Martinel, N., Micheloni, C.: Classification of local eigen-dissimilarities for person re-identification. IEEE Signal Process. Lett. **22**(4), (2015)
34. Liu, K., Zhao, Z., Member, IEEE, Cai, A.: Datum-adaptive local metric learning for person re-identification. IEEE Signal Process. Lett. **22**(9), (2015)
35. Ma, L., Yang, X., Tao, D.: Person re-identification over camera networks using multi-task distance metric learning. IEEE Trans. Image Process. **23**(8), (2014)
36. Ma, A.J., Li, J., Yuen, P.C., Senior Member, IEEE, Li, P.: Cross-domain person re-identification using domain adaptation ranking SVMs. IEEE Trans. Image Process. **24**(5), (2015)

# Machine Translation of Telugu Singular Pronoun Inflections to Sanskrit

T. Kameswara Rao and T.V. Prasad

**Abstract** Inflections in Telugu plays a crucial role in maintaining the meaning of the sentence unchanged though the word order is changed. This is possible since inflections are the internal part of the word. Inflecting nouns and pronouns is customary in Telugu. Main attention is given to inflected pronoun translation in this paper. Various categories of inflections and the ways of translation of pronouns are discussed in detail. For analysis, identification and translation of pronoun inflections, a Morphological Analysis System (MAS) that can perform forward and reverse morphology is developed for Machine Translation (MT) (Telugu (Source Language—SL) to Sanskrit (Target Language—TL)).

**Keywords** Pronoun inflections · Declensions · Machine translation · Morphological analysis

## 1 Introduction

Pronouns are the terms used instead of nouns to designate a person or thing without naming it [1]. In Telugu, pronouns are classified into eight types, viz. relational, adjectival, numerical, ambiguous-numerical, personal, demonstrative, indefinite and interrogative (Table 1) [2].

Scholars greatly differ on the count of exact number of classifications of pronouns in Sanskrit, such as 9 [3], 7 [4], 6 [5], etc. Based on in-depth research, it was found that there are more than 35 pronouns variously classified and used by different scholars [6]. Broadly, these 35 pronouns (Table 2) were classified into six

T. Kameswara Rao (✉) · T.V. Prasad
Department of CSE, Chirala Engineering College, Ramapuram Beach Road, Chirala,
Prakasam Dist 523157, A.P, India
e-mail: tkrphd@gmail.com

T.V. Prasad
e-mail: tvprasad2002@yahoo.com

© Springer India 2016 293
H.S. Behera and D.P. Mohapatra (eds.), *Computational Intelligence in Data Mining—Volume 1*, Advances in Intelligent Systems and Computing 410, DOI 10.1007/978-81-322-2734-2_30

**Table 1** Types of pronouns in Telugu

| Type | Description | Example |
|------|-------------|---------|
| saMbaMdha (relational) | Describes relations | evarYtE (whoever), EdytE |
| viSEshaNa (adjectival) | These are adjectival | aMdarU |
| saMkhyA-vAcaka (numerical) | These are numerical | oka, iddaru, reMDava etc. |
| saMkhyEya vAcaka (ambiguous numerical) | These describe a number without clear information | vArumuggurU veerE (these are the three). Here the gender is unclear |
| purushasaMbaMdha (personal) | First, second and third persons | nEnu (I), neevu (you) ataDu (he), etc. |
| nirdESAtmaka (demonstrative) | These describes precisely | idi (this), avi (they), ivi (these) etc. |
| anirdishT-Arthaka (indefinite) | These pronouns cont give precise information | konni (few), enni (how many), cAla (many), etc. |
| praSnArthaka (interrogative) | These pronouns used to interrogate | evaru (who), EmiTi (what), elA (how) |

**Table 2** Inflections used for nouns in Telugu and Sanskrit

| Case | Telugu affixes | Sanskrit affixes | |
|------|----------------|------------------|---|
| | | M/F | N |
| Nominative | Du, mu, vu | s (-सੁ) | m (-म੍) |
| Accusative | ni, nu, kUrci, guriMci | am (-अम੍) | m (-म੍) |
| Instrumental | cEta, cE, tODa, tO | A (-आ) | A (-आ) |
| Dative | koraku, kY | e (-ए) | e (-ए) |
| Ablative | valana, kaMTe, paTTi | as (-अस੍) | as (-अस੍) |
| Possessive | ki, ku, yokka, lO, lOpala | as (-अस੍) | as (-अस੍) |
| Locative | aMdu, na | i (-इ) | i -(इ) |

types, viz. personal, relative, interrogative, demonstrative indefinite and numerical pronouns [2]. All the pronouns are categorized according to number, gender and case [1]. There are 5 categories of pronouns in usage as given in Sect. 2.

Numbers are singular, dual and plural. Only singular is discussed in this paper. Genders are masculine, feminine and neuter. In Telugu, masculine gender represents men, feminine gender represents women and neuter gender represents objects. But in Sanskrit, the gender is purely grammatical and unpredictable [7]. Masculine gender represents a word that is formed with masculine characteristics, feminine gender represents a word that is formed with feminine characteristics and a neuter gender represents a word that is formed with neuter characteristics that are defined by the Sanskrit grammar [8]. Neuter gender represents sometimes masculine and feminine genders and sometimes neither, e.g. you, lawyer, committee etc. [1].

Cases can be used to express pronoun's relation with other words or with itself. These cases are considered as inflections, viz. accusative, instrumental, dative, ablative, possessive, locative and vocative. The distinction in gender of pronouns appears only in the pronouns in the third person [6, 9].

## 2 Structure of Declensions of Telugu and Sanskrit Pronouns

Declensions are the inflections of a noun or pronoun [10]. The declension of noun or pronoun forms a morpheme that consists of a stem and affix (*pratyaya*). Based on the gender and ending character of the stem, Sanskrit employs the predetermined affix to attach to the stem to form the declension morpheme as a grammatical ordinance. Affixes of Telugu and Sanskrit (singulars) are in Table 2.

Telugu affixes that are described in Table 2, are used to apply for both nouns and pronouns. But, the affixes of Sanskrit are used to affix nouns, but not pronouns. They are described in third section of the paper in detail. There are numerous vernacular declensions in Telugu nouns and pronouns, but not in inflections except in few cases, e.g. '*valla*' is a vogue form of '*valana*'. The colloquialism or other forms in pronouns will also be observed in many cases but they are not considered in this paper.

Formal declensions of singular pronouns that are used popularly are described in Table 15 of Appendix A. Following are the categories of pronouns in regular use [3]:

1. Personal: *asmad* (I), *yushmad* (you), *bhavat* (your honor), *sva* (one's own) are the four pronouns of this category, that are used to represent the human beings of first and second persons. From Telugu, *nEnu* (I), *mEmu* (we), *manamu* (we), *nIvu* (you), *mIru* (you/thou) and their declensions fall under this category [11].
2. Relative pronouns: *yad* (he-who, she-who, that-which) is the pronoun of this category and always occurs in a relative clause. *Datara* and *Datama* are the affixes to apply on *yad* accordingly. From Telugu, *evaDYtE-vADu* (he-who), *evatYtE-Ame* (she-who), *EdYtE-adi* (that-which), etc. and their declensions fall under this category.
3. Interrogative pronoun: *kim* (who, what, which) is the pronoun in this category and are used to interrogate. From Telugu, '*evaru* (who)' and its declensions fall under this category.
4. Demonstrative: Though there are many pronouns in this category, *tad* (he, she, that, it—who or which has been mentioned), *Etad* (this—represents a very near person or thing), *idam* (this—refers a person or thing that is near), *adas* (refers a person or thing at remote) are the pronouns in use. From Telugu, *adi* (that), *ataDu* (he), *Ame* (she), *idi* (this), *itadu* (this), *Ime* (this), *I* (this), *A* (that), etc. and their declensions fall under this category.

5. Indefinite pronouns: These are formed by attaching the affixes *cit, cana, api* or *svit* to the declensions of *kim.* From Telugu, *okAnokaDu* (somebody), *okAnokate* (somebody), *okAnokaTi* (something), etc. and their declensions fall under this category.

Categories of pronouns like Numerical, Directional, Correlative, Reciprocal, Possessive, Emphatic, etc. are the other categories. Some of them are formed by combinations of affixes attached to the primary nouns or pronouns. Indefinite and other categories of pronouns are not discussed in this paper since they are compound (not the base pronouns).

## 3 Morphological Analysis

The morphemes of formal declensions of pronouns in Telugu will have one of the affixes described in Table 1. In MT, affix can be cleaved from the morpheme to identify primary stem (pronoun) of the stem. Its correlated Sanskrit pronoun will be fetched from the data base. Only nominative forms are available in database along with pronoun's gender. Gender is also identified since the formation of Sanskrit morphemes depends purely on the gender of the word. The cases of the affix such as nominative, accusative, etc. have also been identified (Table 1). Correlated affix will be appended to the stem based on the gender. For the convenience of MT, stems are modified so as to generalize the affixes. The generalization is brought in the case of third person pronouns' declensions only since first and second persons' pronouns are very few.

Table 3 of Appendix A describes the stem and affixes of the pronouns *aham* (I) and *tvam* (you), Table 4 *tad* (He) and *Etad* (This), *idam* (This) and *adas* (That)—of Masculine Gender, Table 5 *tad* (She) and *Etad* (This), *idam* (This) and *adas* (That)—of Feminine Gender, Table 6 *tad* (That) and *Etad* (This), *idam* (That) and *adas* (This)—of Neuter Gender, Table 7 *kim* (He/She-Who) and *kim* (That-Which)—of Masculine, Feminine and Neuter genders, Table 8 *kim* (Who) and *kim* (Which)—of Masculine, Feminine and Neuter genders.

## 4 Results and Discussions

The process of MT has made simple and convenient to translate the declensions of singular pronouns, by basing on the similarities in affixes of various pronouns' declensions (Table 12 of Appendix A).

When the declensions of neuter gender viz. *tat* (that), *Etat* (this), *idam* (this), *ayam* (this), *yat* (that-which), *kim* (which) and masculine gender, viz. *saH* (he), *EshaH* (this), *yaH* (he-who), *kaH* (who) of third person are to be created, for all of this pronouns, the affixes are same except for accusative case. For *idam, yaH, saH, ayam, kaH* –*am* is the affix, for *tat, Etat, yat*—*at* is the affix and for *kim*—*im* is the affix. For all these pronouns, corresponding stems and their affixes are given in the Table 9 of Appendix A.

When the declensions of feminine gender viz. *sA* (she), *EshA* (this), *iyam* (this), *yA* (she-who), *kAm* (who) of third person are to be created, for all of these pronouns, the affixes are same. Corresponding stems and their affixes are given in the Table 10 of Appendix A.

When the declensions of masculine gender *asau* (that), and neuter gender *adaH* (that) of third person are to be created, for this pronouns, the affixes are same except for accusative case. For *adaH, aH* is the affix and for *asau, um* is the affix. For all these pronouns, corresponding stems and their affixes are given in the Table 11 of Appendix A.

All the affixes are consolidated and categorized case-wise into following three types for convenient and simple MT:

- Category 1: Affixes of declensions of pronouns of third person Masculine gender like *yaH* (he-who), *saH* (he), *EshaH* (this), *ayam* (this), *kaH* (who) and Neuter gender like *tat* (that), *Etat* (this), *idam* (this), *yat* (that-which), *kim* (which) are same except for accusative case as shown in Table 12 of Appendix A.
- Category 2: Affixes of declensions of pronouns of third person Feminine gender like *sA* (she), *EshA* (this), *iyam* (this), *yA* (she-who), *kAm* (who) as shown in Table 13 of Appendix A.
- Category 3: Affixes of declensions of pronouns of third person Masculine gender *asau* (that) and Neuter gender *adaH* (that) are same except for accusative case as shown in Table 14 of Appendix A.

The gender of the pronoun will be in any one of the categories mentioned above, and its corresponding stem is identified as mentioned in the Tables 9, 10 and 11 of Appendix A. Though some of the terms are context based, MAS is able to translate successfully 95 % of the cases. For instance, the sentence '*I manishiki Astulu lEvu* (This person have no assets)', have two declensions viz. '*I*' and '*manishiki*'. Gender of the '*I*' will not be identified unless the gender 'of the correlated term i.e. *manishiki* is identified. Since the gender of the term '*mahishi*' is treated as masculine, the same gender is applied to '*I*' also. This is the analysis of context, which may not be possible all the time to identify the genders. For gender unidentified issues, the masculine gender is applied.

# 5  Conclusions

Proper/formal singular declensions of pronouns that are mentioned in this paper are machine translated to Sanskrit from Telugu by implementing MAS. There are numerous informal declensions of pronouns are in use in Telugu. They are vernacular, vogue and colloquial.

Only singular declensions are discussed in this paper. It is planned to present implementation of dual and plural declensions' translation also. There is also a need of implementation of an MAS that can machine translate the informal declensions in the same stream.

# Appendix A

See Tables 3, 4, 5, 6, 7, 8, 9, 10, 11, 12, 13, 14 and 15.

**Table 3** Stems and affixes of *aham* and *tvam* pronouns

| Case | *nEnu* (I) | Morphology | | *nIvu* (you) | Morphology | |
|------|------------|------------|-------|--------------|------------|-------|
|      |            | Stem | Affix |              | Stem | Affix |
| Nom. | *aham* | – | – | *tvam* | – | – |
| Acc. | *mAm* | *m* | *Am* | *tvAm* | *tv* | *Am* |
| Instr. | *mayA* | *m* | *ayA* | *tvayA* | *tv* | *ayA* |
| Dat. | *Mahyam* | *m* | *ahyam* | *tubhyam* | *t* | *ubhyam* |
| Abl. | *mat* | *m* | *at* | *tvat* | *tv* | *at* |
| Poss. | *mama* | *m* | *ama* | *tava* | *t* | *ava* |
| Loc. | *mayi* | *M* | *ayi* | *tvayi* | *tv* | *ayi* |

**Table 4** Stems and affixes of *tad—ataDu* (he) and *Etad—itaDu* (this) *idam—I* (this) and *adas– A* (that)—of masculine

| Ca | *ataDu* (he) | Morpgy | | *itaDu* (this) | Morpgy | | *I* (this) | Morpgy | | *A* (that) | Morpgy | |
|----|--------------|--------|-------|----------------|--------|-------|------------|--------|-------|------------|--------|-------|
|    |              | St | Affix |                | St | Affix |            | St | Affix |            | St | Affix |
| N | *saH* | – | – | *EshaH* | – | – | *ayam* | – | – | *asau* | – | – |
| A | *tam* | *t* | *am* | *Etam* | *Et* | *am* | *imam* | *im* | *am* | *amum* | *am* | *Um* |
| In | *tEna* | *t* | *Ena* | *EtEna* | *Et* | *Ena* | *anEna* | *an* | *Ena* | *amunA* | *am* | *unA* |
| D | *tasmY* | *t* | *asmY* | *EtasmY* | *Et* | *asmY* | *asmY* | – | *AsmY* | *amushmY* | *am* | *ushmY* |
| Ab | *tasmAt* | *t* | *asmAt* | *EtasmAt* | *Et* | *asmAt* | *asmAt* | – | *AsmAt* | *amushmAt* | *am* | *ushmAt* |
| P | *tasya* | *t* | *asya* | *Etasya* | *Et* | *asya* | *asya* | – | *Asya* | *amushya* | *am* | *ushya* |
| L | *tasmin* | *t* | *asmin* | *Etasmin* | *Et* | *asmin* | *asmin* | – | *asmin* | *amushmin* | *am* | *ushmin* |

**Table 5** Stems and affixes of *tad—Ame, Etad—Ime, idam—I, adas—A*—feminine

| Ca | *Ame* (she) | Morpgy | | *Ime* (this) | Morpgy | | *I* (this) | Morpgy | | *A* (that) | Morpgy | |
|---|---|---|---|---|---|---|---|---|---|---|---|---|
| | | St | Affix | | St | Affix | | St | Affix | | St | Affix |
| N | *sA* | – | – | *EshA* | – | – | *iyam* | – | – | *asau* | – | – |
| A | *tAm* | *t* | *Am* | *EtAm* | *Et* | *Am* | *Imam* | *im* | *Am* | *amUm* | *am* | *Um* |
| I | *tayA* | *t* | *ayA* | *EtayA* | *Et* | *ayA* | *anayA* | *an* | *AyA* | *amuyA* | *am* | *uya* |
| D | *tasyY* | *t* | *asyY* | *EtasyY* | *Et* | *asyY* | *asyY* | – | *AsyY* | *amushyY* | *am* | *ushyY* |
| Ab | *tasyAH* | *t* | *AsyAH* | *EtasyAH* | *Et* | *asyAH* | *asyAH* | – | *AsyAH* | *amushyAH* | *am* | *ushyAH* |
| P | *tasyAH* | *t* | *AsyAH* | *EtasyAH* | *Et* | *asyAH* | *asyAH* | – | *AsyAH* | *amushyAH* | *am* | *ushyAH* |
| L | *tasyAm* | *t* | *AsyAm* | *EtasyAm* | *Et* | *asyAm* | *asyAm* | – | *AsyAm* | *amushyAm* | *am* | *ushyAm* |

**Table 6** Stems and affixes of *tad—adi*, and *Etad—idi, idam—I, adas—A*—neuter

| Ca | *adi* (that) | Morpgy | | *Idi* (this) | Morpgy | | *I* (this) | Morpgy | | *A* (that) | Morpgy | |
|---|---|---|---|---|---|---|---|---|---|---|---|---|
| | | St | Affix | | St | Affix | | St | Affix | | St | Affix |
| N | *tat* | – | – | *Etat* | – | – | *idam* | – | – | *adaH* | – | – |
| A | *tat* | *t* | *at* | *Etat* | *Et* | *at* | *idam* | *id* | *am* | *adaH* | *ad* | *aH* |
| I | *tEna* | *t* | *Ena* | *EtEna* | *Et* | *Ena* | *anEna* | *an* | *Ena* | *amunA* | *am* | *unA* |
| D | *tasmY* | *t* | *asmY* | *EtasmY* | *Et* | *asmY* | *asmY* | – | *AsmY* | *amushmY* | *am* | *ushmY* |
| Ab | *tasmAt* | *t* | *AsmAt* | *EtasmAt* | *Et* | *asmAt* | *asmAt* | – | *AsmAt* | *amushmAt* | *am* | *ushmAt* |
| P | *tasya* | *t* | *asya* | *Etasya* | *Et* | *asya* | *asya* | – | *Asya* | *amushya* | *am* | *ushya* |
| L | *tasmin* | *t* | *asmin* | *Etasmin* | *Et* | *asmin* | *asmin* | – | *asmin* | *amushmin* | *am* | *ushmin* |

**Table 7** Stems and affixes of *kim—evaru/Edi* masculine, feminine and neuter

| Ca | *evaru* (M) (he-who) | Morphology | | *Evaru* (F) (she-who) | Morphology | | *Edi* (N) (that-which) | Morphology | |
|---|---|---|---|---|---|---|---|---|---|
| | | Stem | Affix | | Stem | Affix | | Stem | Affix |
| N | *yaH* | – | – | *Ya* | – | – | *yat* | – | – |
| Ac | *yam* | *y* | *am* | *yAm* | *y* | *Am* | *yat* | *y* | *at* |
| I | *yEna* | *y* | *Ena* | *yayA* | *y* | *ayA* | *yEna* | *y* | *Ena* |
| D | *yasmY* | *y* | *asmY* | *yasyY* | *y* | *asyY* | *yasmY* | *y* | *asmY* |
| Ab | *yasmAt* | *y* | *asmAt* | *yasyAH* | *y* | *asyAH* | *yasmAt* | *y* | *asmAt* |
| P | *yasya* | *y* | *Asya* | *yasyAH* | *y* | *asyAH* | *Yasya* | *y* | *asya* |
| L | *yasmin* | *y* | *asmin* | *yasyAm* | *y* | *asyAm* | *Yasmin* | *y* | *asmin* |

**Table 8** Stems and affixes of *kim—evaru* masculine, feminine and neuter

| Case | *evaru* (M) (who) | Morphology | | *Evaru* (F) (who) | Morphology | | *Edi* (N) (which) | Morphology | |
|---|---|---|---|---|---|---|---|---|---|
| | | St | Affix | | St | Affix | | St | Affix |
| N | *kaH* | – | – | *kA* | – | – | *kim* | – | – |
| Ac | *kam* | *k* | *am* | *kAm* | *k* | *Am* | *kim* | *k* | *im* |
| I | *kEna* | *k* | *Ena* | *kayA* | *k* | *ayA* | *kEna* | *k* | *Ena* |
| D | *kasmY* | *k* | *asmY* | *kasyY* | *k* | *asyY* | *kasmY* | *k* | *asmY* |
| Ab | *kasmAt* | *k* | *asmAt* | *kasyAH* | *k* | *asyAH* | *kasmAt* | *k* | *asmAt* |
| P | *kasya* | *k* | *asya* | *kasyAH* | *k* | *asyAH* | *Kasya* | *k* | *asya* |
| L | *kasmin* | *k* | *asmin* | *kasyAm* | *k* | *asyAm* | *Kasmin* | *y* | *asmin* |

**Table 9** Stems and affixes of pronoun nominatives *tat*, *Etat*, *idam*, *yaH*, *saH*, *EshaH*, *ayam*, *kaH*, *kim* of masculine and neuter genders and *sA*, *EshA*, *iyam*, *ya*, *kAm* of feminine genders

| Ca | Morphology: that (N—*adi*), this (N—*idi*), this (N—*I*), he-who (M—*evarYtE*), that-which (N—*EdYtE*), He (M—*ataDu*), this (M—*itaDu*), this (M—*I*), who (M—*evaru*), which (M—*evaru*), which (N—*Edi*) | | | | | | | | | | | Morphology: she (F—*Ame*), this (F—*Ime*), this (F—*I*), she-who (F—*evaru*), who (F—*evaru*) | | | | | |
|---|---|---|---|---|---|---|---|---|---|---|---|---|---|---|---|---|---|
| | Stems | | | | | | | | | | Affix | Stems | | | | | Affix |
| | *tat* | *Etat* | *idam* | *yaH* | *yat* | *saH* | *EshaH* | *ayam* | *kaH* | *kim* | – | *sA* | *EshA* | *iyam* | *yA* | *kAm* | – |
| N | t | Et | id* | y* | y | t* | Et* | im* | k* | k# | at/am*/im# | t | Et | im | y | K | Am |
| Ac | t | Et | an | y | y | t | Et | an | k | k | Ena | t | Et | an | y | K | ayA |
| I | t | Et | – | y | y | t | Et | – | k | k | asmY | t | Et | – | y | K | asyY |
| D | t | Et | – | y | y | t | Et | – | k | k | asmAt | t | Et | – | y | K | asyAH |
| Ab | t | Et | – | y | y | t | Et | – | k | k | asya | t | Et | – | y | K | asyAH |
| P | t | Et | – | y | y | t | Et | – | k | k | asmin | t | Et | – | y | K | asyAm |
| L | t | Et | – | y | y | t | Et | – | k | k | | t | Et | – | y | K | |

**Table 10** Stems and affixes of the pronoun nominatives *tat, Etat, idam, yaH, saH, EshaH, ayam, kaH, kim* of masculine and neuter and *sA, EshA, iyam, yA* of feminine gender

| Ca | Morphology: that (N—*adi*), this (N—*idi*), this (N—*I*), he-who (M—*evarYtE*), that-which (N—*EdYtE*), He (M—*ataDu*), this (M—*itaDu*), this (M—*I*), who (M—*evaru*), which (N—*Edi*) | | | | | | | | | | | Morphology: she (F—*Ame*), this (F—*Ime*), this (F—*I*), she-who (F—*evaru*), who (F—*evaru*) | | | | | |
|---|---|---|---|---|---|---|---|---|---|---|---|---|---|---|---|---|---|
| | Stem | | | | | | | | | | Affix | Stems | | | | | Affix |
| | *tat* | *Etat* | *idam* | *yaH* | *yat* | *saH* | *EshaH* | *ayam* | *kaH* | *Kim* | – | *sA* | *EshA* | *iyam* | *yA* | *kAm* | – |
| N | t | Et | id* | y * | y | t* | Et* | im* | k* | k# | at/am*/im# | t | Et | im | y | K | – |
| Ac | t | Et | an | y | y | t | Et | an | k | k | | t | Et | an | y | K | Am |
| I | t | Et | – | y | y | t | Et | – | k | k | Ena | t | Et | – | y | K | ayA |
| D | t | Et | – | y | y | t | Et | – | k | k | asmY | t | Et | – | y | K | asyY |
| Ab | t | Et | – | y | y | t | Et | – | k | k | asmAt | t | Et | – | y | K | asyAH |
| P | t | Et | – | y | y | t | Et | – | k | k | asya | t | Et | – | y | K | asyAH |
| L | t | Et | – | y | y | t | Et | – | k | k | asmin | t | Et | – | y | K | asyAm |

**Table 11** A consolidated affixes for the stems of the pronoun nominatives *asau* and *adaH* of masculine and neuter genders

| Case | Morphology: that (M—A), that (N—A) | | |
|------|------|------|------|
| | Stem | | Affix |
| Nom. | *asau* | *adaH* | – |
| Acc. | *am* | *ad*\* | *um/aH*\* |
| Instr. | *am* | *am* | *unA* |
| Dat. | *am* | *am* | *ushmY* |
| Abl. | *am* | *am* | *ushmAt* |
| Poss. | *am* | *am* | *ushya* |
| Loc. | *am* | *am* | *ushmin* |

**Table 12** Singular personal accusative, instrumental

| Case | Ty | G | San | Accusative | | Instrumental | |
|------|-----|----|------|------------|----------|--------------|----------|
| | | | | Telugu | Sanskrit | Telugu | Sanskrit |
| *nEnu* (I) (FP) | P | A | *Aham* | *nannu, nAguriMci, nannu (guriMci/gUrci)* etc. | *mAm* | *nA (cE/cEta/tO/tODa)* etc. | *mayA* |
| *nIvu* (you) (SP) | P | A | *tvam* | *ninnu, nIguriMci, ninnu (guriMci/gUrci)* etc. | *tvAm* | *nI (cE/cEta/tO/tODa)* etc. | *tvayA* |
| *ataDu* (he) (TP) | D | M | *saH* | *ataDini, ataNNi, ataDini (guriMci/gUrci) ataNNigUrci* etc. | *taM* | *ataDi (cE/tODA/cEta/tO)* etc. | *tEna* |
| *itaDu* (this) (TP) | D | M | *Etad* | *itaNNi (gUrci/guriMci) itaDini (guriMci/gUrci), itaDini, itaNNi,* etc. | *Etam* | *itaDi (cE/cEta/tODA/tO)* etc. | *EtEna* |
| *Ame* (she) (TP) | D | F | *sA* | *Ame (gUrci/guriMci), Amenu (guriMci/gUrci), Ameni, Amenu,* etc. | *tAm* | *Ame (cE/tODa/tO/cEta)* etc. | *tayA* |
| *Ime* (this) (TP) | D | F | *EshA* | *Ime (nu/gUrci/guriMci), Imeni,* etc. | *EtAm* | *Ime (cE/tO/cEta/tODa)* etc. | *EtayA* |
| *idi* (this) (TP) | D | N | *Etat* | *dIni (guriMci/gUrci)* etc. | *Etat* | *dIni (cEta/cE/tODa/tO)* etc. | *EtEna* |
| *adi* (that) (TP) | D | N | *tat* | *dAni (ni/guriMci/gUrci), dAnniguriMci,* etc. | *tat* | *dAni (cEta/cE/tO/tODa/)* etc. | *tEna* |
| *EvaDu* (who) (TP) | I | M | *kaH* | *evaDini (guriMci/gUrci), evaNNi (gUrci/guriMci)* | *kam* | *evaDi (cEta/cE/tODa/tO)* etc. | *kEna* |
| *Evate* (who) (TP) | I | F | *kA* | *evatini (guriMci/gUrci), evateni,* etc. | *kAm* | *Evate (cE/cEta/tO/tODa)* etc. | *kayA* |
| *Edi* (which) (TP) | I | N | *kim* | *dEnini (gUrcni/guriMci)* etc. | *kim* | *dEni (cEta/cE/tODa/tO)* etc. | *kEna* |

**Table 13** Singular personal dative and ablative, possessive, locative

| Case | Ty | G | San | Dative | | Ablative | |
|---|---|---|---|---|---|---|---|
| | | | | Telugu | Sanskrit | Telugu | Sanskrit |
| nEnu (I) (FP) | P | A | aham | nA (korakulkY/korakY) etc. | mahyam | nA (valana/kaMTe), nannubaTTi etc. | mat |
| nIvu (you) (SP) | P | A | tvam | nI (koraku/kY/korakY) etc. | tubhyam | nI (valana/kaMTe) ninnubaTTi etc. | tvat |
| ataDu (he) (TP) | D | M | saH | ataDi (koraku/kY/korakY) etc. | tasmY | ataDi (valana/kaMTe) ataDinibaTTi, ataNNibaTTi, etc. | tasmAt |
| itaDu (this) (TP) | D | M | Etad | itaDi (koraku/kY/korakY) etc. | EtasmY | itaDi (valana/kaMTe), itaDinibaTTi, itaNNibaTTi, etc. | EtasmAt |
| Ame (she) (TP) | D | F | sA | Ame (koraku/kY/korakY) etc. | tasyY | Ame (valana/kaMTe/baTTi), AmenubaTTi | tasyAH |
| Ime (this) (TP) | D | F | EshA | Ime (koraku/korakY/kY) etc. | EtasyY | Imevalana, ImenibaTTi ImenubaTTi ImekaMTe, | EtasyAH |
| idi (this) (TP) | D | N | Etat | dIni (koraku/korakY/kY) etc. | EtasmY | dInivalana, dInikaMTe, dInnibaTTi etc. | EtasmAt |
| adi (that) (TP) | D | N | tat | dAni (koraku/kY/korakY) etc. | tasmY | dAnivalana, dAnikaMTe, dAninibaTTi etc. | tasmAt |
| EvaDu (who) (TP) | I | M | kaH | evaDi (koraku/kY/korakY) etc. | kasmY | evaDi (valana/kaMTe), evaDinibaTTi, evaNNibaTTi etc. | kasmAt |
| Evate (who) (TP) | I | F | kA | Evate (koraku/kY/korakY) etc. | kasyY | evate (valana/kaMTe/baTTi) etc. | kasyAH |
| Edi (which) (TP) | I | N | kim | dEni (koraku/kY/korakY) etc. | kasmY | dEni (valana/kaMTe/baTTi) | kasmAt |

**Table 14** Singular personal possessive and locative

| Case | Ty | G | Sanskrit | Possessive | | Locative | |
|---|---|---|---|---|---|---|---|
| | | | | Telugu | Sanskrit | Telugu | Sanskrit |
| nEnu (I) (FP) | P | A | aham | nA (yokka, nAlO, nAku, nAlOpala, nA), etc. | mama | nA (yaMdu/yaMduna) etc. | mayi |
| nIvu (you) (SP) | P | A | tvam | nI (yokka/lOku/lOpala) etc. | tava | nI (yaMdu/yaMduna) etc. | tvayi |
| ataDu (he) (TP) | D | M | saH | ataDi (ki/yokka/lO/lOpala) etc. | tasya | ataDi (yaMdu/yaMduna) etc. | tasmin |
| itaDu (this) (TP) | D | M | EshaH | itaDi (ki/yokka/lO/lOpala) etc. | Etasya | itaDi (yaMdu/yaMduna) etc. | Etasmin |
| Ame (she) (TP) | D | F | sA | Ame (ki/ku), Ame (yokka/lOpala/lO) etc. | tasyAH | Ame (yaMdu,/yaMduna) etc. | tasyAm |
| Ime (this) (TP) | D | F | EshA | Ime (yokka/lOpala/lO), Imeki, Imeku, etc. | EtasyAH | Ime (yaMdu/yaMduna) etc. | EtasyAm |
| idi (this) (TP) | D | N | Etat | dIni (yokka/lO/lOpala) dIniki, etc. | Etasya | dIni (yaMdu/yaMduna) etc. | Etasmin |
| adi (that) (TP) | D | N | tat | dAni (yokka/lO/lOpala), dAniki etc. | tasya | dAni (yaMdu/yaMduna) etc. | tasmin |
| EvaDu (who) (TP) | I | M | kaH | evaDi (ki/yokka/lO/lOpala) etc. | kasya | evaDi (yaMdu,/yaMduna) etc. | kasmin |
| Evate (who) (TP) | I | F | kA | evate (ku/ki/yokka/lO/lOpala) etc. | kasyAH | evate (yaMdu/yaMduna/yaMdu) etc. | kasyA |
| Edi (which) (TP) | I | N | kim | dEni (ki/yokka/lO ilOpala/lOpala) etc. | kasya | dEni (yaMdu/yaMduna) etc. | kasmin |

**Table 15** Examples for the process of MT of singular pronoun declensions

| Per | G | Declension (T) | Affix (T) | Stem (T) | Morphed Primary stem (T) | Stem (S) | Case | Affix (S) | Final form = Stem (S) + Affix (S) |
|---|---|---|---|---|---|---|---|---|---|
| I | – | 'nAguriMci (of me) | guriMci | nA | nEnu | Id | Acc | Am | id + am = idam |
| II | – | nIcEta (by you) | cEta | nI | nIvu | tv | Instr | ayA | tv + ayA = tvayA |
| III | M | ataDikoraku (for him) | koraku | ataDi | ataDu' | t | Dat | asmY | t + asmY = tasmY |
| III | F | Amevalana (from her) | valana | Ame | Ame | t | Abl | asyAH | t + asyAH = tasyAH |
| III | N | dAniyokka (of it) | Yokka | dAni | adi | t | Poss | asya | t + asya = tasya |

*Legend*: *Per* Person, *G* Gender, *S/San* Sanskrit, *T* Telugu, *M* Masculine, *F* Feminine, *N* Neuter, *Ma* Masculine, *Fe* Feminine; *Ne* Neuter; *A* All, *FP* First; *SP* Second; *TP* Third, *P* Personal; *I* Interrogational; *D* Demonstrative, *Ty* Type, *Morpgy* Morphology

# References

1. Wikner, C.: A Practical Sanskrit Introductory. (1996)
2. Prasad, M.K.: Telugu Vyakaranamu. Victory Publishers (2002)
3. Kale, M.R.: A Higher Sanskrit Grammar. Motilal Banarsidass Publishers, Varanasi (1960)
4. Muller, M.: A Sanskrit Grammar for Beginners. Longmans Green & Co, London (1866)
5. Srinivasacharyulu, K.: Sanskrit in 30 days. Andhra Pradesh Sahithya Academy (1972)
6. Sastry, K.L.V., Ananta Rama Sastri, Pt.L.: Sabda Manjari. R. S. Vadhyar Publishers (2002)
7. Deshpande, M.M.: A Sanskrit Primer. University of Michigan (2007)
8. Kameswara Rao, T., Prasad, T.V.: Machine translation of telugu singular noun inflections to sanskrit. In: International Conference on in Computer Science and Engineering (2015) (submitted)
9. Arden, A.H.: A Progressive Grammar of the Telugu Language, 2nd edn. Simpkin, Marshall, Hamilton, Kent & Co. Ltd., London (1905)
10. Suryanarayana, J.: Sanskrit for Telugu Students. Sri Balaji Printers (1993)
11. McDonell, A.A.: Sanskrit Grammar for Students, 3rd edn. Oxford University Press (1926)
12. Caldwell, R.: A Comparative Grammar of the Dravidian or South-Indian Family of Languages, 2nd edn. Trubner & Co., Ludgate Hill, London (1875) (Rev.)

# Tweet Analyzer: Identifying Interesting Tweets Based on the Polarity of Tweets

M. Arun Manicka Raja and S. Swamynathan

**Abstract** Sentiment analysis is the process of finding the opinions present in the textual content. This paper proposes a tweet analyzer to perform sentiment analysis on twitter data. The work mainly involves the sentiment analysis process using various trained machine learning classifiers applied on large collection of tweets. The classifiers have been trained using maximum number of polarity oriented words for effectively classifying the tweets. The trained classifiers at sentence level outperformed the keyword based classification method. The classified tweets are further analyzed for identifying top N tweets. The experimental results show that the sentiment analyzer system predicted polarities of tweet and effectively identified top N tweets.

**Keywords** Tweets · Sentiment analysis · Opinion polarity · Classification · Top N tweets

## 1 Introduction

Sentiment analysis involves the identification and classification of opinions mentioned in the source content. Social media sites generate large amount of opinionated data such as tweet posts, status messages, blog content etc., The process of knowing the sentiment from these user generated data is essential [1]. The tweets are the data streams since they are generated rapidly and those represent the dynamic behavior of the users present in the entire twitter network. The tweets are generated on various topics representing different domains by many users.

M.A.M. Raja (✉) · S. Swamynathan
Department of Information Science and Technology, College of Engineering Guindy,
Anna University, Chennai 600 025, Tamil Nadu, India
e-mail: arunmanickaraja@gmail.com

S. Swamynathan
e-mail: swamyns@annauniv.edu

© Springer India 2016                                                                                     307
H.S. Behera and D.P. Mohapatra (eds.), *Computational Intelligence in Data Mining—Volume 1*, Advances in Intelligent Systems and Computing 410, DOI 10.1007/978-81-322-2734-2_31

Twitter sentiment analysis mainly involves identification of trending topics and product reputation. Twitter sentiment analysis is particularly different, rather than performing sentiment analysis in any other domain, because of the difficulties in finding the sentiments expressed in tweets. Currently the twitter based recommender system is highly encouraged. The already available recommender systems only consider the relationship between users and items and not considering the relationship among users. In addition, these systems use social graph of users and not considering the messages what they share [2]. In this work, sentiment based tweet analyzer is proposed to process all the tweets by analyzing the opinionated content mentioned in the tweets for providing a better recommendation by identifying top N tweets.

In this paper, Sect. 2 discusses about various tweet analysis related works carried out by different researchers. Section 3 explains various design components of the tweet analyzer system design. Section 4 elaborates the detailed implementation of the tweet analyzer system with classification results. Finally, Sect. 5 provides the conclusion and future work of the sentiment analysis with tweet analyzer.

## 2  Related Work

Social networks evolved as a mandate for sharing information instantly. There is different type of research works have been performed in the literature such as twitter sentiment analysis, trending topic identification, twitter based recommendation system. Rill et al. [3] designed a system called PoliTwi to find the emerging political discussion on various topics. The work investigated the detection of the polarity present in the discussion of different topics represented using hashtags. Sidorov et al. [4] experimented machine learning based approach for opinion mining in tweets. The authors examined how the classifiers perform while doing opinion mining in Spanish twitter data. The experiment mainly performed on three classifiers which are naive bayes, decision trees and support vector machines. Liu et al. [5] proposed semi supervised topic adaptive sentiment classification. The topic adaptive sentiment classification learning algorithm is designed to update topic-adaptive features based on choosing the un-labelled data with collaborative selection method. The topic adaptive sentiment classification algorithm outperformed other well-known supervised classifiers. Gautam et al. [6] performed sentiment analysis using machine learning approaches. The sentiment analysis is performed for customer review classification wherein the information is analyzed in the form of number of positive tweets and number of negative tweets. Pennacchiotti et al. [7] addressed the problem of user classification task in social media. The authors employed machine learning approach which relied on the features derived from the user information. The sentiment analysis task of twitter dataset of laptops incorporates the review mining tasks carried out for online consumer reviews of digital camera datasets in our earlier work [8]. The performance of the different classifiers is evaluated using sentiment based classification. The sentiment analysis performed on digital camera reviews based on the polarity of the opinionated content. Lin et al. [9] presented a large scale machine learning in twitter. It involves the twitter integration of the machine learning tools into Hadoop and Pig analytics platform. Yang et al. [10] proposed a topic detection model known as Trend Sensitive-Latent Dirichlet Allocation as an alternative to understand the tweets and to measure the interestingness of the tweet based on the topics. Aiello et al. [11] discussed trending topic detection in twitter. The authors compared six topic identification methods in three different twitter datasets pertaining to various major events that differ in time scale and the content generation rate. Kim et al. [12] used a probabilistic model for a recommendation system in twitter. The experiments have been done both on real and synthetic data set demonstrate the effectiveness in discovering the interesting and hidden knowledge from the twitter data.

In spite of the tweet mining works carried out by various researchers, the tweets need to be analyzed by finding its polarity which helps to identify top N tweets. In the literature, many works have been done for twitter sentiment analysis. In this work, the sentiment analysis work has been performed using various trained classifiers with large corpuses of sentimental words for training data to improve the accuracy of the classifiers.

## 3 Tweet Analyzer System Design

The proposed system shown in Fig. 1 consists of various components: Tweet Crawler, Tweet pre-processor, Sentimental word extractor and Sentiment Analyzer.

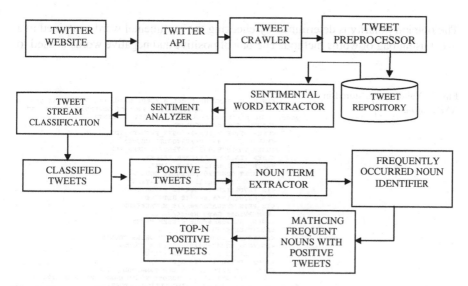

**Fig. 1** Tweet analyzer system design

## 3.1   Tweet Crawler

The tweet crawler consists of various tasks such as URL feed of Twitter website, Twitter API authentication, and tweet collection using REST API.

## 3.2   Tweet Pre-processor

The non-english tweets, tweets with smiley icons, tweets containing only user names, and tweets containing only URLs are all considered as invalid tweets and those tweets are removed. This is achieved by checking the tweets with the use of different regular expressions and pattern matching. Figure 2 shows the valid tweet content with different contents such as language, retweet id, user place, retweet count and etc.

## 3.3   Sentimental Word Extractor

RITA word tagger is used for tagging the tweet. The tagger helps to identify the noun, adjective, adverbs mentioned in the tweet. The tagged adjective terms are used to identify the sentimental words contained in the tweets.

## 3.4   Sentiment Analyzer

The opinion polarity is determined by checking the sentimental word contained in a tweet using bag of words method. The bag of positive and negative words is used to

**Fig. 2** Valid tweets stored in XML format

```
<TWEET>
    <TWEET_CONTENT>HP Spectre x360: Hands On: Video: If you're
after a beefy 2 in 1 laptop for a spot of portable gaming and
your... http://t.co/zTItdOtfan</TWEET_CONTENT>
    <LANGUAGE>en</LANGUAGE>
    <USER_RETWEET_ID>-1</USER_RETWEET_ID>
    <FAVORITE_COUNT>0</FAVORITE_COUNT>
    <USER_ID>592930037436796928</USER_ID>
    <USER_PLACE>null</USER_PLACE>
    <RATE_LIMIT>null</RATE_LIMIT>
    <RETWEET_COUNT>0</RETWEET_COUNT>
    <GEO_LOCATION>null</GEO_LOCATION>
    <IS_FAVORITED>false</IS_FAVORITED>
    <IS_RETWEET>false</IS_RETWEET>
    <IS_RETWEETED>false</IS_RETWEETED>
    <NAME>Video Game News</NAME>
    <USER_NAME>gamerhub_</USER_NAME>
    <LOCATION></LOCATION>
    <FOLLOWERS_COUNT>1928</FOLLOWERS_COUNT>
    <FRIENDS_COUNT>1919</FRIENDS_COUNT>
    <USER_STATUS>1919</USER_STATUS>
    <TIME_ZONE>null</TIME_ZONE>
    <URL>http://topic-hub.com/gamerhub</URL>
    <DESCRIPTION>All the latest video game news and reviews
from around the world.</DESCRIPTION></TWEET>
</TWEET>
```

check the terms tweeted whether it represents positivity or negativity. There are totally 1,907 positive words, 4,750 negative words used for finding the polarity of a tweet. The classifiers are first trained on the collected tweets using these keywords and then tested on tweets.

## 3.5 Machine Learning Classifiers

Some of the classifiers used are J48 classifier, Conjunctive rule based classifier, Naive Bayes classifier, SVM and IBK instance based classifier. The reason for selecting these classifiers is due to the uniqueness of every classifier. The functionalities of the classifiers are described as follows.

J48 is a tree based classification technique. The classification process is modelled by tree construction. The main idea of J48 is to divide the data into various ranges based on the attribute values in the training data [13]. Naive bayes is a simple probabilistic model based classifier. It calculates the probabilities by counting the occurrence and combination of values present in the training data. Conjunctive rule performs well for prediction on numeric and nominal class labels. AdaBoostM1 which helps to boost the nominal class classifier using adaboostM1 method. It highly improves the performance but sometimes overfits to the data. Sequential minimal optimization is a support vector based classifier algorithm. It replaces all missing values and transforms nominal attributes into binary values. It also normalizes all attributes by default. IBK instance based constructs hypotheses directly from the training instances. This means that the hypothesis complexity grows with the data.

The classifiers have been chosen from different categories of classifiers. Thus the learning tendency and the classification accuracy of the classifiers are evaluated meaningfully with the collected twitter data.

## 3.6 Identification of Top N Positive and Negative Tweets

The positively classified tweets are further tagged for extracting the noun terms. The noun terms from all the positive tweets are extracted. The most frequently represented noun terms are identified by counting its number of occurrence and then those identified nouns are matched with tweets. The top N positive tweets are found then the corresponding tweet mentioned laptop brand is also identified for its popularity. Likewise, it is performed for negative tweets.

## 4  Experimental Results and Discussion

Since the machine learning classifiers are better in handling the standard datasets, there is a need for evaluating the classifiers on non-standard datasets. There were totally 50,000 tweets collected using twitter REST API for experimental purposes. The collected tweets were generated in the last 3 months from the date of experiments performed. After pre-processing of the collected tweets, the valid tweets are totally of 43,838. Then the pre-processed tweets are tagged and checked its sentimental polarity using different mining algorithms and its performance measures are evaluated.

The positively classified tweets are further tagged for identifying frequently mentioned nouns in tweets. Then the tweets related to frequently used nouns are identified for finding top N tweets and its corresponding product for which the tweet was posted. Likewise, the negative sentimental words present in the tweets are also identified. The classified tweets using keyword based classification method are shown in Table 1. There are totally 43,838 tweets inputted to the keyword based bag of words method. The classified results indicate that the positive tweets are 17,873, negative tweets are 2,484, neutral tweets are 1,348 and undefined tweets are 22,133. The large amount of tweets is classified as undefined due to the lack of keywords in training data. Table 2 shows some of the common sentiment words presented in the laptop tweets.

The training of classifiers on tweets is shown in Fig. 3. The trained classifiers are further applied on the new tweets for assessing the performance of classifiers. Figure 4 shows the testing of the classifiers on new tweets.

**Table 1**  Classification using keyword based method

| Dataset | Total | Positive | Negative | Neutral | Undefined |
|---------|-------|----------|----------|---------|-----------|
| Laptop tweets | 43,838 | 17,873 | 2484 | 1348 | 22,133 |

**Table 2**  Polarity oriented words in tweets

| Positive polarity | Negative polarity | Neutral polarity | Undefined polarity |
|-------------------|-------------------|------------------|--------------------|
| fantastic, aspire, good, awesome, genuine | bad, not good, poor, drained, less speed, low, expensive, bad quality | but ok, but, however, adjustable, somewhat good | http, laptop, model |

lenovo in it s time to get your hands on the lenovo a   plus make sure you re ready: Positive
hp desktops are better than imacs : Positive
dell laptops are better than apple their marketing isn t : Neutral
new lcd screen with flex cable for hp compaq presario :Undef
  The latest Dell XPS 15 Touch (9530) is a premium desktop-replacement laptop , it doesn't come cheap: Negative

**Fig. 3**  Training of classifiers on tweets

In Testing: deffinitely cool, but faced problem with with Charger For HP Compaq Laptop N193.
In predicted : Both
In Testing: the sony n2: Ac Adapter for TOSHIBA EQUIUM LAPTOP mains lead here.
In predicted : undef (0.0)
In Testing :Finally i'm happy to hear this offer from Dell. I'm aware of the new model.
In predicted : Pos (2.0)
In Testing: In my lenovo since update done, i cant use this wifi. Help pls.
In predicted:  Neg (1.0)

**Fig. 4** Testing of trained classifiers on tweets

**Table 3** Performance evaluation of classifiers on tweets

| Training tweets (4500) test tweets (43,838) | Total number of tweets | Positively classified tweets | Negatively classified tweets | Neutrally classified tweets | Undefined tweets | Accuracy of classifiers |
|---|---|---|---|---|---|---|
| J48 classifier | 43,838 | 32,880 | 350 | 1308 | 9300 | 75.00 |
| Conjunctive rule classifier | 43,838 | 32,172 | 366 | 1277 | 10,023 | 73.38 |
| AdaBoostM1 classifier | 43,838 | 31,079 | 758 | 1708 | 10,293 | 70.86 |
| Naive Bayes classifier | 43,838 | 31,609 | 358 | 711 | 11,160 | 72.10 |
| Sequential minimal optimization | 43,838 | 31,932 | 304 | 1201 | 10,401 | 72.84 |
| IBK instance based classifier | 43,838 | 32,397 | 373 | 1028 | 10,040 | 73.90 |

Table 3 shows the comparison of the classification performed using various classifiers. The classifiers are implemented in Java using WEKA API. The total positively classified tweets are around 30,000 using different classifiers. The total number of nouns extracted from these 30,000 tweets are 66,348. The tweets are collected for 8 laptop brands which include 33 models. The positive sentimental words present with the noun terms in the top 50 positive tweets that include various laptop brands as shown in Table 4. Likewise, the negative sentimental words present with the noun terms in the top 25 negative tweets for different laptop brands are shown in Table 5.

The comparison of various classifiers is shown in Fig. 5. It shows that most of the trained classifiers are relatively equivalent in classifying the positive tweets, negative tweets, neutral tweets and undefined tweets.

# 5 Conclusion

In this paper, a tweet analyzer framework is implemented to perform the sentiment oriented tweet analyzing task. The empirical results show better performance of the trained classifiers better than general keyword based classification methods.

**Table 4** Laptop brands with positive sentimental words

| Laptop brands | Positive sentimental words with noun terms |
|---|---|
| ACER | i. acer aspire 5750z, as5750z-4835 new led wxga hd glossy laptop lcd |
| | ii. high performance battery for acer aspire 7741z-4643 laptop |
| ASUS | i. new asus…just awesome |
| DELL | i. dell inspiron is excellent |
| | ii. the best laptop on the market—the $399 dell |
| | iii. dell backup, finally got a laptop working more than 7 h on single full charge |
| | iv. dell laptops are better than apple. their marketing isn't. hp desktops are better than imacs |
| HP | i. genuine hp pavilion |
| | ii. new genuine oem ac power adapter charger hp sleekbook ultrabook |
| | iii. high performance battery for hp pavilion dv7-3188 cl laptop |
| COMPAQ | i. genuine 65w ac adapter charger Compaq |
| | ii. special edition evo |
| | iii. beautiful bagged burgundy evo viii |
| LENOVO | i. a quick lesson in lenovo's cool history. the future will be even cooler! |
| | ii. you can buy lenovo thinkpad |
| | iii. lenovo made a selfie flash and it is awesome |
| | iv. super lenovo thinkpad tablet *** price as low as $535 *** |
| | v. lenovo actually has the greatest stories to tell |

**Table 5** Laptop brands with negative sentimental words

| Laptop brands | Negative sentimental words with noun terms |
|---|---|
| DELL | i. dell laptops are better but their marketing isn't |
| HP | i. i seriously have no idea why i wanna speedrun hp and the chamber of secrets so bad, i just do |
| | ii. hp you are the king of bloatware and i hate you. |
| | iii. i tried to read hp fanfiction but no. no. the well is too deep. i can't. not yet. i have to be responsible |
| COMPAQ | i. you should make @noobde promise to add this is u win #evo or something that would be tight as hell dawg |
| LENOVO | i. Lenovo thinkpad specifications confirmed—att fans did not have a lot of things |
| LG | i. this does not end here. when we call LG custcare they say the installation is chargeable. our product is not under warranty. |
| | ii. when we call ur custcare back to inform this he says u shud not hv informed the LG people that it is not an online purchase |

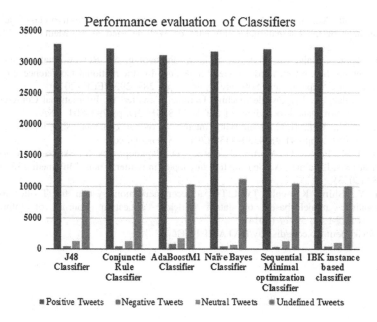

**Fig. 5** Comparison of classifiers

Further, the top N tweets have been identified for finding the most positively tweeted laptop brands. The noun term extraction based top positive tweets identification provides better recommendation. This work can be extended in future by evaluating the adjective level and noun terms mentioned in the individual tweets, with the sentimental score computation for ranking tweets with trending topics.

# References

1. Kim, Y., Shim. K.: TWITOBI a recommendation system for twitter using probabilistic modeling. In: International Conference on Data Mining, pp. 340–349. IEEE (2011)
2. Neethu, M.S., Rajasree, R.: Sentiment analysis in twitter using machine learning techniques, computing. In: International Conference on Communications and Networking Technologies, pp. 1–5. IEEE (2013)
3. Rill, V., Reinel, D., Scheidt, J., Zicari, R. V.: PoliTwi early detection of emerging political topics on twitter and the impact on concept-level sentiment analysis. J. Knowl. Based Syst. **69**, 24–33 (2014). (ScienceDirect)
4. Cagliero, L., Fiori, A.: TweCoM topic and context mining from twitter. In: The Influence of Technology on Social Network Analysis and Mining, vol. 6, pp. 75–100. Springer (2013)
5. Liu, S., Cheng, X., Li, F., Li, F.: TASC: topic-adaptive sentiment classification on dynamic tweets. Trans. Knowl. Data Eng. **27**(6), 1696–1709 (2015). (IEEE)
6. Gautam, G., Yadav, D.: Sentiment analysis of twitter data using machine learning approaches and semantic analysis. In: International Conference Contemporary Computing, pp. 437–442. IEEE (2014)

7. Pennacchiotti, M., Popescu, A.: A machine learning approach to twitter user classification. In: Proceedings of the International Conference on Weblogs and Social Media, pp. 281–288. (2011)

8. Arun Manicka Raja, M., Winster, S.G., Swamynathan, S.: Review analyzer analyzing consumer product reviews from review collection. In: International Conference on Recent Advances in Computing and Software Systems, pp. 287–292. IEEE (2012)

9. Lin, J., Kolcz, A.: Large-scale machine learning at twitter. In: International Conference on Management of Data. Proceedings of the ACM SIGMOD, pp.793–804. USA

10. Yang, M.-C., Rim, H.-C.: Identifying interesting twitter contents using topical analysis. J. Expert Syst. Appl. **41**(9), 4330–4336 (2014). ScienceDirect

11. Aiello, L.M., Petkos, G., Martin, C., Corney, D., Papadopoulos, S., Skraba, R., Goker, A., Kompatsiaris, I., Jaimes, A.: Sensing trending topics in twitter. Trans. Multimedia. **15**, 61268–61282 (2013)

12. Kim, Y., Shim, Kyuseok: TWILITE: a recommendation system for Twitter using a probabilistic model based on latent Dirichlet allocation. Journal of Information Systems ACM **42**, 59–77 (2014)

13. http://wiki.pentaho.com/display/DATAMINING/

# Face Tracking to Detect Dynamic Target in Wireless Sensor Networks

T.J. Reshma and Jucy Vareed

**Abstract** Wireless sensor networks are collection of spatially distributed autonomous actuator devices called sensor nodes. Tracking target under surveillance is one of the main applications of wireless sensor networks. It enables remote monitoring of objects and its environments. In target tracking, sensor nodes are informed when the target under surveillance is discovered. Some nodes detect the target and send a detection message to the nodes on the targets expected moving path. So nodes can wake up earlier. Face tracking is a new tacking framework, in which divides the region into different polygons called Faces. Instead of predicting the target location separately in a face, here estimate the targets movement towards another face. It enables the wireless sensor network to be aware of a target entering the polygon a bit earlier. Face track method failed when target in dynamic motion, i.e. target in random motion or retracing the path again and again. If the target follows a dynamic motion, the polygons are reconstructed repeatedly and energy is wasted in sending messages to create new polygons. Here proposes a framework to track a mobile object in a sensor network dynamically. In this framework, Polygons are created initially in form of clusters to avoid repeated polygon reconstruction. The sensors are programmed in a way that at least one sensor stays active at any instant of time in the polygon to detect the target. Once the target is detected and it entered into the edge of polygon, the nodes of the neighboring polygon is activated. Then, activated polygon keep tracking the target. Thus energy of the sensor nodes are saved because of polygon is not created and deleted, instead activated.

**Keywords** Wireless sensor networks · Target tracking · Face tracking · Node failure

T.J. Reshma (✉) · J. Vareed
Department of Computer Science & Engineering, Vidya Academy of Science & Technology,
Thrissur, India
e-mail: tjreshma75@gmail.com

© Springer India 2016    317
H.S. Behera and D.P. Mohapatra (eds.), *Computational Intelligence
in Data Mining—Volume 1*, Advances in Intelligent Systems
and Computing 410, DOI 10.1007/978-81-322-2734-2_32

# 1 Introduction

Wireless sensor networks are collection of many autonomous actuator devices. These actuator devices are known as sensor nodes. Sensor nodes distributed over an area for the purpose of sensing that environment. They are used in application areas such as environmental monitoring, military, healthcare services, chemical detection and smart homes. A sensor node, also called as a mote, is a node in a wireless sensor network that is capable of gathering information which is sensory in nature, processing this sensed data, and communicating with other connected nodes in the network. A wireless sensor network can effectively senses the data from the environment and communicate with others. A sensor network offers efficient combination of distributed sensing, computing and communication. The basic idea of sensor network is to disperse small sensing devices; which are capable of sensing the changes in parameters of environment. They are used in infinite applications and, at the same time, offer many challenges due to their characteristics, primarily the severe energy constraints to which sensing nodes are subjected.

Target tracking is one of main application of wireless sensor networks. A target tracking system is often required to ensure continuous monitoring. There always exist nodes that can detect the target along its trajectory. In which Sensor nodes get information when the target under surveillance is discovered. Some nodes detect the target and send a detection message to the nodes on the targets expected moving path, so they can wake up earlier. Thus, the nodes in the targets moving path can prepare earlier and remain vigilant as the target moves. To be energy efficient and to accurately track the target under surveillance, only the nodes close to the path can participate in tracking process and providing continuous coverage.

## 1.1 Objective

Tracking a target in an area under surveillance is one of the important applications of wireless sensor networks. When target is detected, the sensor node communicates the detection information regarding the target signature to the neighbor nodes along with the sink in network. The neighboring nodes are activated on reception of the detection message and take part in the tracking responsibility of the target. This is the basic procedure in target tracking. Many existing frameworks failed due to problem of sensor nodes, quick movement of the target, random movement of target etc. Main objective is to propose a new tracking framework which is more energy efficient and tracking more accurately than existing works.

## 2 Related Works

One of the critical issues in WSN research is to track the target under surveillance. There are several target tracking methods [1–4] have been proposed for Wireless sensor networks. Many approaches have been developed to address the issues of target tracking and provide the solution for the same. Some of target tracking methods are given below.

In Tree-Based Target Tracking, nodes in a network may be organized in a hierarchical tree structure or represented as a graph. Dynamic Convoy Tree-Based Collaboration (DCTC) is a tree-based tracking method. Dynamic convey Tree-Based collaboration [5], first detects the target and monitors it by tracking the surrounding area of the target. It relies on a tree structure called convoy tree, that include sensor nodes around the moving object, and the tree is configured to add some nodes and prune some nodes as the target moves. The target first enters the detection region, sensor nodes that can detect the target collaborate with each other to select a root and construct an initial convoy tree. Root node collects information from the sensor nodes and refines this information to obtain more accurate and complete information about the target using some classification algorithms.

Low-Energy Adaptive Clustering Hierarchy [6] is used to reduce energy consumption. In LEACH method sensor nodes are formed as cluster structure and choose one of them as head of cluster. Sensor node first detects the target and sends the information to its cluster-head. Then the cluster head collects and compresses the data collected from all the nodes and sends it to the base station. The Cluster head requires more energy than other nodes in the network. So LEACH uses random rotation of the nodes required to be the cluster-heads to evenly distribute energy consumption in the network.

In prediction based scheme [5], sensors become activated when there are interesting events to report, and only those parts of the network with the most useful information balanced by the communication cost need to be active. The type of network is advantageous for security reasons. The networks could also actively collect information, based on predictions of when and where interesting events will be.

Q Huang suggested a new protocol in name of Mobicast [7]. It is used to estimate the direction of target movement. Another method is polygon tracking [8], which uses polygons to track the target. Main idea of proposed framework based on polygon tracking. These are the main methods currently used for target tracking.

## 3 Overview of Proposed System

To overcome the disadvantages of existing target tracking methods proposes a new tracking method called Enhanced polygon based tracking framework. In which, sensor nodes are deployed and their area divided into different polygons called faces [9]. Each face contains more than three sensor nodes. Sensor nodes are informed when the target under surveillance is discovered. Some nodes detect the target and send detection information to the nodes on the target's expected moving path, so they can wake up earlier. So the nodes in the target's moving path can prepare in advance and remain vigilant in front of it as it moves. To be energy efficient and to accurately track the target, only the nodes close to the path can participate in tracking process and providing continuous coverage. When node failure occurs extend the polygon area so that it do not effect tracking process. The targets movement is random in nature and can follow any path. If the target retracing the same path again and again, the polygons are reconstructed repeatedly and energy is wasted in exchanging messages to create new polygons. It can be overcome by the proposed framework.

In polygon based face tracking method [10], deployed sensors are divided into different polygons. There are mainly five steps. First step is about the system initialization. It includes initial polygon construction in the plane. A node has all of the corresponding polygons information after the wireless sensor network planarization step. This work uses the knowledge of spatial neighborhood defined on a planar graph [8]. Each node find out locally who it's spatial neighbors are by planarization method. When a target is detected by some nodes, that communicates to all of its adjacent neighbors with their detection information. Second step about target detection. Once the target is surrounded by the perimeter of a polygon, it becomes active polygon Pc. Step 3 about brink detection. Here the term brink represents the crossing edge of polygon. Brink detection algorithm mainly used to generate the forward polygon. So that the nodes of forward polygon can wake up and prepare before the target moves to it. Couple nodes are the nodes which send target detection information to neighbor polygon and also to sink node. Couple nodes are selected by optimal selection algorithm. A joint-request message [8] is sent to forward polygon when target approach to the crossing edge, saying that the target is approaching. All nodes in forward polygon receive the request, change their state to an awakening, and then start sensing. When the target crosses the brink, another joint-request message is sent to the nodes in forward polygon, saying that the target is crossing the brink. After the target crosses over the brink, another message is sent to the NNs in the previous Pc. After receiving the message, all NNs, except the previous CNs, return to the inactive state. When target in dynamic motion, it retracing the path again and again, polygon creation and deletion message frequently pass between active polygon and forward polygon. So all nodes involved in tracking process run out of energy quickly. To overcome this energy wastage, only one node is sending information to forwarding polygon. Once target is detected, polygons are not created and deleted instead activated. When a node in active polygon failed, one of the nearest neighbour takes the role of failed node by extending the polygon area (Fig. 1).

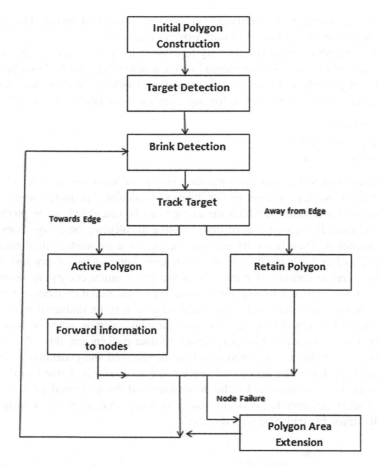

**Fig. 1** Proposed tracking framework

## 4 Experimental Results

The proposed Enhanced polygon tracking scheme was applied on a number of sensor nodes of simulations on the ns-2 network simulator. The deployment of sensor nodes in the region of interest is carried out using NS-2. Sensor nodes are given statically by the user. The current sensing are consist of 50 mobile sensor nodes with one node as target. After the initial sensor deployment the nodes are communicated to each other and they are ready to send data among them. After neighbor discovery, face discovery takes place. First, each node in network collects the neighbor information by hello message. Then, each node computes its face neighbor nodes with help of a Gabriel Graph (GG). The face neighbor node means the link between this node and its neighbor conforms to Gabriel Graph. Then target node start to move with the speed 10–30 m/s into the deployed sensor network area.

Nodes in the polygons are activated with the movement of target. The current method tested on the basis of previous work in the field.

After implementation, it is necessary to analyze the performance of the model. Performance analysis can be represented with the help of graphs. Ns2 tool provides necessary graphs whose values are taken put from the trace file. For analyzing the performances of the mechanism following parameters are taken into consideration:

- Throughput
- Energy consumption
- Packet delivery ratio

It is important to find out the remaining energy of the nodes in order to check whether the network is efficient or not. Energy is calculated is using energy model of the network simulator in which the energy can be calculated as the product of power and time. In each interval of time we have to calculate the energy consumed for the operation. Then find outs average energy. As a parameter Initial energy is given as 200 J at the beginning of the simulation. For each and every task it will consume a certain amount of energy from the given initial energy. In communication networks, network throughput is the average successful delivery of message over a communication channel. This gathered data may be delivered over a link may be logical or physical, or can be pass through a certain network node. The throughput is measured in bits per second or data packets per time slot. Packet Delivery Ratio (PDR) is calculated based on the received and generated packets as recorded in the trace file. In general, packet delivery ratio is defined as the ratio between the received packets by the destination and the generated packets at the source. Packet Delivery Ratio is calculated by using awk script. Awk script processes the trace file and produces the result.

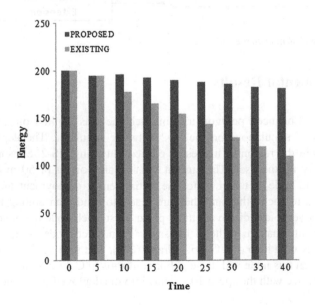

# 5  Conclusion

Target tracking is a key application of wireless sensor networks. It helps to detect target entering into the area of surveillance. It is a complex task to accurately achieve in practice. Main existing methods are cluster based and tree based methods. They result in network failure due to node failure. Here uses a new tracking framework that detects the movements of a target using polygon tracking. This tracking framework can estimate a targets positioning area, achieve tracking ability with high accuracy, and reduce the energy cost of WSNs. It detects arrival of target a bit time earlier and overcome node failure. This proposed framework is better in energy efficiency and throughput than other existing solutions.

# 6  Future Works

A future research direction is to varying the brink length of polygons and make wireless sensor network more energy effective.

**Acknowledgments** I express my deepest thanks to Mrs. Jucy Vareed (Asst. Professor, CSE Department) for taking part in discussions and giving necessary advices and guidance, facilitating preparation of this paper.

# References

1. Zhong, Z., Zhu, T., Wang, D., He, T.: Tracking with unreliable node sequence. In: Proceedings of IEEE INFOCOM, pp. 1215–1223. (2009)
2. Wang, Z., Lou, W., Wang, Z., Ma, J., Chen, H.: A novel mobility management scheme for target tracking in cluster-based sensor networks. In: Proceedings of IEEE DCOSS, pp. 172–186. (2010)
3. Zhou, Y., Li, J., Wang, D.: Posterior cramer-rao lower bounds for target tracking in sensor networks with quantized range-only measurements. IEEE Signal Process. Lett. **17**(2), 377–388 (2010)
4. Kaltiokallio, O., Bocca, M., Eriksson, L.M.: Distributed RSSI processing for intrusion detection in indoor environments. In: Proceedings of the Ninth ACM/IEEE International Conference on Information Processing in Sensor Networks (IPSN), pp. 404–405. (2010)
5. Zhang, W., Cao, G.: Dynamic convoy tree-based collaboration for target tracking in sensor networks. IEEE Trans. Wireless Commun. **12**(4), 1689–1701 (2004)
6. Kaplan, L.M.: Global node selection for localization in a distributed sensor network. IEEE Trans. Aerosp. Electron. Syst. **42**(1), 113–135 (2006)
7. Chan, H., Luk, M., Perrig, A.: Using clustering information for sensor network localization. In: Proceedings of the International Conference on Distributed Computing in Sensor Systems (DCOSS05). (2005)
8. Bhuiyan, M.Z.A., Wang, G., Wu, J.: Polygon-based tracking framework in surveillance wireless sensor networks. In: Proceedings of IEEE International Conference on Parallel and Distributed Systems (ICPADS), pp. 174–181. (2009)

9. Berg, M.D., Kerveid, M.V., Overmars, M., Schwarzkof, O.: Computational Geometry. Springer (1998)
10. Bhuiyan, M.Z.A., Wang, G., Wu, J.: Detecting movements of a target using face tracking in wireless sensor networks. In: Proceedings of IEEE International Conference on Parallel and Distributed Systems (ICPADS), pp. 174–181. (2010)

# Abnormal Gait Detection Using Lean and Ramp Angle Features

Rumesh Krishnan, M. Sivarathinabala and S. Abirami

**Abstract** Gait Recognition plays a major role in Person Recognition in remote monitoring applications. In this paper, a new gait recognition approach has been proposed to recognize the normal/abnormal gait of the person with improved feature extraction techniques. In order to identify the abnormality of the person lean angle and ramp angle has been considered as features. The novelty of the paper lies in the feature extraction phase where the walking abnormality is measured based on foot movement and lean angle which is measured between the head and the hip. In this paper, the feature vector is composed of measured motion cues information such as lean angle between the head and the hip and ramp angle between the heel and the toe for two legs. These features provide the information about postural stability and heel strike stability respectively. In the training phase, classifier has been trained using normal gait sequences by exemplar calculation for each video and distance metric between the sequences separately for samples. Based on the distance metric, the maximum distance has been calculated and considered as threshold value of the classifier. In the testing phase, distance vector between training samples and testing samples has been calculated to classify normal or abnormal gait by using threshold based classifier. Performance of this system has been tested over different data and the results seem to be promising.

**Keywords** Abnormal gait detection · Person identification · Gait recognition

R. Krishnan (✉) · M. Sivarathinabala · S. Abirami
Department of Information Science and Technology, College of Engineering,
Anna University, Chennai, India
e-mail: svp.research14@gmail.com

M. Sivarathinabala
e-mail: sivarathinabala@gmail.com

S. Abirami
e-mail: abirami_mr@yahoo.com

© Springer India 2016                                                    325
H.S. Behera and D.P. Mohapatra (eds.), *Computational Intelligence
in Data Mining—Volume 1*, Advances in Intelligent Systems
and Computing 410, DOI 10.1007/978-81-322-2734-2_33

# 1 Introduction

Video Surveillance is one of the important research fields and this has been mainly used for security purpose. In Video Surveillance, Gait Recognition becomes a major research towards identifying and also in understanding human behavior with respect to the environment. Gait [1, 2] refers to the walking style of the person which provides the unique identity of each person from the videos. Many biometrics such as face, iris, palm print and finger print has been used in the previous literatures. This biometrics has been extracted only with the cooperation of the users. For the security purpose, identification of the person is necessary even without proper user's cooperation. In this respect, if we consider remote monitoring applications, patients are monitored continuously through gait videos. The patient's neurological status has been recognized and the clinical treatment has been performed based on the abnormality of gait.

Human gait abnormality has been predicted from the posture instability and heel to toe behavior. Postural Instability refers to a neck drop that is relative to the flexed posture of the trunk and limbs. Heel to Toe walking behavior is one of the abnormalities that the person has the inability to make heel contact with the ground during the initial stance phase of the gait cycle and the absence of full foot contact with the ground during the remainder of the gait cycle. Toe walking may be due to alcohol consumption, weakness and leg tremors. Most elderly patients have difficulty with heel to toe walking due to general neural loss impairing a combination of strength and co-ordination. To recognize a normal/abnormal gait, body postures and the body parts movement during various situations is still an open challenge. Thus we are motivated to propose a new model to identify the normal/abnormal gait with respect to postural changes and heel strike ability of humans in gait videos.

In this work, our Contribution lies in threefold: a new Gait recognition approach has been introduced to recognize the normal/abnormal gait of the person in various postures. In feature extraction phase, lean angle and ramp angles are extracted as features to analyze the relationships between normal/abnormal of the person. The classifier has been used to classify the gait pattern either normal or abnormal.

This paper has been organized as follows: Sect. 2 introduces the previous works that has been carried out in this research area. Section 3 describes the proposed methodology, Sect. 3.1 deals about preprocessing step, Sect. 3.2 details the way in which features are extracted and Sect. 3.3 details about the classification procedure. Section 4 gives the implementation details and results. Finally, Sect. 5 concludes the paper.

## 2 Related Works

Various works in the field of video analytics have been devoted for gait recognition and they are addressed in this section in detail with respect to the abnormal gait detection. Andriacchi et al. [3] presented the method for observing normal person gait and patients gait with knee disabilities. They analyzed time distance measurements and ground reaction force parameters based on walking speed and used regression analysis to derive the functional relations between ground reaction force amplitudes and walking speed. The basic time distance measurements observed over a range of walking speeds can be useful indicators of gait abnormalities associated with knee disabilities.

Tafazzoli et al. [4] proposed an approach for human gait recognition to analyze the leg and arm movements. A model has been created using active contour models and the Hough transform. Fourier analysis has been used here to describe the motion patterns. The classification has been performed using k-nearest neighbour rule. The authors discussed their model with the extra features produced from the motion of the arms which leads a solution for self occlusion. Detection of this gait characteristics detection has gained more interest in field of biomechanics and rehabilitation sciences. Mostayedl et al. [5] proposed a method to detect the abnormal gait by employing Discrete Fourier Transform (DFT) analysis and the joint angle characteristics in frequency domain. Here, abnormal gait has been recognized using harmonic coefficient recognition.

From the clinical gait theory [6] it is well known that, a normal gait exhibits periodic stride-cycles with stride-angle around 45° between the legs, whereas neurological patients walk with shortened stride-angle with high variability between the stride-cycles. According to the above fact, Khan et al. [7] analyzed the features such as stride-cycles which are based on the cyclic leg motion and posture lean which is based on the angle between leaned torso and axis of gravity. In contrast to the existing approaches, the main focus of this work is on abnormal gait recognition through predominant features called lean angle and ramp angle. This technique has also been proved to reduce the false recognition rates, which is very common in gait recognition system.

## 3 Proposed Methodology

In this research, a special attempt has been made to identify/differentiate the abnormal from the normal gait from the tracked blobs. This work attempts to detect the abnormality in gait sequences. Gait abnormality [8, 9] occurs mainly due to the variations in posture and heel strike ability. Motivated by this, our approach relies on the exploration of two different features lean angle and ramp angle to describe the pose of the person. In this framework, abnormality in posture (or) posture instability is determined the variation of lean angle in a complete gait cycle.

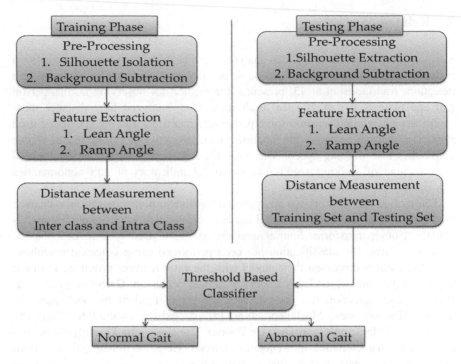

**Fig. 1** Framework for abnormal gait detection

Also abnormality in toe and heel behaviour is determined through the variations that occur over the ramp angle in the gait cycle. Distance vector has been measured between inter class and intra class variations in the training phase. In the testing phase, distance has been measured between the training and testing set. Threshold has been fixed based on the distance measures and the classification has been performed using Threshold based classifier. The abnormality can be determined from the pose and heel strike actions of the gait cycle. Abnormal gait recognition framework has been shown in Fig. 1.

## 3.1 Pre-processing

Video Sequence has been represented by gait patterns and their relationships. In videos, gait features have been extracted widely that involves background modelling, foreground detection and foreground extraction. Pre-processing [10–13] is the technique that generally removes low-frequency background noise, normalizes the intensity of the individual particles images, removes reflections, and masking portions of images. To represent gait features, the silhouette images are extracted and detected using foreground segmentation. The dark pixel values of background

**Fig. 2** Lean angle
measurement. **a** Normal gait.
**b** Abnormal gait

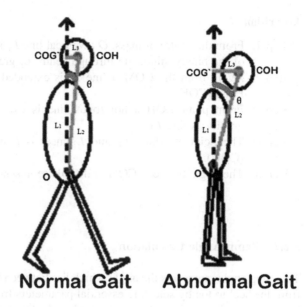

**Normal Gait    Abnormal Gait**

are eliminated and the foreground pixels which represent the human body are
preserved in the image. The binary image is filtered in order to remove small blobs.
Dilation and erosion operations have been performed in order to remove the
anomalies in the filtered image. Then the obtained silhouette is isolated using a
bounding rectangle as shown in Fig. 2a, b.

## 3.2 Feature Extraction

Here feature extraction, includes the extraction of lean angle and ramp angle from
the pre-processed images.

### 3.2.1 Lean Angle Calculation

Lean angle is the angle which lies between the center of gravity and center of Head.
Three steps have been proposed to calculate lean angle. As a first step, head has
been segmented from silhouette image. The original silhouette height through taken
20 % of top of the image consists of head. The contour detection method is used
here to detect the centre of head from the segmented head region and Centre of
Mass (COM) calculated from the silhouette image. Finally, plotting the head and
hip into the silhouette image and the line from head to centre of mass by which this
line to calculating lean angle between this line and centre of gravity. Lean angle
measurement has been shown in Fig. 2.

**Algorithm**

Step 1: From the centre of mass, $O$ a vertical line $L_1$ is extended towards the top of the binary silhouette through centre of gravity.

Step 2: Again from the COM a line $L_2$ is extended in the angle of Centre of Head (COH).

Step 3: From point COH a horizontal line is extened towards the centre of gravity as line $L_3$.

Step 4: The point at which $L_1$ and $L_3$ meet is referred as Center of Gravity (COG).

Step 5: The angle between COG and COH represents the lean angle.

### 3.2.2 Ramp Angle Calculation

The ramp angle represents the angle which lies between the ground surface and line from the heel to toe by selecting essential parameters from heel to toe as explained by the steps given below. The ramp angle calculation of both legs separately for every half gait cycle.

**Algorithm**

Step 1: Segment the binary silhouette image using contour detection method.

Step 2: Segment the edge image in to 10 parts in column wise.

Step 2.1: Each column of edge image is scanned in the bottom up approach and consider first white pixel as edge of foot.

Step 2.2: Extract the valid points which contains more distances between adjacent point from average of distance each point.

Step 2.3: End points are considered as heel and toe.

Step 3: Draw line $L_1$ from heel $H_1$ to toe $T_1$ of left side leg of binary silhouette, similarly line $L_2$ from heel $H_2$ to toe $T_2$ of right side leg of binary silhouette.

Ramp angles are calculated by measuring the angle which lies between L1 and L3 and L2 and L4 as shown in Fig. 3.

## 3.3 Classification

The classification method used here to classify normal or abnormal gait is an unsupervised learning method. The classification parameter such as maximum lean angle, minimum lean angle, average lean angle, maximum ramp angle, minimum ramp angle, and average ramp angle are considered as feature vector. In the training

**Fig. 3** Ramp angle measurement

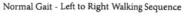

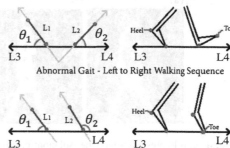

phase, the feature vector extracted from a gait sequence of a normal walking behavior of a person is used to train the classifier. Also this proposed classifier has been used to classify abnormality in gait walking sequence such as posture instability or heel strike. The exemplar points are calculated by using centroid calculation of feature points such as lean angle and ramp angle separately. Euclidean distance has been measured for inter and intra classes and maximum distance has been considered as threshold value. The classifier uses this threshold [14] to classify abnormality in gait in testing phase.

The testing phase input videos are taken and binary silhouette images are extracted by using pre-processing steps. This is followed by silhouette isolation and proposed feature vector has been extracted using feature extraction methods. The distance vector is calculated between training samples and testing samples by using exemplar points calculated in the training phase. Finally, the minimum distance has been considered to classify the given input video whether it belongs to normal or abnormal video. If it is abnormal gait, abnormality has been further labeled as posture instability and heal strike instability problem.

## 4 Results and Discussion

The implementation of this gait recognition system has been done using MATLAB (Version2013a), Python and Numpy. The input videos are taken from existing dataset CASIA [15] and KTH [16] for normal walking and synthetic dataset has been taken for abnormal walking both heel to toe and posture instability. The proposed abnormality detection framework has been applied and tested over many different test cases and the results are shown here. In our experiments all sequences were divided with respect to the subjects into a testing set and a validation set. There are 8 video files for each combination of 2 subjects, 2 actions and 2 scenarios. Each file contains about four sub sequences used as a sequence in our experiments. The recognition results were obtained on the test set.

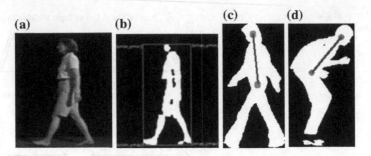

**Fig. 4  a, b** Represents silhouette isolation; **c, d** represents lean angle measurements

**Table 1** Lean angle for normal and abnormal gait

|  | Maximum lean angle (in degrees) | Average lean angle (in degrees) | Minimum lean angle (in degrees) |
|---|---|---|---|
| Normal gait | 6 | 4 | 2 |
| Abnormal gait | 52 | 45 | 40 |

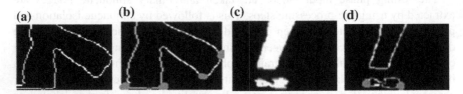

**Fig. 5  a, b** Represents the Ramp angle calculations for normal gait and **c, d** represents the abnormal gait

Silhouette isolation and lean angle measurements has been shown in Fig. 4a–d. From this head and hip (centre of mass) has been detected which is followed by lean angle for every silhouette image in a gait cycle. The extracted lean angle represents the gait cycle by maximum, average and minimum lean angle in the feature vector for normal and abnormal gait to be shown in Table 1.

The ramp angle calculations for both normal and abnormal gait by considering maximum and minimum ramp angle of both the legs separately for each half gait cycle. The ramp angle calculation for the normal gait has been shown in Fig. 5a, b. The ramp angle calculation for abnormal gait has been shown in Fig. 5c, d.

The results of the ramp angle calculated are shown in the Table 2 for both normal and abnormal gait with taking maximum and minimum ramp angle both legs separately for each half gait cycle.

**Table 2** Ramp angle for normal and abnormal gait

| | Normal gait | Abnormal gait | |
| --- | --- | --- | --- |
| | | Person P1 | Person P2 |
| Maximum ramp angle of left leg for first half gait cycle (in degrees) | 58 | 22 | 25 |
| Minimum ramp angle of left leg for first half gait cycle (in degrees) | 2 | 6 | 2 |
| Maximum ramp angle of right leg for first half gait cycle (in degrees) | 21 | 20 | −5 |
| Minimum ramp angle of right leg for first half gait cycle (in degrees) | 0 | 6 | −22 |

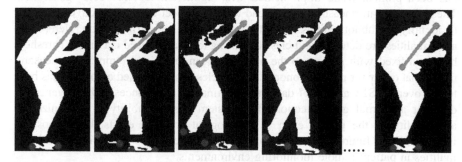

**Fig. 6** Posture instability—abnormal gait

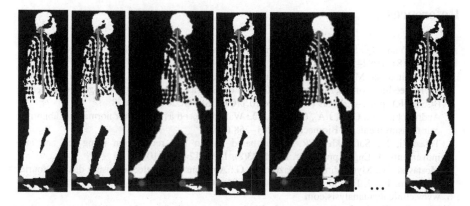

**Fig. 7** Abnormal gait based on Heel and Toe

Figure 6 represents the posture variations in the abnormal gait and Fig. 7 represents the abnormal heel strike actions in the gait cycle. The Classification results have been tabulated in Table 3.

**Table 3** Result of gait classification

| Testing videos | Gait | Abnormality |
| --- | --- | --- |
| Video 1 | Normal gait | – |
| Video 2 | Abnormal gait | Posture instability |
| Video 3 | Normal gait | – |
| Video 4 | Normal gait | – |
| Video 5 | Abnormal gait | Heel strike instability |

## 5 Conclusion

In this research, an automated gait recognition system is proposed to recognize the abnormality of the person using ramp angle and lean angle as features. This system has been proposed and implemented with significant feature vector generation. These features are measured from head to hip and heel to toe respectively in order to identify the abnormalities in posture and in the heel strike action. Here, these abnormalities are detected and classified using threshold based classifier. Threshold has been fixed with respect to the minimum lean angle and ramp angle measurements that occur for normal/abnormal gait cycles. The proposed algorithm has been tested over CASIA and KTH dataset for normal gait sequences and self generated data for abnormal gait detection. This system has the ability to recognize the abnormality of the person even if neurological problem exists due to postural instabilities. In future, this work could be implemented to recognize gait abnormalities in patient remote monitoring environments.

## References

1. Yin, J., Yang, Q.: Sensor-based abnormal human activity detection. IEEE Trans. Knowl. Data Eng. **20**(8), 1082–1090 (2008)
2. Sivarathinabala, M., Abirami, S.: Automatic identification of person using fusion of features. In: Proceedings of International Conference on Science Engineering Management Research (ICSEMR), pp. 1–5. IEEE (2014). doi:10.1109/ICSEMR.2014.7043628
3. Andriacchi, T.P., Ogle, J.A., Galante, J.O.: Walking speed as a basis for normal and abnormal gait measurements. J. Biomech. **10**, 261–268 (1977)
4. Tafazzoli, F., Safabakhsh, R.: Model-based human gait recognition using leg and arm movements. J. Eng. Appl. Artif. Intell. **23**(8), 1237–1246 (2010)
5. Mostayed, A., Mynuddin, M., Mazumder, G.: Abnormal gait detection using discrete fourier transform. Int. J. Hybrid Inf. Technol. **3**(2), 36–40 (2010)
6. www.clinicalgaitanalysis.com
7. Khan, T., Westin, J., Dougherty, M.: Motion cue analysis for parkinsonian gait recognition. Biomed. Eng. **7**, 1–8
8. Kong, K., Bae, J., Tomizukaa, M.: Detection of abnormalities in a human gait using smart shoes. In: International Society for Optical Engineering, Proceeding of SPIE Digital Library, vol. 6932. (2008)
9. Bauckhage, C., Tsotsos, J.K., Bunn, F.E.: Automatic detection of abnormal gait. J. Image Vis. Comput. **27**(2), 108–115 (2009)

10. Abd el Azeem Marzouk, M.: Modified background subtraction algorithm for motion detection in surveillance systems. J. Am. Arabic Acad. Sci. Technol. **1**(2), 112–123 (2010)
11. Gowshikaa, D., Abirami, S., Baskaran, R.: Automated Human Behaviour Analysis from Surveillance Videos: A Survey Artificial Intelligence Review. (2012). doi:10.1007/s2-012-9341-3
12. Gowsikhaa, D., Manjunath, Abirami, S.: Suspicious human activity detection from surveillance videos. Int. J. Internet Distrib. Comput. Syst. **2**(2), 141–149 (2012)
13. Gowsikhaa, D., Abirami, S., Baskaran, R.: Construction of image ontology using low level features for image retrieval. In: Proceedings of the International Conference on Computer Communication and Informatics, pp. 129–134. (2012)
14. Mohammadi, L., van de Geer, S.: On Threshold-Based Classification Rules. Mathematical Institute, University of Leiden (2000)
15. Zheng, S., Zhang, J., Huang, K., He, R., Tan, T.: Robust view transformation model for gait recognition. In: Proceedings of the IEEE International Conference on Image Processing. (2011)
16. Schuldt, C., Laptev, I., Caputo, B.: Recognizing human actions: a local SVM approach. In: Proceedings of ICPR'04, Cambridge, UK

# Fuzzy Logic and AHP-Based Ranking of Cloud Service Providers

Rajanpreet Kaur Chahal and Sarbjeet Singh

**Abstract** The paper presents an approach to rank various Cloud-Service providers (CSPs) on the basis of four parameters, namely, Reliability, Performance, Security and Usability, using a combination of Likert scale, Fuzzy logic and Analytic Hierarchy Process (AHP). The CSPs have been ranked using data collected through a survey of cloud computing users. The purpose of this paper is to help provide a technique through which various CSPs can be ranked on the basis of a number of parameters keeping in mind the users' perceptions. We have chosen fuzzy logic because it has the capability to model the vague and imprecise information provided by the users and can handle the uncertainty involved. Similarly, AHP has been chosen because of its ability to make complex decisions on the basis of mathematics and psychology. The proposed model derives its strength from the fact that it takes into account the uncertainty and dilemma involved in decision-making.

**Keywords** Fuzzy logic · Analytic hierarchy process (AHP) · Ranking of CSPs

## 1 Introduction

Cloud computing is the next big trend in Information Technology. The main idea behind cloud computing is "why buy when you can rent". The platform provides on-demand services on pay per use basis. It provides the consumers with an opportunity to rent the resources provided by the Cloud Service Provider and use them for their own purpose.

National Institute of Standards and Technology has defined cloud computing as "a model for enabling convenient, on-demand network access to a shared pool of

R.K. Chahal (✉) · S. Singh
Panjab University, Chandigarh, India
e-mail: rajan_chahal@yahoo.com

S. Singh
e-mail: sarbjeet@pu.ac.in

© Springer India 2016
H.S. Behera and D.P. Mohapatra (eds.), *Computational Intelligence in Data Mining—Volume 1*, Advances in Intelligent Systems and Computing 410, DOI 10.1007/978-81-322-2734-2_34

configurable computing resources that can be rapidly provisioned and released with minimal management effort or service provider interaction" [1]. Cloud computing provides three service delivery models, namely, Software as a Service (SaaS), Platform as a Service (PaaS) and Infrastructure as a Service (IaaS). SaaS provides an opportunity to clients to access a software application over Internet on rent basis rather than buying it. PaaS defines an environment and platform for developers to develop services and applications over the Internet. IaaS provides access to computing resources across Internet. The model proposed in this paper caters to the SaaS delivery model. However, it can also be applied to the PaaS as well as IaaS models since the parameters considered for ranking the CSPs can be measured for all the three delivery models.

Recognizing the potential of cloud computing, many companies have started offering the cloud services. Having a multitude of vendors at their disposal, the users are often faced with the dilemma of choosing the best CSP. This paper proposes a technique to help users deal with this dilemma.

Section 2 of the paper contains information about the related work, Sect. 3 contains the detailed explanation of the proposed method along with results and Sect. 4 presents the conclusion.

## 2 Related Work

Saurabh et al. have proposed an AHP based mechanism to rank the Cloud Service Providers. The mechanism is divided into three phases: breaking up of problem, judging the user's priorities and aggregating the priorities. They have used the SMICloud framework and the mechanism provides the cloud providers an idea of their performance as compared to their competitors [2]. Zibin et al. have proposed a ranking method called QoS Ranking Prediction which employs the past experiences of other users. They have given two ranking algorithms known as CloudRank1 and CloudRank2 which presents the QoS ranking directly without invoking the QoS values. The experiments have proved this approach to perform better than other rating approaches and the famous greedy method [3]. To decrease the computational demands, Smitha et al. have proposed a broker-based approach where the cloud brokers maintain the data of the cloud service providers and rank them accordingly, keeping in view the request made by the customer. The CSS algorithm designed by them has shown significant improvements over other approaches [4]. Saurabh et al. have suggested a framework to prioritise the cloud services on the basis of quality measurement of each of the distinctive cloud services. The framework helps to develop a sense of healthy competition between various service providers in conformance to their Service Level Agreements (SLA). The feasibility of such a framework has been shown with the help of a case study [5]. Ioannis et al. have addressed sheer uncertainty and vagueness in comparing various cloud services against the user requirements and specifications. The authors have proposed a framework for ranking cloud services on the basis of varying level of fuzziness and evaluated each service on the heterogeneous model of usage characteristics. The framework allows for a unified method of multi objective assessment of cloud services with precise and imprecise metrics [6]. Arezoo et al. have proposed a Weight Service Rank approach for ranking the cloud computing services. The approach uses real-world QoS features and the authors have conducted comprehensive experiments to prove that their approach is more flexible, scalable and takes less time for execution than AHP and SVD techniques [7]. Zia et al. have proposed a parallel cloud service selection technique wherein QoS history of cloud services is evaluated using MCDM process during different periods of time. The proposed approach uses TOPSIS and ELECTRE as the MCDM technique. Using these techniques, the cloud services are ranked for each time period and then these results are combined to arrive at an overall ranking of cloud services [8]. Praveen and Morarjee have proposed a personalized ranking prediction structural framework known as cloud rank. It has been developed to forecast quality of service ranking without requiring any need of real time feeds or real world services invocation from the consumers or users [9].

## 3 Explanation of the Process with Results

The proposed process ranks the CSPs on the basis of data collected from the users. Users were asked to rate five CSPs on the basis of Reliability, Performance, Security and Usability on a scale of 1–5, where 1 stands for very low and 5 stands for very high. The ratings given by the users represent their perception or opinion formed after accessing multiple services provided by the five CSPs and are not based on the use of any particular service. Moreover, no dataset could be found which measured the CSPs for the parameters needed in our research. Hence, we conducted a survey of the cloud computing users for the agreed upon parameters after explaining the meaning of each parameter.

### 3.1 Dataset

The data collected through the survey is of the following form:

Table 1 shows the data containing ratings given by a single user to the five CSPs. Similar data is collected from all the users. The parameters have the following meanings:

Reliability: It is measured in the form of opinion of the user regarding the number of failures or outages experienced while using the service. Lower the number of failures, greater is the reliability.

Performance: It is measured on the basis of the speed of the service as perceived by the user. Higher the speed, greater is the performance.

**Table 1** Data collected
through survey from one user

| | Reliability | Performance | Security | Usability |
|---|---|---|---|---|
| CSP1 | 4 | 4 | 4 | 4 |
| CSP2 | 4 | 5 | 4 | 3 |
| CSP3 | 3 | 4 | 3 | 4 |
| CSP4 | 3 | 3 | 4 | 3 |
| CSP5 | 3 | 3 | 5 | 5 |

*1* very low, *2* low, *3* average, *4* high, *5* very high

Security: It represents a measure of the confidentiality of user data as perceived by the user. Greater the confidentiality, higher is the security.

Usability: It is a measure of the ease of use of the service as opined by the user.

The ratings given by all the users are taken into account and a value known as Percent$_{rating}$ (Parameter) is calculated which signifies the percentage of users who have given a particular rating to the CSPs. This is the basis of Likert scale. For example, Percent$_{rating=1}$ (Reliability) = 13 for CSP1 means that 13% users have given rating 1 i.e. very low for the reliability of CSP1. These values are shown in Table 2.

$$Percent_{rating=i}(Parameter) = \frac{Number\ of\ users\ rating\ Parameter\ as\ i}{Total\ number\ of\ users} \quad (1)$$

where i = rating given by users (i.e. 1, 2, 3, 4, 5)

Table 2 shows the values of Percent$_{rating}$ (Parameter) for CSP1. Columns 2,3,4,5 represent Percent$_{rating=1, 2, 3, 4, 5}$ (Reliability), Percent$_{rating=1, 2, 3, 4, 5}$ (Performance), Percent$_{rating=1, 2, 3, 4, 5}$ (Security) and Percent$_{rating=1, 2, 3, 4, 5}$ (Usability) respectively for CSP1.

Similar datasets are created for the other four CSPs as well. The above values are then fed into four Fuzzy Inference Systems (FIS), one for each parameter. The Percent$_{rating=1, 2, 3, 4, 5}$ (Reliability) values for each CSP are fed into Reliability FIS one by one to arrive at a single Reliability value. Similarly, the Percent$_{rating}$ (Parameter) values for the other three parameters are fed into Performance FIS, Security FIS and Usability FIS respectively, to arrive at single Performance, Security and Usability values.

Figure 1 shows the Reliability FIS. The inputs to the FIS are the Percent$_{rating=1, 2, 3, 4, 5}$ (Reliability) values and the output is a single Reliability value.

**Table 2** Percent$_{rating}$
(parameter) values based on
Likert scale

| | Reliability | Performance | Security | Usability |
|---|---|---|---|---|
| Rating = 1 | 13 | 5 | 8 | 7 |
| Rating = 2 | 13 | 16 | 11 | 10 |
| Rating = 3 | 24 | 17 | 23 | 16 |
| Rating = 4 | 24 | 40 | 34 | 36 |
| Rating = 5 | 26 | 22 | 24 | 31 |

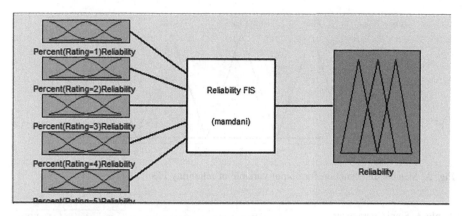

**Fig. 1** Reliability FIS

Performance FIS, Security FIS and Usability FIS have the inputs Percent$_{rating=1, 2, 3, 4, 5}$ (Performance), Percent$_{rating=1, 2, 3, 4, 5}$ (Security) and Percent$_{rating=1, 2, 3, 4, 5}$ (Usability) respectively., with outputs single Performance, Security and Usability values.

The membership functions for the inputs and output of Reliability FIS are shown in Figs. 2 and 3 respectively.

The other three FIS also have the same membership functions for their input and output variables. The single values of Reliability, Performance, Security and Usability obtained from the four FIS are shown in Table 3.

These values are used in the further analysis. For reference purposes, we term these values as Reliability(CSP1), Performance(CSP1) and so on.

## 3.2 Steps Using AHP

After obtaining the Reliability, Performance, Security and Usability values as explained in the previous part, the further analysis is carried out using the Analytic

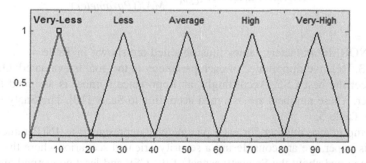

**Fig. 2** Membership function for the input variables of reliability FIS

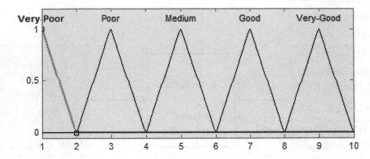

**Fig. 3** Membership function for output variable of reliability FIS

**Table 3** Single parameter values received from FIS

|  | Reliability | Performance | Security | Usability |
|---|---|---|---|---|
| CSP1 | 5 | 5.5 | 5 | 7 |
| CSP2 | 7 | 5.5 | 5.5 | 7 |
| CSP3 | 5 | 5 | 7 | 5 |
| CSP4 | 5 | 5 | 5 | 5 |
| CSP5 | 5.5 | 5.5 | 5 | 5 |

Hierarchy Process (AHP) technique. The whole process is explained in the steps below:

**Step 1**: From the values given in Table 3, find the maximum value in each column and term it as MAX(Parameter). So the values obtained are:

MAX(Reliability) = 7
MAX(Performance) = 5.5
MAX(Security) = 7
MAX(Usability) = 7

**Step 2**: A normalized matrix P is created using the values of Table 3 and MAX (Parameter). The normalized values thus obtained are termed as NORM(Parameter).

$$NORM(Parameter)_{CSPx} = \frac{Parameter(CSPx)}{MAX(Parameter)} \qquad (2)$$

where x = 1, 2, 3, 4, 5

The NORM(Parameter) values thus obtained are shown in Table 4.

**Step 3**: Relative importance of each parameter to the goal is evaluated. Our goal is to select the best CSP. Accordingly, an importance number is assigned to each parameter. These numbers are assigned according to Saaty [10]. The Saaty scale is given in Table 5.

The importance number for each parameter is represented as IN(Parameter).

In this paper, we are considering a security oriented scenario, where the user is most concerned about the Security aspect of the CSP and least concerned about the

**Table 4** Normalized matrix P

|  | NORM (reliability) | NORM (performance) | NORM (security) | NORM (usability) |
|---|---|---|---|---|
| CSP1 | 0.71428571 | 1 | 0.714286 | 1 |
| CSP2 | 1 | 1 | 0.785714 | 1 |
| CSP3 | 0.71428571 | 0.909090909 | 1 | 0.714286 |
| CSP4 | 0.71428571 | 0.909090909 | 0.714286 | 0.714286 |
| CSP5 | 0.78571429 | 1 | 0.714286 | 0.714286 |

**Table 5** Saaty scale for importance of parameters [10]

| Importance number | Definition |
|---|---|
| 1 | Equal importance |
| 2 | Weak |
| 3 | Moderate importance |
| 4 | Above moderate |
| 5 | Strong importance |
| 6 | Above strong |
| 7 | Very strong |
| 8 | Very, very strong |
| 9 | Extreme importance |

Usability. So the importance numbers assigned to the parameters in this scenario are given in Table 6.

Depending on the user's requirements, other scenarios can also be considered such as performance oriented, where the user is most concerned about the performance and least concerned about the usability. The importance numbers can be changed according to the scenarios.

**Step 4**: Using the importance numbers assigned to each parameter, an importance matrix M is created. The values in the matrix M are referred to as IMP (Parameter$_{mn}$).

$$IMP(Parameter_{mn}) = \frac{IN(Parameter_m)}{IN(Parameter_n)} \qquad (3)$$

where Parameter$_m$, Parameter$_n$ = Reliability, Performance, Security, Usability
IMP(Parameter$_{mn}$) = Importance of Parameter m compared to Parameter n.

**Table 6** Importance numbers

| Parameter | Importance number |
|---|---|
| Reliability | 7 |
| Performance | 8 |
| Security | 9 |
| Usability | 3 |

**Table 7** Importance matrix M

|  | Reliability | Performance | Security | Usability |
|---|---|---|---|---|
| Reliability | 1 | 0.875 | 0.777778 | 2.333333 |
| Performance | 1.1428571 | 1 | 0.888889 | 2.666667 |
| Security | 1.2857143 | 1.125 | 1 | 3 |
| Usability | 0.4285714 | 0.375 | 0.333333 | 1 |
| Sum | 3.8571429 | 3.375 | 3 | 9 |

Thus, the importance matrix M is shown in Table 7.

After calculating the IMP(Parameter$_{mn}$) for all the parameters, a column-wise Sum is calculated for each parameter in matrix M. This value is referred to as SUM (IMP(Parameter$_{mn}$)).

**Step 5**: A weight matrix W is constructed using the IMP(Parameter$_{mn}$) and SUM (IMP(Parameter$_{mn}$)) values, according to (4).

$$W(Parameter_{mn}) = \frac{IMP(Parameter_{mn})}{SUM(IMP(Parameter_{mn}))} \tag{4}$$

where W(Parameter$_{mn}$) = weight of Parameter$_m$ compared to Parameter$_n$.

The final weights for each parameter, represented as W(Parameter), are then obtained using (5).

$$W(Parameter) = \frac{\sum W(Parameter_{mn})}{x} \tag{5}$$

where x = number of parameters (i.e. 4)

The weights obtained are shown in Table 8.

As can be verified from the matrix given in Table 8, the sum of the weights of the parameters equals to 1.

**Step 6**: A rank matrix R is constructed using the weights of the parameters and the normalized values calculated in matrix P. The value in each cell of matrix R is known as Rank(Parameter).

$$Rank(Parameter) = W(Parameter) * NORM(Parameter) \tag{6}$$

Matrix R is shown in Table 9.

**Table 8** Weight matrix W

|  | Reliability | Performance | Security | Usability | Weights |
|---|---|---|---|---|---|
| Reliability | 0.2592593 | 0.259259259 | 0.259259 | 0.259259 | 0.259259 |
| Performance | 0.2962963 | 0.296296296 | 0.296296 | 0.296296 | 0.296296 |
| Security | 0.3333333 | 0.333333333 | 0.333333 | 0.333333 | 0.333333 |
| Usability | 0.1111111 | 0.111111111 | 0.111111 | 0.111111 | 0.111111 |

**Table 9** Rank matrix R

|      | Reliability | Performance | Security | Usability |
|------|-------------|-------------|----------|-----------|
| CSP1 | 0.18518519  | 0.296296296 | 0.238095 | 0.111111  |
| CSP2 | 0.25925926  | 0.296296296 | 0.261905 | 0.111111  |
| CSP3 | 0.18518519  | 0.269360269 | 0.333333 | 0.079365  |
| CSP4 | 0.18518519  | 0.269360269 | 0.238095 | 0.079365  |
| CSP5 | 0.2037037   | 0.296296296 | 0.238095 | 0.079365  |

**Table 10** Ranks of the CSPs for Scenario 1

| CSPs | Rank     |
|------|----------|
| CSP1 | 0.830688 |
| CSP2 | 0.928571 |
| CSP3 | 0.867244 |
| CSP4 | 0.772006 |
| CSP5 | 0.81746  |

**Step 7**: Finally, the CSPs are ranked by summation of Rank(Parameter) values.

$$\text{Rank(CSP)} = \sum_{\substack{\text{Parameter}_m = \text{Reliability,} \\ \text{Performance, Security,} \\ \text{Usability}}} \text{Rank(Parameter}_m) \qquad (7)$$

The final ranks of the CSPs thus obtained are shown in Table 10.
The CSPs are ranked in the order CSP2 > CSP3 > CSP1 > CSP5 > CSP4.

## 4 Conclusion and Future Scope

In this paper, we have proposed a Likert, Fuzzy and AHP-based ranking of cloud service providers according to the users themselves who decide on the importance of the parameters considered for ranking the CSPs based on their preferences. This approach has the advantage of being flexible since the users can mention the degree of importance of the parameters to their requirements and how these parameters affect the ranking of providers. Moreover, the approach is scalable as the number of parameters and the CSPs can be increased according to the demand of the situation.

In future, we would like to compare the performance of our approach with other mechanisms.

**Acknowledgments** I sincerely thank all those people who helped me in completing this research paper.

# References

1. Liu, F., Tong, J., Mao, J., Bohn, R., Messina, J., Badger, L., Leaf, D.: NIST Cloud Computing Reference Architecture. Recommendations of the National Institute of Standards and Technology. Special Publication 500-292, Gaithersburg (2011)
2. Garg, S.K., Versteeg, S., Buyya, R.: SMICloud: a framework for comparing and ranking cloud services. In: Proceedings of the Fourth IEEE International Conference on Utility and Cloud Computing, pp. 210–218. (2011). doi:10.1109/UCC.2011.36
3. Zheng, Z., Wu, X., Zhang, Y., Lyu, M.R., Wang, J.: QoS ranking prediction for cloud services. IEEE Trans. Parallel Distrib. Syst. 24(6), 1213–1222 (2013). doi:10.1109/TPDS.2012.285
4. Sundareswaran, S., Squicciarini, A., Lin, D.: A brokerage-based approach for cloud service selection. In: Proceedings of the IEEE Fifth International Conference on Cloud Computing, pp. 558–565. (2012). doi:10.1109/CLOUD.2012.119
5. Garg, S.K., Versteeg, S., Buyya, R.: A framework for ranking of cloud computing services. Future Gener. Comput. Syst. 29(4), 1012–1023 (2013). doi:10.1016/j.future.2012.06.006
6. Patiniotakis, I., Rizou, S., Verginadis, Y., Mentzas, G.: Managing imprecise criteria in cloud service ranking with a fuzzy multi-criteria decision making method. In: Service-Oriented and Cloud Computing, pp. 34–48. (2013). doi:10.1007/978-3-642-40651-5_4
7. Jahani, A., Khanli, L.M., Razavi, S.N.:W_SR: A QoS based ranking approach for cloud computing service. Comput. Eng. Appl. J. 3(2), 55–62 (2014). ISSN:2252-5459
8. Rehman, Z., Hussain, O.K., Hussain, F.K.: Parallel cloud service selection and ranking based on QoS history. Int. J. Parallel Prog. 42(5), 820–852 (2014). doi:10.1007/s10766-013-0276-3
9. Kumar, G.P., Morarjee, K.: Ranking prediction for cloud services from the past usages. Int. J. Comput. Sci. Eng. 2(9), 22–25 (2014). E-ISSN:2347-2693
10. Saaty, T.L.: Decision making with the analytic hierarchy process. Int. J. Serv. Sci. 1(1), 83–98 (2008). doi:10.1504/IJSSCI.2008.017590

# Classification of Tabla Strokes Using Neural Network

**Subodh Deolekar and Siby Abraham**

**Abstract** The paper proposes classification of tabla strokes using multilayer feed forward artificial neural network. It uses 62 features extracted from the audio file as input units to the input layer and 13 tabla strokes as output units in the output layer. The classification has been done using dimension reduction and without using dimension reduction. The dimension reduction has been performed using Principal Component Analysis (PCA) which reduced the number of features from 62 to 28. The experiments have been performed on two sets of tabla strokes, which are played by professional tabla players, each comprises of 650 tabla strokes. The results demonstrate that correct classification of instances is more than 98 % in both the cases.

**Keywords** Computer music · Neural network · Tabla · Classification · Multilayer perceptron

## 1 Introduction

Musical instruments are normally categorized into four families: brass, woodwind, percussion and strings. Brass instrument produces sound from vibration of player lips. It uses a set of valves and slides for producing sound. Trumpet and cornet are examples of brass instruments. Woodwind instruments generate sound with the help of wind. Flute and bagpipe are instances of such instruments. Percussion instruments are played by striking by hand or using sticks. Tabla and drums are such instruments. String instruments produce sound using vibration of strings.

S. Deolekar (✉)
Department of Computer Science, University of Mumbai, Mumbai 400098, India
e-mail: subodhdeolekar@gmail.com

S. Abraham
Department of Maths & Stats, G. N. Khalsa College, University of Mumbai, Mumbai, India
e-mail: sibyam@gmail.com

© Springer India 2016

347

H.S. Behera and D.P. Mohapatra (eds.), *Computational Intelligence in Data Mining—Volume 1*, Advances in Intelligent Systems and Computing 410, DOI 10.1007/978-81-322-2734-2_35

Strings are normally bowed or plucked against the instrument body. Violin and guitar belongs to this category of instruments. The family of percussion instruments is believed to include the oldest musical instruments and tabla is considered to be the most popular and universally played percussion instrument. Tabla is a Hindustani percussion instrument which is used for solo performance as well as accompany for the vocal and other instruments. Tabla is comprised of two drums: generally right side drum is called as "daayaan" (also called as tabla), which is made up of wood and the left side drum is called "baayaan" (also called as Dagga), which is made up of metal. (When we refer to tabla in the remaining part of the paper, it refers to the 'tabla set').

The tabla has an interesting construction. The daayaan (right hand drum) is almost always made of wood, which is hollow tapered cylindrical block of wood. The bottom is solid and tapered in reverse direction. The diameter at the membrane may run from just under five inches to over six inches. The top is mounted by a goat-skin. The baayaan (left hand drum) may be made of iron, aluminium, copper, steel, or clay; yet brass with a nickel or chrome plate is the most common material. Undoubtedly the most striking characteristic of the tabla is the large black spot on each of the playing surfaces. These black spots, which are Shyahee, are a mixture of gum, soot, and iron filings. Shyahee gives the weight on the skin, that in turn, controls the overall vibrations of the skin. Their function is to create the bell-like timbre that is characteristic of the instrument, baayaan provides the bass sound and daayaan gives treble [1, 2]. The strokes which are played on tabla are called *bols*. These are the mnemonics which are used to define different rhythmic patterns that are played on tabla. Table 1 gives the list of basic strokes which are played on *daayaan*, *baayaan* separately and together:

**Table 1** List of basic tabla strokes

| *DaayaanBols* | *BaayaanBols* | *Both Together* |
| --- | --- | --- |
| Ge (गे / घे) | Na (ना) | Dha (धा) |
| Ka (क / की) | Tu (तू) | Dhin (धीं) |
|  | Ti (ति) | Tin (तीं) |
|  | Ta (ट) |  |
|  | N (न) |  |
|  | T (त / र) |  |
|  | Tra (त्र) |  |
|  | Din (दिं) |  |

**Fig. 1** Tabla set

Recognition of tabla strokes is considered to be an expert job which requires years of experience. Though the language of tabla is well defined, it differs from one musical tradition to another. Most of the *bols* have similar name but different timbral property. It is difficult to identify the correct *bols* even for experienced tabla players [3]. Classification will help us to recognize and characterize these *bols*. We use Multilayer Perceptron for this purpose. The paper attempts to recognize different strokes of the tabla instrument, which is further expected to generate creative compositions through machine intelligence as shown in Fig. 1.

## 2 Related Work

Classification of musical instruments has been addressed by many researchers [4–9] but very few have worked in the field of tabla classification. Gillet and Richard [10] tried for automatic labeling of tabla strokes by segmenting them from audio file. Learning and classification of the strokes is done using trigram (3-grams) and Hidden Markov Model (HMM). Results are compared with k-NN, Naïve Bayes and Kernel density estimator classifiers, where HMM approach offers 93.6 % recognition rate. Parag [3], who uses the recordings from Gillet and Richard and with some additional dataset, tries to segment and recognize the tabla strokes. In his study, the focus is on four classifiers, namely Multivariate Gaussian, Feed forward Neural Network, Probabilistic neural network and Binary tree. Features like temporal centroid, attack time, zero-crossing, spectral centroid, skewness, kurtosis and MFCC are extracted and principal component analysis is used to reduce the dimensionality of the same. The average recognition rate with 10-fold cross validation is 89, 92, 92 and 78 % respectively. Sarkar [11] developed a system called TablaNet which recognizes the tabla strokes over the network and helps in utilizing the maximum network bandwidth with less delay. He uses kNN classifier and PCA reduction technique in order to classify the tabla Strokes.

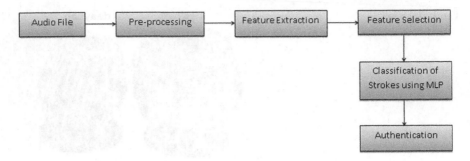

**Fig. 2** System architecture

# 3 Methodology

The overall architecture of the classification system proposed is described in Fig. 2.

## 3.1 Audio File

The System takes input as an audio file which comprises of different tabla strokes. The number of samples of audio carried per second, also called as sample rate, is kept as 44,100 Hz.

## 3.2 Pre-processing

These strokes are not of same duration. Most often, only some initial part of the strokes contribute to the melody or pitch. The remaining part is just a humming, which does not play much role for defining the properties of the strokes. In order to have standard duration of time and to have only contributing part of the strokes, they are clipped to specific time duration. This preprocessing helps to have the same length for each audio file. Thus, preprocessing involves the segmentation of individual strokes.

## 3.3 Feature Extraction

Different sets of spectral and temporal features are considered for analyzing different content of the tabla strokes for the classification purpose. Features like Zero-crossing, Centroid, Roll off, Flux, MFCC0-MFCC12, Peak Ratio of Chroma scales are extracted from the audio file of individual strokes.

## 3.4 Feature Selection

In order to reduce the dimensionality, we have used Principal Component Analysis (PCA) feature selection method. PCA works with the eigenvalue decomposition and perform well. The generation of feature vector in PCA depends on the dataset being used [11]. Total 28 features were reduced out of 62 features.

## 3.5 Classification of Strokes

We have used Multilayer Perceptron (MLP) as a Neural Network [12] in order to classify tabla strokes. MLP Classifier uses back propagation to classify instances. The multilayer perceptron network that we have used here has 62 input nodes, 37 hidden nodes ([input nodes + output nodes]/2) and 13 output nodes. The learning rate is set to 0.3 and so the weights of the synapse get updated accordingly. The choice of Multilayer Perceptron is deliberate since it works well with multiple classes and linear or non-linear activation functions [13].

## 3.6 Authentication

We have used correctly classified instances and confusion matrix for authentication purpose. The confusion matrix helps us to drive the performance measures like sensitivity, specificity and accuracy which gives us better statistical foundation along with other measures.

# 4 Experimental Set Up and Results

We have used the Marsyas (Music Analysis, Retrieval and Synthesis for Audio Signals) open source software framework for retrieving the features from audio files. Using Marsyas, which is a popular Music Information Retrieval (MIR) [14] tool, we have extracted 21361 and 18802 instances of dataset 1 and dataset 2 respectively. The selection of the feature set is based on various audio descriptors which contribute for retrieving different features like temporal and spectral. These include Zero-crossing, Centroid, Roll off, Flux, MFCC0 to MFCC12 and Peak Ratio of Chroma scales. Total 62 features were used for the classification purpose. Later these features were reduced using PCA. The data extracted using Marsyas are stored in ARFF file and are supplied to Weka machine learning environment [15] for classification.

| | S. no. | Tabla stroke | Duration (s) | Duration after clipping (s) |
|---|---|---|---|---|
| **Table 2** Strokes and their duration | 1 | Dha | 0.6–0.9 | 0.5 |
| | 2 | Dhin | 0.5–0.6 | 0.5 |
| | 3 | Din | 1.5–2 | 0.5 |
| | 4 | Ge | 0.5–1 | 0.5 |
| | 5 | Ka | 0.5–0.6 | 0.5 |
| | 6 | N | 0.5–0.7 | 0.5 |
| | 7 | Na | 0.7–1 | 0.5 |
| | 8 | T | 0.2–0.3 | 0.2 |
| | 9 | Ta | 0.2–0.3 | 0.2 |
| | 10 | Ti | 0.25–0.3 | 0.3 |
| | 11 | Tin | 0.5–0.7 | 0.5 |
| | 12 | Tra | 0.1–0.3 | 0.2 |
| | 13 | Tu | 0.2–0.5 | 0.5 |

## 4.1 Dataset Used

We have considered basic tabla strokes which are recorded from the professional tabla players,[1] The tabla set is tuned to C# scale. Table 2 shows the timings after clipping of these strokes.

## 4.2 Performance Measures Used

Table 3 shows the confusion matrix which is obtained from the classifier by 10-fold cross validation. This matrix is used for calculating the values of various performance measures.

The percentages of correct classification without PCA are 99.73 and 99.60 % for dataset 1 and dataset 2 respectively. These are higher than the results obtained with the PCA as a feature reduction technique.

### 4.2.1 Sensitivity

Table 4 shows the sensitivity of the MLP with PCA and without using PCA. It is used to measure the performance of the classifier, and defined as follows.

---

[1]Tabla players were Mr. Arun Kundekar and Mr. Kshitij Patil, both of them are Sangeet Visharad in Tabla.

**Table 3** Confusion matrix for tabla strokes

| | Dha | Dhin | Din | Ge | Ka | N | Na | T | Ta | Ti | Tin | Tra | Tu |
|---|---|---|---|---|---|---|---|---|---|---|---|---|---|
| Dha | 2195 | 2 | 1 | 0 | 2 | 0 | 0 | 0 | 0 | 0 | 0 | 0 | 0 |
| Dhin | 0 | 2198 | 0 | 1 | 0 | 0 | 0 | 1 | 0 | 0 | 0 | 0 | 0 |
| Din | 1 | 0 | 2197 | 0 | 1 | 0 | 1 | 0 | 0 | 0 | 0 | 0 | 0 |
| Ge | 0 | 0 | 0 | 2198 | 1 | 0 | 0 | 0 | 0 | 0 | 1 | 0 | 0 |
| Ka | 0 | 1 | 1 | 1 | 1047 | 0 | 0 | 0 | 0 | 0 | 2 | 0 | 0 |
| N | 0 | 0 | 0 | 0 | 0 | 2197 | 0 | 0 | 0 | 0 | 0 | 0 | 0 |
| Na | 1 | 0 | 0 | 0 | 0 | 0 | 2199 | 0 | 0 | 0 | 0 | 0 | 0 |
| T | 0 | 1 | 0 | 1 | 0 | 0 | 0 | 862 | 5 | 3 | 0 | 0 | 1 |
| Ta | 0 | 0 | 0 | 0 | 0 | 1 | 0 | 8 | 890 | 0 | 0 | 1 | 0 |
| Ti | 0 | 0 | 0 | 0 | 0 | 0 | 0 | 2 | 7 | 1263 | 3 | 0 | 0 |
| Tin | 0 | 0 | 0 | 0 | 0 | 0 | 0 | 0 | 0 | 5 | 2193 | 0 | 0 |
| Tra | 0 | 0 | 0 | 0 | 0 | 0 | 0 | 0 | 0 | 0 | 0 | 450 | 0 |
| Tu | 0 | 0 | 0 | 0 | 0 | 0 | 0 | 0 | 0 | 0 | 0 | 1 | 1415 |

**Table 4** Sensitivity (true positive rate)

| Dataset # | Sensitivity (true positive rate) | |
| --- | --- | --- |
| | MLP without PCA (%) | MLP with PCA (%) |
| 1 | 99.64 | 97.24 |
| 2 | 99.60 | 97.93 |

**Table 5** Specificity (true negative rate)

| Dataset # | Specificity (true negative rate) | |
| --- | --- | --- |
| | MLP without PCA (%) | MLP with PCA (%) |
| 1 | 99.97 | 99.82 |
| 2 | 99.96 | 99.79 |

$$Sensitivity = number\ of\ true\ positive\ instances /$$
$$(number\ of\ true\ positive\ instances + number\ of\ false\ negative\ instances)$$
$$(1)$$

Thus, sensitivity describes how well a classifier classifies these instances that belong to the specified classes. The experimental results show that the sensitivity approaches 100 %, which means that even with feature reduction, the methodology is quite effective.

### 4.2.2 Specificity

It describes how well a classification task classifies those observations that do not belong to a particular class. Table 5 shows the specificity, which is defined as follows:

$$Specificity = number\ of\ true\ negative\ instances /$$
$$(number\ of\ true\ negative\ instances + number\ of\ false\ positive\ instances)$$
$$(2)$$

### 4.2.3 Accuracy

Table 6 shows the Accuracy and it is defined as the proportion of the total number of test tabla strokes that are correctly classified.

$$Accuracy = (number\ of\ true\ positive\ instances$$
$$+ number\ of\ true\ negative\ tinstances)/(Total\ number\ of\ instances)$$
$$(3)$$

**Table 6** Accuracy

| Dataset # | Accuracy | |
|---|---|---|
| | MLP without PCA (%) | MLP with PCA (%) |
| 1 | 99.95 | 99.68 |
| 2 | 99.92 | 99.62 |

# 5 Discussion and Conclusion

The paper discusses the effectiveness of Multilayer Feed Forward Neural Network as an effective classification tool to classify tabla strokes. Classification accuracies above 99 %, achieved through the experimental results using feature reduction and without using feature reduction, show that the proposed methodology performs well. The minimal difference in the accuracies with and without using PCA as a feature reduction technique reveals that the PCA could successfully extract almost all the features which actually contributed to the correct classification. Though this work is restricted to the classification of tabla stokes, it can also be extended to other percussion instruments like dholki, drum set etc. even though they are played in a different context.

# References

1. An Unquiet Mind: Western Classical Versus Indian Classical Music. http://skeptic. skepticgeek.com/2011/08/31/western-classical-vs-indian-classical-music/
2. David Courtney: Learning the Tabla, vol. 2. M. Bay Publications (2001). ISBN:0786607815
3. Chordia, P.: Segmentation and recognition of tabla strokes. In: International Conference on Music Information Retrieval, pp. 107–114. (2005)
4. Herrera, P., Dehamel, A., Gouyon, F.: Automatic labeling of unpitched percussion sounds. In: Audio Engineering Society Convention Paper. (2003)
5. Herrera, P., Yeterian, A., Gouyon, F.: Automatic classification of drum sounds: a comparison of feature selection methods and classification techniques. In: International Conference on Music and Artificial Intelligence (ICMAI). (2002)
6. Kursa, M., Rudnicki, W., Wieczorkowska, A., Kubera, E., Kubik-Komar, A.: Musical instruments in random forest. Found. Intell. Syst. **5722**, 281–290 (2009)
7. Kostek, B.: Automatic classification of musical instrument sounds. J. Acoust. Soc. Am. **107** (5), 2818 (2000)
8. Cemgil, A.T., Gürgen, F.: Classification of musical instrument sounds using neural networks. Proc. SIU97 (1), 1–10 (1997)
9. Marques, J., Moreno, P.J.: A study of musical instrument classification using Gaussian mixture models and support vector machines. Cambridge Research Laboratory Technical Report Series, CRL 99/4(June). (1999)
10. Gillet, O., Richard, G.: Automatic labelling of tabla signals. In: ISMIR 2003, 4th International Conference on Music Information Retrieval, Baltimore, Maryland, USA, October 27–30. (2003)
11. Sarkar, M.: TablaNet: A Real-Time Online Musical Collaboration System for Indian Percussion. (2007)
12. Haykin, S.: Neural Networks. Macmillan, New York (1994)

13. Ruck, D.W., Rogers, S.K., Kabrisky, M., Oxley, M.E., Suter, B.W.: The multilayer perceptron as an approximation to a Bayes optimal discriminant function. IEEE Trans. Neural Netw. **1**(4), 296–298 (1990)
14. Tzanetakis, G., Cook, P.: MARSYAS: a framework for audio analysis. Org. Sound **4**(3), 169–175 (1999)
15. Witten, I.H., Frank, E.: Data Mining: Practical Machine Learning Tools and Techniques, 2nd edn. Morgan Kaufmann, San Francisco (2005)

# Automatic Headline Generation for News Article

K.R. Rajalakshmy and P.C. Remya

**Abstract** Newspaper plays a significant role in our day to day life. In a news article, readers are attracted towards headline. Headline creation is very important while preparing news. It is a tedious work for journalists. Headline needs maximum text content in short length also it should preserve grammar. The proposed system is an automatic headline generation for news article. There are several methods to generate a headline for news. Here the headline is generated using n-gram model. After pre-processing, extract all possible bi-grams from the news. Frequency of bi-grams in the document is found out and sentence is identified. Select all sentences in which the most frequent bi-gram appears. After that rank the sentences based on the locality of the sentence in the document. Then highly ranked sentence will be selected. Finally using some phrase structural rules, new headlines are generated.

**Keywords** Bi-grams · Pos tag · Tokens

## 1 Introduction

Text summarization is the process of automatically summarizes the text documents. The objective of summarization is to produce a paragraph or abstract that summarizes a large collection of data. Here the headline generation is the task of producing a very short summary. Headline of a news article is defined as a short statement that gives reader a general idea about the main contents of the news. Headline generation is an important task, journalists creates headline for an article by taking lots of time. It is both interesting and difficult process because headline is

K.R. Rajalakshmy (✉) · P.C. Remya
Department of Computer Science & Engineering, Vidya Academy
of Science and Technology, Thalakkotukara, India
e-mail: rajalakshmy2009@gmail.com

P.C. Remya
e-mail: remyapc@vidyaacademy.ac.in

© Springer India 2016

357

H.S. Behera and D.P. Mohapatra (eds.), *Computational Intelligence
in Data Mining—Volume 1*, Advances in Intelligent Systems
and Computing 410, DOI 10.1007/978-81-322-2734-2_36

the first thing that catches everyone. Different news agencies have different style of writing. The news model of various countries may change. Headline should focus on most relevant theme expressed in the input article. It can be used for a reader to quickly identify information which they have interest. Automatic headline generation for news article is beneficial for journalists.

## 1.1  Syntax

Syntax is basically the structure of sentence. Sentences have to follow certain structural rules in order to make meaningful. Computer language syntax is generally classified into three levels.

- Words—it is the smallest unit on the syntax level.
- Phrases—it is the group of words in grammar level.
- Context—determining what objects refer to.

## 1.2  Semantics

It is mathematical study of the meaning of a programming language. It focuses on the relation between signifiers like words, phrases, signs and symbols also what they stand for.

## 1.3  N-gram Model

For Linguistic analysis it is helpful if we split the sentence into words. Each individual word is called a token. N-gram is an adjoining sequence of n items from a given sequence of text or speech. An n-gram of size one is called as a "unigram", size two is "bigram" and size three is "trigram". It estimates probability of each word given prior context.

## 1.4  Parts of Speech Tagging

It is the assignment of part of speech tag to each of the token in a document. English Parts of Speech consists of Noun, Verb, Adjective, Adverb, Preposition, Determiner, Coordinating Conjunction etc. Here it is done by a tool called Stanford parser.

Eg: he is going to college.
  PRN VBZ VBG TO NN

## 2  Related Work

In this section we briefly describe the existing works of headline generation. Soricut et al. [1] presented a model in which relevant sentence is extracted from the input document and compressed to required length. Input document is transformed into an intermediate representation called WIDL expression. A wide NLG engine, which operates on WIDL expression, is used as a back-end module for headline formation. It begins from an important, individual words and phrases, that are formed together to create a fluent sentence. Algorithm is used to found out keywords. Then using the returned keywords new phrases are extracted from the document which are linked together to create headlines. Here it uses keywords to identify the words that carry most of the information in an input document. It also use syntactic information extracted from parse tree.

Banko et al. [2] presented Extractive summarization approach cannot generate document summaries shorter than a single sentence. Summarization has two steps: (1) Content selection (2) Surface realization. At first the input documents undergoes some preprocessing such as formatting and removing mark-up data such as font and html tags. After that punctuation and apostrophes were detached. System will try to determine a model of the similarity between the representation of some aspects in a document and representation of similar aspects in summary. Then Estimate the reasonableness of some token present in summary and some tokens appeared in the document to be summarized. A conditional probability of a word occlude in the summary with the word present in the document is found out. In surface realization probability of word sequence is calculated. In a bigram language model the probability of a word ordering is calculated by the product of probability it's next left content. Probabilities for sequence that were not seen in the training data are calculated by back off weights.

Zhou et al. [3] presented different Bag of words Models. First is Sentence position model. Sentence position is very useful for recognition of text. This theory is used to select headline words. When a sentence with its position is considered its possibility to contain in headline words is estimated. Next is unigram model, here unigram probabilities were considered. After considering the models keywords will be selected. The keywords that appeared in the first sentence will be selected. Then a clustering algorithm is performed to found the neighborhood of each word. Thus a set of phrases will generate and using some handwritten rules new Headline is created.

In our headline generation, phrase based sentence selection method is used. If phrase is used to identify a sentence the context of the word will preserves. So the most important sentence can be selected. Table 1 shows different sentence selection models. In existing methods most of them used keywords to select sentence. Our method is better than the existing one because more accurate result can be generated. After sentence selection rules are applied to generate new headlines. Rule based

**Table 1** Different sentence selection models

| Techniques | Keyword based sentence selection | Phrase based sentence selection |
|---|---|---|
| WIDL expression | ✓ | × |
| Statistical translation | ✓ | × |
| Summarization at ISI | ✓ | × |
| Template based sentence generation | ✓ | ✓ |
| Statistical paraphrase summary generation | ✓ | × |

sentence generation is better than template based and corpus based. In template based sentence generation predefined templates are used. It needs a lot of templates, to generate a sentence, while in our method no predefined templates are used. In corpus based, large corpus of data is required. So our method is very simple and efficient.

## 3 Architecture

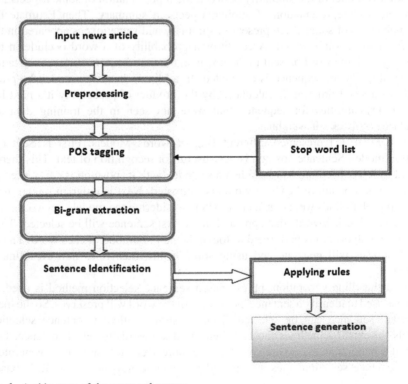

**Fig. 1** Architecture of the proposed system

**Pseudo code**

**Input: news article (text).**
**Output: headline for the news.**

1 Preprocessing
2 Tokenize the input text.
3 Matrix[i][j]=void
4 For i starts at 0, i<length-1, increment i
{
    For j starts at 0, j<length-1, increment j
{
5 Read tokens from the input.
6 Calculate the occurrence of one token preceding the other.
7 If Matrix[i][j]>0 then it is valid bi-grams and Matrix[i][j]<0 will
  be invalid bigrams.
8   End for.
9 Initialize count[i] =0;
10 count[i] ++;
11 Find out the most frequent bigram
12 Extract all possible sentences in which these bi-grams appeared.
13 Select one sentence from that based on position of sentences in the document.
14 Finally phrase structural rules are used to generate a new sentence.

## 3.1   Preprocessing

In preprocessing the input text is cleaned. The comma, semicolons, exclamation marks and other special characters will be removed. The full stop is retained to know the end of a sentence. All letters is reduced to lowercase. For instance of 'Automobile' to 'automobile'. If case folding is not done then the tokens will be distinct each other. This will lead to duplication of tokens.

### 3.1.1   Stop Words Removal

Stop word is a frequently used word in a document. Stop words may be a set of determiners for example "the", "a", "an" or it may be a set of adjectives like "any", "bad", "good" or prepositions like "above", "across", "before". These words conveys less information compared to other words in the document. The most commonly used stop words are maintained as a list that was shown in the Fig. 2.

- Start with a list of stop words.
- Look up the table.
- Match each token against the table.
- Remove the list of stop words.

"a", "new", "told", "about", "above", "above", "across", "after", "afterwards", "again", "against", "all",
"almost", "alone", "along", "already", "also", "although", "always", "am", "among", "amongst",
"amongst", "amount", "an", "and", "another", "any", "anyhow", "anyone", "anything", "anyway",
"anywhere", "are", "around", "as", "at", "back", "be", "became", "because", "become", "becomes",
"becoming", "been", "before", "beforehand", "behind", "being", "below", "beside", "besides",
"between", "beyond", "bill", "both", "bottom", "but", "by", "call", "can", "cannot", "cant", "co", "con",
"could", "cry", "de", "describe", "detail", "do", "done", "down", "due", "during", "each", "eg", "eight",
"either", "eleven", "else", "elsewhere", "empty", "enough", "etc", "even", "ever", "every", "everyone",
"everything", "everywhere", "except", "few", "fifteen", "fifty", "fill", "find", "fire", "first", "five", "for",
"former", "formerly", "forty", "found", "four", "from", "front", "full", "further", "get", "give", "go",
"had", "has", "hasn't", "have", "he", "hence", "her", "here", "hereafter", "hereby", "herein", "hereupon",
"hers", "herself", "him", "himself", "his", "how", "however", "hundred", "ie", "if", "in", "inc", "indeed",
"interest", "into", "is", "it", "its", "itself", "keep", "last", "latter", "latterly", "least", "less", "ltd", "made",
"many", "may", "me", "meanwhile", "might", "mill", "mine", "more", "moreover", "most", "mostly",
"move", "much", "must", "my", "myself", "name", "namely", "neither", "never", "nevertheless", "next",
"nine", "no", "nobody", "none", "no one", "nor", "not", "nothing", "now", "nowhere", "of", "off",
"often", "on", "once", "one", "only", "onto", "Mr.", "or", "other", "others", "otherwise", "our", "ours",
"ourselves", "out", "over", "own", "part", "per", "perhaps", "please", "put", "rather", "re", "same", "see",
"seem", "seemed", "seeming", "seems", "serious", "several", "she", "he".

**Fig. 2** List of stop words

## 3.2 Tagging

In tagging all the tokens in the document is interpreted with its part of speech
information. This task is performed by Stanford parser. A natural language parser is
a program that accomplishes the grammatical structure of sentences. It will identify
all the nouns, verbs, determiners etc. There is a tag set associated with each parser.
Stanford parser is a Probabilistic Part of speech tagger. The Probabilistic Parser is
based on the first order or second order Markov models. In English language there
exists an ambiguity among certain words like 'store' which may be a noun or a
verb. This ambiguity is resolved by the context of the word in a sentence. Figure 3
shows the Parts Of Speech tagging method.

## 3.3 Tokenization

Tokenization is splitting the text into sequence of tokens which is "words". The
token is the individual units in a text. It may be a word, number or punctuation
mark. Tokens are separated by identifying the boundaries.

## 3.4 Bigram Generations

A bigram is a sequence of two adjacent elements in a string of tokens. Bigram
probability matrix for tokens is calculated. All the tokens after stop word removal

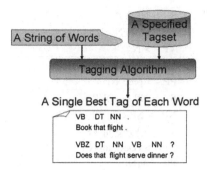

**Fig. 3** POS tagging method

are taken into rows and columns of matrix. Then occurrence of one token after the other will be calculated by referring to the position in the sentence. An increment to count is assigned to all possible bi-gram. After that valid bi-grams will be extracted. 0 values will assign for all non bigrams and 1 for all possible bigram. Matrix[i][j] > 0 will be all possible bi-grams.

## 3.5 Topic Identification

Topic identification is the process of choosing a set of sentences that best expresses the summary of the document. Here a simple statistics of the bigram is used to find out the set of topic sentences. To determine the ranking of sentences, a relative "significance" score is calculated and then top scoring sentence are selected. The frequency of bi-gram existence in a document gives a valuable significance of the topic words that occur often in the document are likely.

Matrix[i][j] < 1 will be ignored and Matrix[i][j] > 0 will be stored in an array. These bi-grams will be sorted. Most frequent bi-gram will be found out. The sentences which have this bi-gram will be extracted. Then we rank these sentences based on the count of bi-grams in that sentences. After that the highly ranked sentence is selected. Some phrase structural rules are applied on that sentence and new headline is generated.

## 3.6 Rules for Sentence Generation

The English sentences have a main clause, which consists of a subject and a verb also there may be an object depending upon the verb. All sentences are about someone or something, which is called a subject. For example 'Raji go to college'

**Table 2** Different rules for sentence generation

| |
|---|
| S - > NP VP |
| NP - > det N |
| VP - > N |
| VP - > tv NP |
| PP - > prep NP |
| S - > NP (aux)VP |
| NP- > NP conj NP |
| VP - > VP conj VP |
| NP - > (Det)(adj)N PP |
| PP- > P(NP) |
| NP- > (Det)N(PP) |
| PP- > P(NP) |
| VP- > V(NP)(PP)(Adv) |

here Raji is the subject. Object follows the verb. Table 2 shows different sentence generation rules.

- A sentence consists of NP (noun phrase) succeed by a VP (verb phrase).
- A noun phrase consists of an optional (determiner) succeed by an N (noun).
- Noun can be preceded by an optional AP (adjective phrase) and succeed by an optional PP (prepositional phrase).

# 4 Results and Discussions

In this system we can give news form a news article as input, the output will be a headline. The headline is grammatically correct and meaningful. The system is simple and efficient. It is better than the template based sentence generation. Here phrase structural rules are used to make the sentence.

Table 3 shows a comparison between existing summary generation tool available in Google, our system and headline in the news article. Here gold standard headline are taken from Hindu newspaper. Semantic similarity between two Headlines is done using a tool "Mechaglot: Calculate Semantic Similarity" which is free and open source publicly available. It is an enhancement of the Vector-Space analysis found within the Classifier4j, which does considered the linguistic meanings of the words. The similarity measures included in the evaluation are cosine and Euclidian distance and can be applied in different ways.

Figure 4 shows that the similarity to original headline is high for our system compared to the Google summary tool.

**Table 3** Headline generated by humans (G), our system (H) and google summary tool (S)

| S | For the time, soft tissue like structures that appear to be red blood cells and collagen fibers have been found in 75 million year old (cretaceous) dinosaur specimen that are not well preserved |
|---|---|
| H | A dinosaur specimen appears under the structure |
| G | Now, soft tissue found in ancient dinosaur sample |
| S | Below-normal rainfall in June and in the first week of July has affected cultivation of paddy, the chief crop, in Udupi district even as farmers keep hoping that the monsoon will pick up in the coming weeks |
| H | The rainfall affected on a paddy cultivation |
| G | Deficient rainfall hits paddy cultivation in Udupi district |
| S | State-run BSNL will launch free roaming, starting today, which will allow all its mobile customers across the country to receive incoming calls at no cost |
| H | The communication has about the BSNL roaming scheme |
| G | BSNL to launch free roaming today |

**Fig. 4** This graph shows semantic similarity measurement of different headline generation system

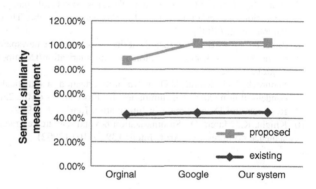

## 5 Conclusion

N gram model has a diversity of applications in NLP tasks. In this work bi-gram is used in the content selection phase. Bi-gram preserves the context of the word. If we take only the keyword for sentence scoring the context of the news may change. Bigram is helpful for meaningful generation of sentence. After generating the bi-grams phrase structural rules are applied to generate a sentence. Here the headline shorter than a sentence is obtained. The generated headline is tested with human generated headline using semantic similarity tool. In future, the semantics of the words can be included. Each sense of the words can be understood by using Lesk algorithm. If the semantics of the words are understand, more accurate result can be found out also word disambiguation can eliminated. In future the wordnet can be used to find the synonyms of the words. Wordnet is a large dictionary which contains the synonyms, hypernonyms of the words, which will enhance the result. The synonyms can be used instead of some words which will be a better one.

# References

1. Soricut, R., Marcu, D.: Stochastic language generation using WIDL-expressions and its application in machine translation and summarization. In: Proceedings of the 21st International Conference on Computational Linguistics and the 44th Annual Meeting Association for Computational Linguistics, pp. 1105–1112. (2006)
2. Banko, M., Mittal, V., Witbrock, M.: Headline generation based on statistical translation. In: Proceedings of the 38th Annual Meeting Association for Computational Linguistics, pp. 318–325. (2000)
3. Zhou, L., Hovy, E.: Headline summarization at ISI. In: Proceedings of HLT-NAACL Text Summarization Workshop and Document Understanding Conference, pp. 174–178. (2003)
4. Feng, Y., Lapata, M.: Automatic Caption generation for news images. IEEE Trans. Pattern Anal. Mach. Intell. **35**(4), (2013)
5. Yadav, A.K., Borgohain, S.K.: Sentence generation from a bag of words using N-gram model. In: IEEE International Conference on Advanced Communication Control and Computing Technologies (ICAC-CCT). (2014)
6. Zhou, L., Hovy, E.: Template filtered headline summarization. In: Proceedings of the ACL Workshop on Text Summarization, Barcelona, July 2004
7. Wan, S., Dras, M., Dale, R.: Towards statistical paraphrase generation: preliminary evaluations of grammaticality. In: Proceedings of the Third International Workshop on Paraphrasing. (2005)
8. Zajic, D., Dorr, B.: Automatic headline generation for newspaper stories. In: The Proceedings of the ACL Workshop on Automatic Summarization/Document Understanding Conference. (2002)
9. Witbrock, M.J., Mittal, V.O.: Ultra summarization: a statistical approach to generating highly condensed non extractive summaries. In: Proceedings of the 22nd International Conference on Research and Development in Information Retrieval. (1999)
10. Knight, K., Marcu, D.: Summarization beyond sentence extraction: a probabilistic approach to sentence compression. Artif. Intell. **139**, 91107 (2002)

# Identifying Natural Communities in Social Networks Using Modularity Coupled with Self Organizing Maps

Raju Enugala, Lakshmi Rajamani, Kadampur Ali
and Sravanthi Kurapati

**Abstract** Community detection in social networks plays a vital role. The understanding and detection of communities in social networks is a challenging research problem. There exist many methods for detecting communities in large scale networks. Most of these methods presume the predefined number of communities and apply detection methods to exactly find out the predefined number of communities. However, there may not be the predefined number of communities naturally occurring in the social networks. Application of brute force inorder to predefine the number of communities goes against the natural occurrence of communities in the networks. In this paper, we propose a method for community detection which explores Self Organizing Maps for natural cluster selection and modularity measure for community strength identification. Experimental results on the real world network datasets show the effectiveness of the proposed approach.

**Keywords** Community detection · SOM · Modularity · Social networks

R. Enugala (✉) · K. Ali
Department of Computer Science and Engineering, SR Engineering College,
Warangal, Telangana State, India
e-mail: raju.cse999@gmail.com

K. Ali
e-mail: ali.kadampur@gmail.com

L. Rajamani
Department of Computer Science and Engineering, Osmania University,
Hyderabad, India
e-mail: drlakshmiraja@gmail.com

S. Kurapati
Department of Computer Science and Engineering, University College
of Engineering, Kakatiya University, Warangal, Telangana State, India
e-mail: sravanthi.k1982@gmail.com

© Springer India 2016 367
H.S. Behera and D.P. Mohapatra (eds.), *Computational Intelligence
in Data Mining—Volume 1*, Advances in Intelligent Systems
and Computing 410, DOI 10.1007/978-81-322-2734-2_37

# 1  Introduction

A social network is a combination of set of nodes and edges. The edges are used to connect nodes in various relationships such as friendship, kinship, etc. Community in this network is defined as a group of subsets of nodes such that the connections within the group are denser whereas the connections are sparse between groups [1, 2]. The communities, when detected, reveal interrelationship, associations, and behavioral trends among the members [3]. For example, the research community, when detected, in a social network may reveal domain specific Special Interest Groups (SIGs) which will be further used for effective research interactions among the members. However, the challenge is to detect the exact number of naturally occurring communities in such networks. There is substantial work in community detection, but very little in finding natural communities [4] in the networks.

In this paper, a novel method is devised to know the number of natural clusters in a network and characterize their strengths using a parameter called modularity. The work becomes important considering the fact that knowledge discovery from such social networks is a buzz activity and has broad range of applications.

## 1.1  Preliminaries

**Social Network** A social network is represented by a graph $G = (V, E)$, where V is a set of nodes from 1 to $n$ and E is a set of edges or links. In this graph nodes are used to represent individuals or organizations as entities. Similarly, links represents the interactions between entities. The above defined social network can be represented by an adjacency matrix $(A = (a_{ij})_{n \times n})$, where n is the number of entities. The values in the adjacency matrix can be assigned either 1 or 0. Where the value 1 $(a_{ij} = 1)$ represents a link between two entities and 0 represents no link.

**Modularity** The Modularity (Q) of a social network can be defined as the ratio between number of nodes inside a community to the number of nodes outside the community. This can be represented using the following equation.

$$Q = \frac{1}{2m} \sum_{\ell=1}^{k} \sum_{i,j \in C_\ell} (A_{ij} - deg_i * deg_j / 2m) \tag{1}$$

where the quantity Q directly reflects the strength of connectivity, m is the number of edges, and $deg_i$ is number of nodes adjacent to a node $v_i$ called as the degree of a node $v$.

## 2 Literature Review

Communities have a long history in social sciences. In recent times different algorithms have been proposed by different disciplines such as mathematics, machine learning, statistics, data mining for detecting communities in social and other complex networks [5–8].

Girvan and Newman first stated the idea of community structure in social networks [9] and modularity [2]. According to Girvan and Newman a *community structure* in a social network is a sub graph of nodes appearing as a group such that the density of internal connections between nodes of sub graph is larger than the connections with the remaining nodes in the network.

In [4], Hopcroft presented an agglomerative hierarchical clustering method to identify stable or natural communities in large linked networks. According to Hopcroft, "a cluster is assumed as a natural cluster if it appears in the clustering process when a given percentage of links are removed".

Li et al. [10] has proposed a technique for identifying communities in complex networks using SOM. In this approach a complex network is divided into dense sub graphs based on the topological connection of each node which uses weight updating scheme. This method can also be used for directed networks and bipartite networks.

## 3 Problem Statement

The community detection problem in a social network is to find a partition $C = \{c_1, c_2, \ldots, c_k\}$ of a simple graph G = (V, E), where V is set of vertices and E is set of edges such that $\forall_i, c_i \subseteq V$ and $\forall_{i,j}, c_i \cap c_j = \phi$ and each $c_i i = 1, \ldots, k$ is a sub-graph (community) of G with intra-cluster density $(\rho_i)$ being higher than inter-cluster density $(\rho_o)$ of edges.

Consider the graph G = (V, E) in Fig. 1, where V = {1,2,3,4,5,6,7}, E = {(1,2), (1,3), (2,3), (2,4), (4,5), (4,7), (5,6), (5,7), (6,7)}.

If we partition the graph along the plane PP', inter cluster density $\rho_o = 0.083$ and intra cluster densities $(\rho_i \text{ s}')$ are 1 (for Left side of partition) and 0.8333 (for Right side of partition). Since $\rho_i = 1 > \rho_o = 0.083$ and $\rho_i = 0.833 > \rho_o = 0.083$, there are two communities, $c_1 = \{1, 2, 3\}$ and $c_2 = \{4, 5, 6, 7\}$, hence the community detection.

**Fig. 1** A simple network graph with two community structures

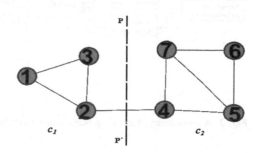

Given the number of partitions there are many algorithms that find the communities. However, the availability of natural communities in the network remains undiscovered. In this paper, a novel method based on Kohonen's Maps (SOM) is devised to know the number of natural communities. The paper also explores modularity for strength assessment and Eigen vector formation.

## 4 Proposed Approach

The general methodology of the proposed work on community detection in social networks is represented as shown in the schematic block diagram of Fig. 2.

The adjacency matrix A, basically represents the real world graphs where in nodes represent entities and edges represent some type of interaction between pairs of those entities. For example, in a social network, nodes may represent individual people and edges may represent friendships. The adjacency matrix which is the true representative of a real time graph (network) acts as input to the system in Fig. 2.

In order to find modularity matrix and identify community strengths we need to compute the degree matrix (**deg**) of A and its transpose (**deg**$^T$). The modularity matrix $B = A - \mathbf{deg} * \mathbf{deg}^T / 2 * m$ is computed, where $m$ denotes the number of edges in the given network (half the count of 1's in A). The modularity matrix leads to find out eigenvectors which in turn represent direction of strong community links in the network. In other words, modularity matrix is significant in identifying denser communities instead of hopping on random communities in the network. This improves the quality of processing and eigenvector evaluation brings in inherent characteristic of processing with reduced dimensionality.

De-noising phase basically involves selecting top eigenvectors corresponding to largest eigenvalues. The selected eigenvectors capture the prominent interaction patterns, representing approximate community partitions indicative of likely community structures in the given network.

The next phase involves grouping of likely community structures to identify natural communities. In this phase there is possibility of using any of the well known clustering methods including k-means clustering. Majority of these clustering

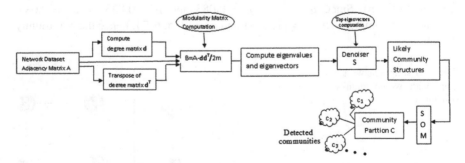

**Fig. 2** A community detection process using modularity coupled with SOM

methods need to know the number of clusters beforehand the execution of the algorithm. Particularly the k-means algorithm which is widely used in the literature suffers from this drawback of brute force selection of number of clusters (k), and fails to report the naturally available clusters in the network. Alternatively, Kohonen's Self Organizing Maps (SOM) use a neighborhood function to preserve the topological properties of the input space and continue clustering by attaching weights to each phase of the evolving grouping. This results in formation of natural groups in the data space. In Fig. 2 SOM is applied on likely community structures and natural community structures are computed. The post processing activity on the outcome of SOM results in community partition C containing natural communities of the network.

## 4.1  The Algorithm

```
Algorithm 1: Natural Community Detection

Input: filename f_in //network dataset
Output: community partition C; // C = {c_1, c_2, ...}
{
  A= load_adjacency_matrix (f_in);
  [rows, cols] = size (A);
  e =0;    //edge count variable
  for i = 1 to rows
  {
   count =0;
    for j = 1 to cols
       {
           if (A[i,j] == 1)   count++;
       }
       deg[i] =count; //degree vector of nodes
        e = e + count;
  }
  m = e/2;  //total number of edges in the graph
  deg^T = transpose(deg); // transposed vector
  B = (A - deg*deg^T)/(2*m);   //modularity matrix
  [V,D] = eig(B); // eigenvectors and eigenvalues
  S = denoise(V,D);// likely community structures
  C = SOM(S); // natural communities of the network

}
```

## 5 Experimental Results

In this section, the step by step procedure of community detection on synthetic network shown in Fig. 1 is demonstrated. Then the following subsections demonstrates results obtained on the real time datasets, namely, Zachary Karate Club dataset [11], American College Football network dataset [9] and Bottlenose Dolphin network dataset [12].

Referring the network in Fig. 1:

**Step 1**: The adjacency matrix of the network (Fig. 1) is

$$
A = \begin{bmatrix}
0 & 1 & 1 & 0 & 0 & 0 & 0 \\
1 & 0 & 1 & 1 & 0 & 0 & 0 \\
1 & 1 & 0 & 0 & 0 & 0 & 0 \\
0 & 1 & 0 & 0 & 1 & 0 & 1 \\
0 & 0 & 0 & 1 & 0 & 1 & 1 \\
0 & 0 & 0 & 0 & 1 & 0 & 1 \\
0 & 0 & 0 & 1 & 1 & 1 & 0
\end{bmatrix}
$$

**Step 2**: The degree matrix **deg** of the synthetic network is

$$
deg = \begin{bmatrix}
2 \\
3 \\
2 \\
3 \\
3 \\
2 \\
3
\end{bmatrix}
$$

**Step 3**: The transpose $\mathbf{deg}^T$ of the degree matrix **deg** is

$$
\mathbf{deg}^T = \begin{bmatrix} 2 & 3 & 2 & 3 & 3 & 2 & 3 \end{bmatrix}
$$

**Step 4**: The modularity matrix $B = (A - \mathbf{deg} * \mathbf{deg}^T)/2m$ is

$$
B = \begin{bmatrix}
-0.22 & 0.67 & 0.78 & -0.33 & -0.33 & -0.22 & -0.33 \\
0.67 & -0.50 & 0.67 & 0.50 & -0.50 & -0.33 & -0.50 \\
0.78 & 0.67 & -0.22 & -0.33 & -0.33 & -0.22 & -0.33 \\
-0.33 & 0.50 & -0.33 & -0.50 & 0.50 & -0.33 & 0.50 \\
-0.33 & -0.50 & -0.33 & 0.50 & -0.50 & 0.67 & 0.50 \\
-0.22 & -0.33 & -0.22 & -0.33 & 0.67 & -0.22 & 0.67 \\
-0.33 & -0.50 & -0.330 & 0.50 & 0.50 & 0.67 & -0.50
\end{bmatrix}
$$

**Step 5**: Eigenvectors of the modularity matrix are:

$$\text{eigenvectors} = \begin{bmatrix} 0.15 & 0.11 & 0.71 & -0.27 & -0.37 & 0.25 & -0.44 \\ -0.52 & -0.22 & 0 & 0.54 & -0.38 & -0.30 & -0.40 \\ 0.15 & 0.11 & -0.71 & -0.27 & -0.37 & 0.25 & 0.14 \\ 0.58 & 0 & 0 & 0 & -0.39 & -0.70 & 0.39 \\ -0.35 & -0.55 & 0 & -0.53 & -0.38 & 0.01 & 0.39 \\ 0.33 & -0.22 & 0 & 0.54 & -0.37 & 0.55 & 0.35 \\ -0.35 & 0.76 & 0 & 0 & -0.38 & 0.01 & 0.39 \end{bmatrix}$$

**Step 6**: The top two eigenvectors are

$$S = \begin{bmatrix} -0.44 & 0.15 \\ -0.40 & -0.52 \\ -0.44 & 0.15 \\ 0.14 & 0.58 \\ 0.39 & -0.35 \\ 0.35 & 0.33 \\ 0.39 & -0.35 \end{bmatrix}$$

**Step 7**: The community partition C is

$$C = \{\{1, 2, 3\}, \{4, 5, 6, 7\}\}$$

where $c_1 = \{1, 2, 3\}$ and $c_2 = \{4, 5, 6, 7\}$. Hence $c_1$ and $c_2$ are detected communities which obviously match with the onlookers' observation of Fig. 1.

## 5.1 Real Time Datasets

Experiments were conducted with three real time datasets, namely, Zachary Karate Club, American College Football Network and Bottlenose Dolphin Network dataset.

### 5.1.1 Zachary Karate Club Dataset

Zachary Karate Club dataset consist 34 vertices and 78 edges. Basically it represents interactions among the members of a karate club. In this dataset, the students are represented as vertices and edges represent linkage between two students if they are good friends. The original division of the club into 2 communities is taken as ground truth.

### 5.1.2 American College Football Dataset

The American College Football network dataset represents the schedule of games between Division IA colleges of United States during the season Fall 2000 [5].

Teams are represented by vertices in the network and the regular season games between two teams are represented by edges. This dataset consist 115 vertices and 616 edges. The teams are divided into 12 different groups.

### 5.1.3 Bottlenose Dolphin Dataset

The network of bottlenose dolphins contains 62 vertices and 159 edges. Vertices represent dolphins and edges represent frequent association between animals.

In order to confirm that we are getting the natural clusters, we tested the proposed method with small (5 %) perturbation in the network. Initially on the selected real time datasets we applied the method and noted the number of clusters produced, purity of clusters and f-measure. We measure quality of clustering in two instances, in original instance and in the perturbed instance and confirm with the results that clustering remains consistent. This consistent clustering confirms detection of natural clusters. Another important quality measure for clustering is f-measure. In the experiments conducted, the f-measure values are also evaluated

**Table 1** Results of proposed method on three real world datasets with and without perturbation

| Dataset | Without perturbation (ground truth) | | | | | With 5 % perturbation | | | | |
|---|---|---|---|---|---|---|---|---|---|---|
| | No. of clusters | Purity of clusters | P | R | F | No. of clusters | Purity of clusters | P | R | F |
| Zachary karate club | 2 | 0.974 | 0.94 | 0.95 | 0.941 | 2 | 0.937 | 0.87 | 0.88 | 0.875 |
| Dolphin network | 2 | 0.962 | 0.92 | 0.87 | 0.902 | 2 | 0.924 | 0.88 | 0.85 | 0.865 |
| College football network | 12 | 0.942 | 0.90 | 0.84 | 0.868 | 12 | 0.89 | 0.86 | 0.78 | 0.818 |

*P* Precision, *R* Recall, *F* f measure

**Fig. 3** Comparison of f-measure values for both proposed and k-means algorithm

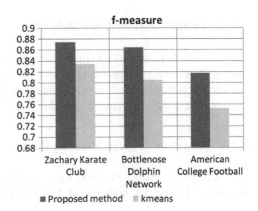

and reported (Table 1). In Fig. 3, the results of f-measure values (with 5 % perturbed data) are compared with k-means algorithm and show the effectiveness of the proposed method.

# 6 Conclusion and Future Work

A method is proposed for detecting natural communities in social networks in this paper. The critical aspect of the paper is the usage of Self Organizing Maps (SOM) for clustering which presumably work on the principle of neighborhood relationships and produce natural clusters. This approach is in contrast to brute force methods of clustering such as k-means clustering. The paper also emphasizes on the usage of modularity definition to identify strengths of communities. Modularity and subsequent eigenvector usage helps to operate on the data with reduced dimensionality. The detection of natural communities by this method is proved on the synthetic data as well as on real time data.

In future, the work can be improved by devising parallel and scalable architectures to deal with large datasets. Instead of eigenvectors and SOM, methods such as SVD and improved clustering algorithms can be used as part of the proposed method. The work can be extended to dynamically changing networks.

# References

1. Watts, D.J., Strogatz, S.H.: Collective dynamics of small-world networks. Nature **393**, 440–442 (1998)
2. Newman, M.E.J., Girvan, M.: Finding and evaluating community structure in networks. Phys. Rev. E **69**, 026113 (2004)
3. Ting, I.H.: Web mining techniques for online social networks analysis. In: Proceedings of International Conference on Service Systems and Service Management, pp. 1–5. Melbourne, June 30–July 2 (2008)
4. Hopcroft, J.E., Khan, O., Kulis, B., Selman, B.: Natural communities in large linked networks. In: Proceedings of International Conference on Knowledge Discovery and Data Mining, pp. 541–546 (2003)
5. Newman, M.E.J.: Modularity and community structure in networks. Proc. Natl. Acad. Sci. **103** (23), 8577–8582 (2006)
6. Newman, M.E.J.: Fast algorithm for detecting community structure in networks. Phys. Rev. E **69**, 066133 (2004)
7. Clauset, A., Newman, M.E.J., Moore, C.: Finding community structure in very large networks. Phys. Rev. E **70**, 066111 (2004)
8. White, S., Smyth, P.: A spectral clustering approach to finding communities in graph. In: SDM (2005)
9. Girvan, M., Newman, M.E.J.: Community structure in social and biological networks. Proc. Natl. Acad. Sci. **99**(12), 7821–7826 (2002)
10. Li, Z., Wang, R., Zhang, X.-S., Chen, L.: Self organizing map of complex networks for community detection. J. Syst. Sci. Complexity. **23**(5), 931–941 Springer-Verlag (2010)

11. Zachary, W.W.: An information flow model for conflict and fission in small groups. J. Anthropol. Res. **33**(4), 452–473 (1977)
12. Lusseau, D.: The emergent properties of a dolphin social network. Proc. R. Soc. B: Biol. Sci. **270**(2), S186–S188 (2003)

# A Low Complexity FLANN Architecture for Forecasting Stock Time Series Data Training with Meta-Heuristic Firefly Algorithm

D.K. Bebarta and G. Venkatesh

**Abstract** Prediction of future trends in financial time-series data related to the stock market is very important for making decisions to make high profit in the stock market trading. Typically the economic time-series data are non-linear, volatile and many other factors crash the market for local or global issues. Because of these factors investors find difficult to predict consistently and efficiently. The motive of designing a framework for predicting time series data is by using a low complexity, adaptive functional link artificial neural network (FLANN). The FLANN is basically a single layer structure in which non-linearity is introduced by enhancing the input pattern. The architecture of FLANN is trained with Meta-Heuristic Firefly Algorithm to achieve the excellent forecasting to increase the accurateness of prediction and lessen in training time. The projected framework is compared by using FLANN training with conventional back propagation learning method to examine the accuracy of the model.

**Keywords** Stock forecasting · Trading point prediction · FLANN · Meta heuristic firefly algorithm

## 1 Introduction

In an optimistic scenario gain of investing in stocks reflects an enormous growth in economy in a short duration. Even though it poses an obstacle of high risk the investors never stops taking risks. Stock market is dynamically changing and unpredictable environment. The major part is to know when to buy and when to

D.K. Bebarta (✉) · G. Venkatesh
Department of CSE, GMR Institute of Technology, Rajam, India
e-mail: bebarta.dk@gmrit.org

G. Venkatesh
e-mail: venkateshganena@gmail.com

© Springer India 2016
H.S. Behera and D.P. Mohapatra (eds.), *Computational Intelligence in Data Mining—Volume 1*, Advances in Intelligent Systems and Computing 410, DOI 10.1007/978-81-322-2734-2_38

sell. So, it is become a challenging task in the field and crucial research area in stock investment.

As the technology developed rapidly, it offers leverage of using many tools to solve problems in many fields which includes forecasting and prediction. Artificial Neural Network (ANN) [1] successfully applied to solve the currency exchange, stock market, and other areas of forecasting problems. ANN's in particular the Multilayer Perceptron (MLP) [2] are capable of producing multipart mapping between the input and the output in nonlinear decision boundaries. A major benefit of using ANNs is that it incorporates prior knowledge to improve the performance of stock market prediction. It also allows the adaptive adjustment to the model and nonlinear description of the problems. ANN [3–5] based models can handle a large number and variety of input variables, and the results are more robust. But ANN's are not efficient for producing effective result because of to choose the multiple layers input, output and hidden layers. This makes the computations complex in terms of time and space. An alternative approach is introduced called FLANN to avoid these type of problems. This structure removes the hidden layer as compared to MLP architecture to help in sinking the neural architectural complexity and provides a better representation of network without degrading the capable of executing a non-linear separable assignment.

The high-order neural network architecture without hidden units was introduced by expanding the input space into a higher dimensional space. The structure of FLANN, formerly established by Pao and Takefji [6] includes a single layer neural network without any hidden layers in 1992. By using FLANN architecture a single input can be enhanced with a set of polynomial functions across n-dimensional spaces, which are constructed from the original input patterns. This popular FLANN model is used in many applications like stock market forecasting [7], for prediction of foreign exchange rate [8], Dual Junction solar cells [9], wireless sensor networks [10], and etc. it has demonstrated its viability and robustness, and ease of computation in these fields.

In this piece of work, we have used the low complexity FLANN structure [11–14] for forecasting the stock data is trained by using both conventional approach i.e. back propagation learning method and meta-heuristic firefly optimization algorithm [15–17] inspired by the flashing behavior of fireflies. We have used three varieties of stock historical data IBM, ORACLE, and Gold Corp. as input.

## 2  Proposed Framework

### 2.1  FLANN Architecture

FLANN is a category of higher order neural networks and successfully used in many applications. In FLANN, the input vector is expanded with a rightfully enhanced representation of the input nodes, thereby synthetically escalating the

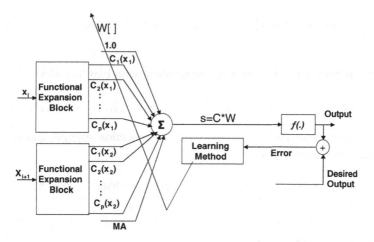

**Fig. 1** The FLANN structure

dimension of the input space. With the help of this enhanced input features are used as input in the input layer. Then the weighted sum of inputs is calculated and pass it through activation function to produce network output. Figure 1 depicts the FLANN structure with 2 inputs.

The normalized input is considered from stock data on closing price is stored in X as given in Eq. 1

$$X = [x_1, x_2, \ldots\ldots\ldots, x_{i-1}, x_i, x_{i+1}, \ldots\ldots\ldots, x_n] \tag{1}$$

Each input window considered from $X_k$ is $[x_i, x_{i+1}]$ is expanded by using Chebyshev polynomial basis function.

## 2.2 Chebyshev Polynomial Functions

Chebyshev Polynomials are given in Eqs. (2) and (3)

$$\begin{aligned} c_1(x) &= x, & c_2(x) &= 2x^2 - 1 \\ c_3(x) &= 4x^3 - 3x, & c_4(x) &= 8x^4 - 8x^2 + 1 \end{aligned} \tag{2}$$

The recursive advanced order Chebyshev Polynomial is

$$c_{t+1} = 2 * x * c_t(x) - c_{t-1}(x). \tag{3}$$

For input pattern given in Eq. 1
Calculation of Moving average is as follows

$$MA = (1.0 + c_1(x_1) + c_2(x_1) + \cdots + c_p(x_1) + c_1(x_2) + c_2(x_2) + \cdots + c_p(x_2))/13 \tag{4}$$

Output pattern derived from given input along with bias and MA

$$S = [1.0, c_1(x_i), c_2(x_i), \ldots, c_p(x_i), c_1(x_{i+1}), c_2(x_{i+1}), \ldots, c_p(x_{i+1}), MA]^T \tag{5}$$

The weight vector [W] in Eq. (6) used as connection strength and is initially initialized with values from −1.0 to +1.0.

$$W = [w_0, w_1, w_2, w_3, w_4, w_5, w_6, w_7, w_8, w_9, w_{10}, w_{11}, w_{12}, w_{13}] \tag{6}$$

## 2.3 Learning Methods

There are two learning methods used to update the weight vector [W] of the FLANN.

I. **The back propagation (BP) algorithm.**

$$o = \tanh(A) \tag{7}$$

The linear activation 'A' is as follows

$$A = S * W \tag{8}$$

The rule to revise the weight $W_{k+1}$ is as follows

$$w_{t+1} = w_t + e_t \frac{\partial o_t}{dw} \tag{9}$$

The $o = \tanh(A)$ function is used at the output node. So, the revised rule become

$$w_{t+1} = w_t + \alpha * e_t * (1 - o_t)^2 * x \tag{10}$$

II. **Meta-Heuristic Firefly Algorithm**

Input: Non-linear input pattern $S$ in Eq. 5 and its corresponding target, learningparameter, light absorption coefficient.

Output: Graph plotted for target and predicted values.

Begin: Generate a list of different weights. Eq. 6

Calculate error for weights. Here list of errors is considered as a performance index and each error value is treated as one firefly.

While True:

Find minimum error from error list and assign it to $f_j$ (brighter firefly)

While k < n

Find any value other than $f_j$ and assign it to $f_i$ (less brighter).

If ($f_j < f_i$):

Calculate the distance between $f_i$ and $f_j$ using below Eq.

$$d_{ij} = \|f_i - f_j\| = \sqrt{\sum (f_i - f_j)^2} \tag{11}$$

Move the firefly $f_i$ towards $f_j$ using below eq.

$$\nabla f_i = f_i + (L_0 e^{(-\eta d^2)})(f_j - f_i) + \alpha(rand - \tfrac{1}{2}) \tag{12}$$

Modify corresponding weight values using below eq.

$$w_{t+1} = w_t - \nabla f_i \tag{13}$$

End If

Now recalculate the error with new set of modified weights

Increase the value of K by one.

End While

Repeat the loop until optimized values are obtained.

End While

End

## 2.4 Time Series Indian Stock Data

The stock data like Oracle, Gold corp and IBM for analysis are used to validate our proposed model. There are 3000 trading days stock data used to analyze our planned model. The simulation results and performance measures are discussed to validate our model. Our forecasting system takes the normalized input from 0.0 to 1.0 which is calculated by using the equation given in Eq. (14).

$$u = (a - a_{\min})/(a_{\max} - a_{\min}). \tag{14}$$

where 'u' and 'a' are the normalized and actual data; $a_{\max}$ and $a_{\min}$ represents the high and low values among the actual data.

### I. Performance Measures

This accuracy evaluation is computed by using popular mean absolute percentage error (MAPE) method is defined in Eq. 15.

$$MAPE = \frac{1}{N}\sum_{j=1}^{N}[abs(e)/\bar{y}] \times 100, \quad \text{where} \quad \bar{y} = \frac{1}{N}\sum_{j=1}^{N}[y]. \quad (15)$$

where $e = y_f - y_a$, $y_f$ and $y_a$ are the forecast and actual values; $\bar{y}$ is the mean value of forecasting period; and N is the number of forecasted period.

## 2.5  Results and Analysis

The testing result of our planned forecasting model based on Back-propagation and Meta-Heuristic Firefly learning algorithm is revealed in below figures (Figs. 2, 3, 4, 5, 6 and 7). Performance comparison shown in Table 1.

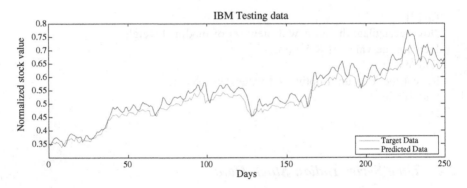

**Fig. 2**  IBM testing result using back-propagation learning algorithm

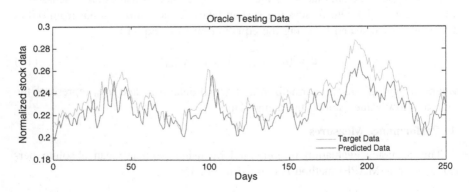

**Fig. 3**  Oracle testing result using back-propagation learning algorithm

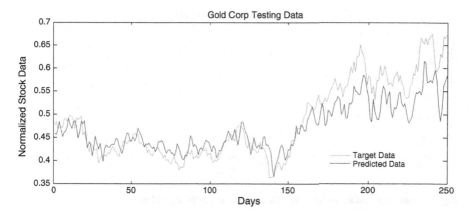

**Fig. 4** Gold corp testing result using back-propagation learning algorithm

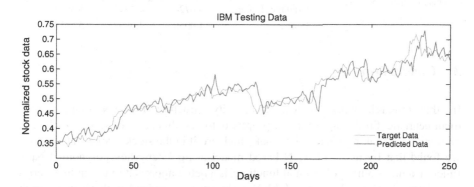

**Fig. 5** IBM testing result using meta-heuristic firefly algorithm

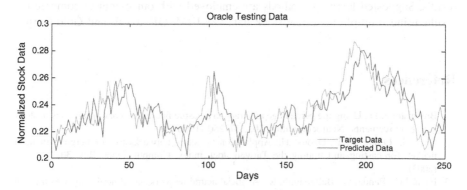

**Fig. 6** Oracle testing result using meta-heuristic firefly algorithm

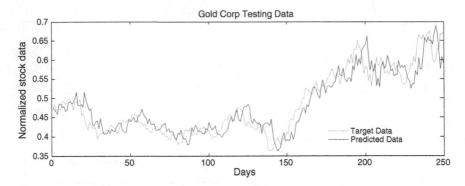

**Fig. 7** Gold corp testing result using meta-heuristic firefly algorithm

**Table 1** Performance comparison using MAPE

| Methods | IBM | Oracle | Gold corp. |
|---|---|---|---|
| Gradient descent | 4.6724 | 4.4491 | 5.8914 |
| Firefly | 4.0050 | 4.0891 | 5.1973 |

## 3   Conclusion

In this research work, a low complexity structure FLANN training with meta-heuristic firefly algorithm is proposed for nonlinear time series stock data prediction for making decisions to make high profit in the stock market. The studies we found that by using FLANN based model using Chebyshev polynomial basis function among other polynomial functions is a better approach to use in time series data forecasting. Further this FLANN architecture is trained with by using both conventional approach i.e. backpropagation learning method and meta-heuristic optimization algorithm inspired by the flashing behavior of fireflies to defend our results. Suggested learning methods are employed with our model to compare the results using multiple historical stock data like IBM, ORACLE, and Gold Corp.

## References

1. Brownstone, D.: Using the percentage accuracy to measure neural network predictions in stock market movements. Neurocomputing **10**, 237–250 (1996)
2. Chen, S., Leung, M.T., Daouk, H.: Application of neural networks to an emerging financial market: forecasting and trading the Taiwan stock index. Comput. Oper. Res. **30**, 901–923 (2003)
3. Patra, J.C., Panda, G., Baliarsingh, R.: Artificial neural network based nonlinearity estimation of pressure sensors. IEEE Trans. Instrum. Meas. **43**(6), 874–881 (1994)

4. Hao, H.-N.: Short-term forecasting of stock price based on genetic-neural network. In: 2010 Sixth International Conference on Natural Computation (ICNC 2010), IEEE Conference Publications

5. Bebarta, D.K., Biswal, B., Rout, A.K., Dash, P.K.: Efficient prediction of stock market indices using adaptive neural network. In: International Conference on Soft Computing for Problem Solving ScoPros-2011, Vol. 131, pp. 287–294. Springer (2012)

6. Pao, Y.H., Takefji, Y.: Functional-link net computing. IEEE Comput. J. 76–79 (1992)

7. Majhi, R., Panda, G., Sahoo, G.: Development and performance evaluation of FLANN based model for forecasting of stock markets. Experts Syst. Appl. Int. J. (2008). (Elsvier)

8. Sahu, K.K., Biswal, G.R., Sahu, P.K., Sahu, S.R., Behera, H.S.: A CRO based FLANN for forecasting foreign exchange rates using FLANN. In: Proceedings of the International Conference on CIDM, Springer (2014). doi:10.1007/978-81-322-2205-7_60

9. Patra, J.C.: Chebyshev neural network-based model for dual-junction sollar cells. IEEE Trans. Energy Convers. 26(1), (2011)

10. Patra, J.C., Bornand, C., Meher, P.K.: Laguerre neural network-based smart sensors for wireless sensor networks. In: I2MTC-2009, IEEE (2009)

11. Bebarta, D.K., Biswal, B., Dash, P.K.: Comparative study of stock market forecasting using different functional link artificial neural networks. Int. J. Data Anal. Tech. Strat. 4(4), 398–427 (2012)

12. Bebarta, D.K., Biswal, B., Rout, A.K., Dash, P.K.: Forecasting and classification of Indian stocks using different polynomial functional link artificial neural networks. In: INDCON, IEEE Conference, pp. 178–182. (2012). doi:10.1109/INDCON.2012.6420611

13. Bebarta, D.K., Biswal, B., Rout, A.K., Dash, P.K.: Dynamic recurrent FLANN based adaptive model for forecasting of stock indices. In: Emerging ICT for Bridging the Future-Proceedings of the 49th Annual Convention of the Computer Society of India CSI, vol. 2, pp. 509–516. Springer International Publishing (2015)

14. Bebarta, D.K., Sudha, T.E., Bisoyi, R.: An intelligent stock forecasting system using a unify model of CEFLANN, HMM and GA for stock time series phenomena. In: Emerging ICT for Bridging the Future-Proceedings of the 49th Annual Convention of the Computer Society of India CSI, vol. 2, pp. 485–496. Springer International Publishing (2015)

15. Kazem, A., Sharifi, E., Hussain, F.K., Saberi, M., Hussain, O.M.: Support vector regression with chaos-based firefly algorithm for stock market price forecasting. Appl. Soft Comput. 13, 947–958 (2013). (Elsevier)

16. Yafei, T., Weiming, G., Shi, Y.: An improved inertia weight firefly optimization algorithm and application. In: IEEE International Conference on Control Engineering and Communication Technology (2012). doi:10.1109/ICCECT.2012.38

17. dos Santos Coelho, L., Klein, C.E., Luvizotto, L.J., Mariani, V.C.: Firefly approach optimized wavenets applied to multivariable Identification of a thermal process. In: IEEE EuroCon, Zagreb, Croatia (2013)

# A Novel Scheduling Algorithm for Cloud Computing Environment

Sagnika Saha, Souvik Pal and Prasant Kumar Pattnaik

**Abstract** Cloud computing is the most recent computing paradigm, in the Information Technology where the resources and information are provided on-demand and accessed over the Internet. An essential factor in the cloud computing system is Task Scheduling that relates to the efficiency of the entire cloud computing environment. Mostly in a cloud environment, the issue of scheduling is to apportion the tasks of the requesting users to the available resources. This paper aims to offer a genetic based scheduling algorithm that reduces the waiting time of the overall system. However the tasks enter the cloud environment and the users have to wait until the resources are available that leads to more queue length and increased waiting time. This paper introduces a Task Scheduling algorithm based on genetic algorithm using a queuing model to minimize the waiting time and queue length of the system.

**Keywords** Cloud computing · Scheduling · Genetic algorithm · Queuing model · Waiting length

## 1 Introduction

The scheduling of tasks successfully has turned out to be one of the problem areas in the field of Computer Science. The aim of the scheduler in a cloud computing environment is to determine a proper assignment of resources to the tasks to cease all the tasks received from the users. Vast numbers of users submit their tasks to the

S. Saha (✉) · S. Pal · P.K. Pattnaik
School of Computer Engineering, KIIT University, Bhubaneswar, India
e-mail: sagnika10@gmail.com

S. Pal
e-mail: souvikpal22@gmail.com

P.K. Pattnaik
e-mail: patnaikprasantfcs@kiit.ac.in

© Springer India 2016                                                                387
H.S. Behera and D.P. Mohapatra (eds.), *Computational Intelligence
in Data Mining—Volume 1*, Advances in Intelligent Systems
and Computing 410, DOI 10.1007/978-81-322-2734-2_39

cloud system by sharing Cloud resources. Subsequently, scheduling these large numbers of tasks turns into a challenging issue in the environment of cloud computing. The principle target of Cloud Computing is to execute the user needs as per Quality of Service (QoS) and to enhance the cloud provider's profit. To accomplish these, better algorithms for task scheduling are expected to schedule different user tasks since a good scheduling algorithm minimizes total computation time and the entire cost associated with it. An efficient scheduling algorithm is one that improves the overall system performance.

Genetic Algorithm (GA) is a heuristic search algorithm based on the principle of natural selection and evaluation that gives an optimal solution. The above problem may be solved using Genetic Algorithm. GAs can figure out the optimal task sequence that is to be designated to the resources.

In this paper, a genetic based scheduling algorithm has been developed that minimizes the waiting time and furthermore reduces the queue length of the overall system. The rest of the paper is organized as follows. Section 2 spotlights on related work; in Sect. 3 the proposed model is depicted; in Sect. 4 the performance analysis of the problem is presented. The last part contains the conclusion and future work.

## 2  Related Work

The Scheduling of task in the cloud has been a well known issue in both academic and industrial spheres. A good scheduling algorithm won't just raise the utilization of resources additionally satisfy the requirements of the users. It is important to deal with these resources in such a way that resources are properly used and the waiting time for resources decreases. For proper scheduling of tasks many algorithms are available as well as methods in cloud computing. The following identifies some of the related works done with scheduling and queuing model:

Snehal Kamalapur, Neeta Deshpande in paper [1] proposed a GA based algorithm for process scheduling. GA is used as a function of process scheduling to produce effective results. The proposed technique gives better results against other traditional algorithms.

Luqun Li in paper [2] presented a non pre-emptive priority M/G/1 queuing model after analysing QoS requirements of Cloud Computing user's jobs. The goal is to find the optimal result for each job with different priority.

Chenhong Zhao, Shanshan Zhang, Qingfeng Liu, Jian Xe, Jicheng Hu in paper [3] focused on an optimization algorithm in light of Genetic Algorithm which will schedule tasks in adaptation to memory constraints and performance.

Yujia Ge, Guiyi Wei in paper [4] displayed a new task scheduler taking into account Genetic algorithm for the Cloud Computing Systems in Hadoop MapReduce. After evaluation of the entire tasks in the queue, the proposed technique makes a new scheduling decision. Genetic Algorithm is applied as an optimization method for the new scheduler. The performance analysis demonstrates that the new scheduler attains a better make span for tasks against FIFO scheduler.

S. Selvarani, Dr. G. Sudha Sadhasivam in paper [5] proposed an improved cost based scheduling algorithm to schedule tasks in a productive way. This algorithm doesn't just measure the computation power and resource cost additionally upgrades the computation ratio by grouping the tasks of the users.

Gan Guo-ning, Huang Ting-Iei, GAO Shuai in paper [6] developed a task scheduling algorithm based on genetic simulated annealing algorithm considering Quality of Service (QoS) requirements of different tasks.

Eleonora Maria Mocanu, Mihai Florea in paper [7] proposed a scheduler in view of genetic algorithm that improves Hadoop's functionality. Hadoop has several task schedulers as FIFO, FAIR, and Capacity Schedulers however; none of them reduces the global execution time. The goal of this report is to improve Hadoop's functionality that prompts a better throughput.

Hamzeh Khazaei, Jelena Misic, Vojislav B. Misic in paper [8] built up a model on a M/G/m/m + r queuing system where single task arrives and the task buffer has a finite capacity. This model obtains a probability distribution of waiting and response time and no. of tasks in the system.

Jyotirmay Patel, A.K. Solanki in paper [9] suggested a hybrid scheduling algorithm using genetic approaches for CPU scheduling since the genetic algorithm gives efficient results. Then it is compared with other algorithms and finds out the minimum waiting time.

Pardeep Kumar, Amandeep Verma in paper [10] proposed a scheduling algorithm in which Min-Min and Max-Min algorithm is combined with Genetic algorithm. How to allocate the requests to the resources is a difficult issue in scheduling of the user's tasks and this algorithm finds out the minimum time required by the requested tasks to complete.

Hu Baofang, Sun Xiuli, Li Ying, Sun Hongfeng in paper [11] proposed an improved scheduling algorithm on adapted genetic algorithm PAGA based on priority. This model brings down the execution time and guarantees Qos requirements of users. Here the fitness function is projected in an idealistic way that reduces several iterations.

H. Kamal Idrjssi, A. Quartet, M. El Marraki [12] studied the underlying ideas of cloud computing that incorporates cloud service models, cloud deployment models, subject area of cloud products and cloud protection and secrecy.

Xiaonian Wu, Mengqing Deng, Runlian Zhang, Bing Zeng, Shengyuan Zhou in paper [13] proposed an optimizing algorithm based on QoS in Cloud Computing systems (TS-QoS). In this method, the tasks are arranged by their precedence. The tasks are mapped on the resources with minimum completion time.

Randeep in paper [14] produced a genetic algorithm for efficient process scheduling. This algorithm finds out minimum waiting time is using genetic algorithm and afterward with other algorithms as FCFS and SRTF.

R. Vijayalakshmi, Soma Prathibha in paper [15] presented a scheduling algorithm where the Virtual Machines (VMs) are allocated to tasks based on priority. The tasks are mapped to VM after the tasks are organized by their priority. With the help of CloudSim toolkit, this entire model is simulated. The test result indicates that the projects are assigned efficiently and the execution time also minimizes.

Ge Junwei, Yuan Yongsheng in paper [16] presented a Genetic Algorithm that considers 3 constraints, i.e. total task completion time, average task completion time and cost. The algorithm enhances task scheduling and resource allocation and maximizes efficiency of the system.

S. Sindhu, Dr. Saswati Mukherjee in paper [17] proposed a scheduling algorithm that is in view of Genetic algorithm that is applicable for application centric and resource centric. The proposed procedure tries to improve make span and average processor utilization.

S. Devipriya, C. Ramesh in paper [18] enhanced Max-Min algorithm in light of RASA algorithm. The primary aim of this algorithm is to allocate the tasks to the resources with maximum execution time that will result in minimum completion time against the original Max-Min algorithm.

# 3 Proposed Model

The focus of the system is to have a maximum usage of resources and to decrease the waiting time and queue length of the entire system. The proposed model of scheduling environment is demonstrated in Fig. 1. Assume Cloud users send n number of tasks $\{T_1, T_2, T_3...T_n\}$ for the resources and these requests from various users are at first stored into the buffer. The controller then apportions these tasks to the proper resources. The task queue is structured by mapping the tasks to the resources. In this paper, FCFS and GA are used as the scheduling algorithms and these algorithms are applied over the task queue. The aim is to discover the right scheduling order that lessens the waiting time of the system. Next the scheduling orders are recovered both for FCFS and GA that minimize the waiting time. The queuing model is then applied over the scheduling orders that are retrieved through FCFS and GA algorithms. It is used to minimize the queue length as well as waiting time of the tasks. It is found that GA offers better results against FCFS.

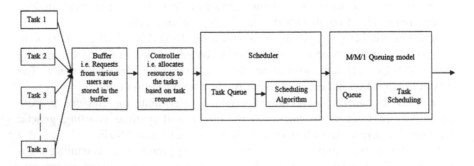

**Fig. 1** A scenario of task to scheduler

Presently, the proposed algorithm is discussed step by step:

(a) Cloud users send n number of tasks to the buffer for resources.
(b) Keep the record of the Burst time range of the tasks.
(c) Then, permute the burst time of the tasks to the number of possible ways.
(d) Now, find the minimum waiting time by applying both FCFS and GA algorithm to each of the permuted sequence.
(e) Next, choose the sequence with minimum waiting time that is discovered using FCFS and GA.
(f) Apply queuing model on the sequences with minimum waiting time.

The input here is the n number of tasks sent by the cloud users and output provides the comparative analysis between FCFS and GA using queuing model that reduces the waiting time of the overall system.

## 4  Tools for Experimental Environments and Result Analysis

GA was initially developed by John Holland in 1975. GA is a search heuristic method, taking into account the process of natural selection and evaluation. This heuristic method is used to generate optimized solutions. A genetic algorithm first begins with a set of tasks that are known as initial population to find out an optimal solution. The tasks are chosen from the initial population and certain operations are performed to form the next generation. A fitness function is used to find an optimal solution for the problem under consideration. In this paper, the fitness of tasks finds the minimum average waiting time, and the one with the minimum value is thought to be the fittest as compared to the others.

The fitness function of a solution $S_r$ is given by,

$$\text{Fitness } (S_r) = \frac{\sum_{i=1}^{N} Wti}{N} \tag{1}$$

(i = 1, 2, 3… N) where $Wt_i$ is the waiting time of the task Sr and N is the total no of tasks.

Roulette wheel is used as a random selection process. Each task is assigned a slot size in proportion to its fitness of the roulette wheel.

The probability of each task is calculated as:

$$P[i] = \frac{Fitness(S_r)}{TotalFitness(S_r)} \tag{2}$$

where Fitness($S_r$) is the fitness function of a solution and TotalFitness($S_r$) is the summation of all fitness functions.

The ordered crossover is applied in this case. Two random crossover points are chosen for partitioning from two parent tasks and divided into left, middle and right portions. The ordered crossover is carried out in the following way. The left and right portions remain unchanged and the middle portion's strings interchange.

Mutation is a process of swapping the position of two genes. Two points are selected from the given tasks and are swapped to get the new child. After applying all the genetic operators on the selected parent, one new child is created. At that point this new child is added to the existing population.

Queuing model is a mathematical theory that deals with managing and providing a service on a queue or on a waiting time. It happens when enough service capacity is not provided that causes the users to wait. The queuing model is recommended by specifying the arrival process of users, service process, no of servers and server capacity. Here, queuing model is used to reduce queue length and waiting time. Poison distribution is taken into consideration as arrival patterns of the users. $\lambda$ is taken as an estimated value for this distribution. The time taken between the start of a service and to its completion is known as service time.

Let $S_i$ be the service time of the ith user. So, the mean or average service time will be

$$E(S) = \frac{\sum_{i=0}^{n} Si}{n} \tag{3}$$

where n is the number of users.

The service rate will be calculated as

$$\mu = \frac{1}{E(S)} \tag{4}$$

The condition provided for making a system stable is that the Utilization factor should be

$$\rho = \frac{\lambda}{\mu} \leq 1. \tag{5}$$

Individual solutions are generated arbitrarily to form an initial population. Crossover creates new population. The fittest solutions are chosen by the parents to reproduce the offspring for the new population. The fitness function is characterized by taking into FCFS to achieve minimum waiting time.

N, no of tasks are sent by the Cloud users for the resources to the request queue, for example $T_1$, $T_2$,...Tn. Consider n no of tasks that are ready to execute, the possible no of ways of performing tasks are n!. In this paper, we have taken 4 tasks that are ready to execute, the possible no of ways are 4! or 24 ways. Let the burst time of the processes are $T_1 = 0.015$, $T_2 = 0.008$, $T_3 = 0.019$, $T_4 = 0.002$.

Table 1 demonstrates the calculation of minimum waiting time by FCFS and GA. The result shows that GA can reduce the waiting time of the system.

**Table 1** Calculation of minimum waiting time for FCFS and GA

| Serial no. | Tasks (T₁, T₂, T₃, T₄) | F(i) of FCFS | P(i) | CP(i) | New chromosome | Crossover | Mutation | F(i) of GA |
|---|---|---|---|---|---|---|---|---|
| 1 | 1, 2, 3, 4 | 0.020 | 0.050 | 0.050 | 3, 4, 1, 2 | 3, 1, 4, 2 | 3, 2, 4, 1 | 0.019 |
| 2 | 2, 1, 3, 4 | 0.018 | 0.045 | 0.095 | 3, 2, 4, 1 | 3, 4, 2, 1 | 3, 1, 2, 4 | 0.024 |
| 3 | 3, 1, 2, 4 | 0.024 | 0.060 | 0.155 | 4, 3, 1, 2 | 4, 1, 3, 2 | 4, 2, 3, 1 | 0.010 |
| 4 | 4, 1, 2, 3 | 0.011 | 0.027 | 0.182 | 2, 3, 4, 1 | 2, 4, 3, 1 | 2, 1, 3, 4 | 0.018 |
| 5 | 1, 3, 2, 4 | 0.023 | 0.057 | 0.239 | 1, 2, 3, 4 | 1, 3, 2, 4 | 1, 4, 2, 3 | 0.014 |
| 6 | 1, 4, 3, 2 | 0.017 | 0.042 | 0.281 | 2, 4, 1, 3 | 2, 1, 4, 3 | 2, 3, 4, 1 | 0.016 |
| 7 | 1, 2, 4, 3 | 0.016 | 0.040 | 0.321 | 2, 3, 4, 1 | 2, 4, 3, 1 | 2, 1, 3, 4 | 0.018 |
| 8 | 4, 3, 2, 1 | 0.018 | 0.045 | 0.366 | 3, 1, 2, 4 | 3, 2, 1, 4 | 3, 4, 1, 2 | 0.019 |
| 9 | 3, 2, 1, 4 | 0.022 | 0.055 | 0.421 | 3, 1, 4, 2 | 3, 4, 1, 2 | 3, 2, 1, 4 | 0.022 |
| 10 | 3, 4, 2, 1 | 0.017 | 0.042 | 0.463 | 1, 4, 3, 2 | 1, 3, 4, 2 | 1, 2, 4, 3 | 0.016 |
| 11 | 3, 4, 1, 2 | 0.019 | 0.047 | 0.51 | 1, 3, 4, 2 | 1, 4, 3, 2 | 1, 2, 3, 4 | 0.020 |
| 12 | 2, 3, 1, 4 | 0.019 | 0.047 | 0.557 | 4, 2, 1, 3 | 4, 1, 2, 3 | 4, 3, 2, 1 | 0.018 |
| 13 | 2, 3, 4, 1 | 0.016 | 0.040 | 0.597 | 1, 3, 2, 4 | 1, 2, 3, 4 | 1, 4, 3, 2 | 0.017 |
| 14 | 1, 3, 4, 2 | 0.021 | 0.052 | 0.649 | 1, 4, 2, 3 | 1, 2, 4, 3 | 1, 3, 4, 2 | 0.021 |
| 15 | 3, 2, 4, 1 | 0.019 | 0.047 | 0.696 | 3, 2, 1, 4 | 3, 1, 2, 4 | 3, 4, 2, 1 | 0.017 |
| 16 | 4, 1, 3, 2 | 0.014 | 0.035 | 0.731 | 2, 1, 4, 3 | 2, 4, 1, 3 | 2, 3, 1, 4 | 0.019 |
| 17 | 4, 2, 3, 1 | 0.010 | 0.025 | 0.756 | 4, 1, 2, 3 | 4, 2, 1, 3 | 4, 3, 1, 2 | 0.015 |
| 18 | 4, 2, 1, 3 | 0.009 | 0.022 | 0.778 | 3, 4, 2, 1 | 3, 2, 4, 1 | 3, 1, 4, 2 | 0.022 |
| 19 | 2, 4, 3, 1 | 0.012 | 0.030 | 0.808 | 2, 1, 3, 4 | 2, 3, 1, 4 | 2, 4, 1, 3 | 0.011 |
| 20 | 2, 4, 1, 3 | 0.011 | 0.027 | 0.835 | 4, 1, 3, 2 | 4, 3, 1, 2 | 4, 2, 1, 3 | 0.009 |
| 21 | 1, 4, 2, 3 | 0.014 | 0.035 | 0.87 | 1, 2, 4, 3 | 1, 4, 2, 3 | 1, 3, 2, 4 | 0.023 |
| 22 | 3, 1, 4, 2 | 0.022 | 0.055 | 0.925 | 2, 4, 3, 1 | 2, 3, 4, 1 | 2, 1, 4, 3 | 0.014 |
| 23 | 2, 1, 4, 3 | 0.014 | 0.035 | 0.96 | 4, 3, 2, 1 | 4, 2, 3, 1 | 4, 1, 3, 2 | 0.014 |
| 24 | 4, 3, 1, 2 | 0.015 | 0.037 | 1 | 4, 2, 3, 1 | 4, 3, 2, 1 | 4, 1, 2, 3 | 0.011 |

Accordingly the particular sequence that minimizes the waiting time must be stored into the buffer queue. The sequence that reduces the waiting time of the overall system is now used as a part of queuing model to find the service rate. In GA the sequence 4, 1, 3, 2 give the minimum waiting time. Furthermore, in case of FCFS we have taken the sequence 1, 2, 3, 4 since it is basically a first come first served algorithm.

For queuing system, the server has two parts, i.e. S1 and S2, and these two parts are sequentially arranged. It is to be noted that when one task is executing in one part then that same task cannot execute in another part. We assume that one task is an entity, i.e. one task can be executed in one and only part at a same time. The two data centres will be executed alternatively (Figs. 2, 3).

We approve our queuing model by using a different stream of arrival rates, $\lambda = 6$, 1, 22, 25, 34 and service rates, $\mu = 38.09$, 40.40 which are arranged in Tables 2 and 3. Here the M/M/1 queuing model is used.

The graphical representations of the outcomes are presented in Figs. 4, 5, 6 and 7.

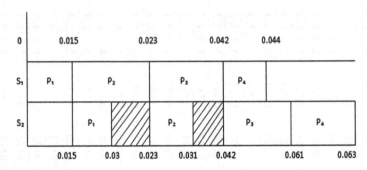

**Fig. 2** The gantt chart of FCFS algorithm with mean service time E(S) = 0.02625 and service rate $\mu = 38.09$

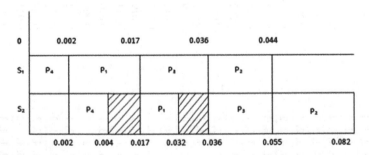

**Fig. 3** The gantt chart of genetic algorithm (GA) with mean service time E(S) = 0.02475 and service rate $\mu = 40.40$

**Table 2** The queue length and waiting time using FCFS $\mu = 38.09$

|  | Lq | Ls | Wq | Ws |
|---|---|---|---|---|
| $\lambda = 6$ | 0.03 | 0.19 | 0 | 0.03 |
| $\lambda = 14$ | 0.21 | 0.58 | 0.02 | 0.04 |
| $\lambda = 22$ | 0.79 | 1.37 | 0.04 | 0.06 |
| $\lambda = 25$ | 1.25 | 1.91 | 0.05 | 0.08 |
| $\lambda = 34$ | 7.42 | 8.31 | 0.22 | 0.24 |

**Table 3** The queue length and waiting time using GA $\mu = 40.40$

|  | Lq | Ls | Wq | Ws |
|---|---|---|---|---|
| $\lambda = 6$ | 0.03 | 0.17 | 0 | 0.03 |
| $\lambda = 14$ | 0.18 | 0.53 | 0.01 | 0.04 |
| $\lambda = 22$ | 0.65 | 1.2 | 0.03 | 0.05 |
| $\lambda = 25$ | 1 | 1.62 | 0.04 | 0.06 |
| $\lambda = 34$ | 4.47 | 5.31 | 0.13 | 0.16 |

**Fig. 4** The average number of customers in the queue(Lq) using FCFS and GA

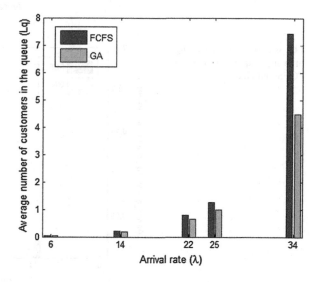

**Fig. 5** The average number of customers in the system (Ls) using FCFS and GA

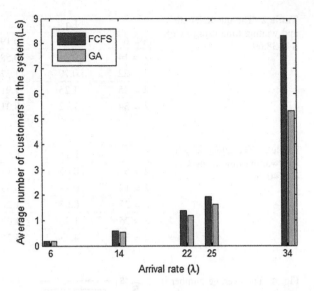

**Fig. 6** The average waiting time in the queue(Wq) using FCFS and GA

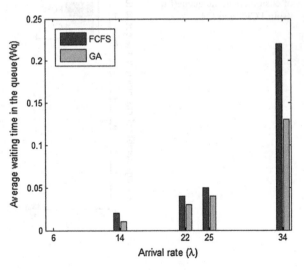

**Fig. 7** The average waiting time in the system(Ws) using FCFS and GA

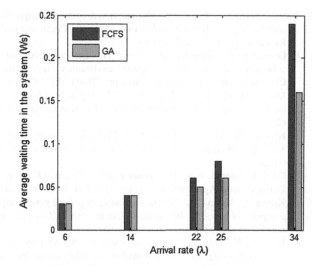

## 5 Conclusion

This paper proposes a hybrid approach for task scheduling algorithm for the cloud environment with the combination of Genetic Algorithm (GA) and Queuing model as a tool. This algorithm reduces the waiting time and queue length for satisfying user requirements where GA is used to minimize the waiting time and the queuing model is used to reduce both the queue length and waiting time. A comparative analysis between the FCFS and GA algorithm is introduced taking into account simulation. The simulation outcomes show that the Genetic Algorithm approach gives 20 % better results against FCFS. Genetic Algorithm and Queuing model approaches has been conveyed for reducing both queue length and waiting time. For future work, this algorithm can be deployed on batch processing that may prompt to good scheduling decisions.

## References

1. Kamalapur, S., Deshpande, N.: Efficient CPU scheduling: a genetic algorithm based approach. In: International Symposium on Ad Hoc and Ubiquitous Computing, pp. 206–207. IEEE (2006)
2. Li, L.: An optimistic differentiated service job scheduling system for cloud computing service users and providers. In: Third International Conference on Multimedia and Ubiquitous Engineering, pp. 295–299. IEEE (2009)
3. Zhao, C., Zhang, S., Liu, Q., Xie, J., Hu, J.: Independent tasks scheduling based on genetic algorithm in cloud computing. In: 5th International Conference on Wireless Communications, Networking and Mobile Computing, pp. 1–4. IEEE (2009)
4. Ge, Y., Wei, G.: GA-based task scheduler for the cloud computing systems. In: International Conference on Web Information Systems and Mining, vol. 2, pp. 181–186. IEEE (2010)

5. Selvarani, S., Sadhasivam, G.S.: Improved cost-based algorithm for task scheduling in cloud computing. In: International Conference on Computational Intelligence and Computing Research, pp. 1–5. IEEE (2010)
6. Guo-ning, G., Ting-Iei, H., Shuai, G.: Genetic simulated annealing algorithm for task scheduling based on cloud computing environment. In: International Conference on Intelligent Computing and Integrated Systems, pp. 60–63. IEEE (2010)
7. Mocanu, E.M., Florea, M., Andreica, M., Tapus, N.: Cloud computing—task scheduling based on genetic algorithms. In: International Conference on System Conference, pp. 1–6. IEEE (2012)
8. Khazaei, H., Misic, J., Misic, V.B.: Performance analysis of cloud computing centers using M/G/M/M+R queuing systems. IEEE Trans. Parallel Distrib. Syst. **23**, 936–943 (2012). (IEEE)
9. Patel, J., Solanki, A.K.: Performance Enhancement of CPU Scheduling by Hybrid Algorithms Using Genetic Approach, vol. 1, pp. 142–144. IJARCET (2012)
10. Kumar, P., Verma, A.: Scheduling using improved genetic algorithm in cloud computing for independent tasks. In: International Conference on Advances in Computing, Communications and Informatics, pp. 137–142. ACM (2012)
11. Baofang, H., Xiuli, S., Ying, L., Hongfeng, S.: An improved adaptive genetic algorithm in cloud computing. In: 13th International Conference on Parallel and Distributed Computing, Applications and Technologies, pp. 294–297. IEEE (2012)
12. Idrissi, H.K., Kartit, A., Marraki, M.: A taxonomy and survey of cloud computing. In: National Security Days, pp. 1–5. IEEE (2013)
13. Wu, X., Deng, M., Zhang, R., Zeng, B., Zhou, S.: A task scheduling algorithm based on QoS driven in cloud computing. In: First Conference on Information Technology and Quantitative Management, vol. 17, pp. 1162–1169. Elsevier (2013)
14. Randeep.: Processor scheduling algorithms in environment of genetics. Int. J. Adv. Res. Eng. Technol. **1**, 14–19 (2013). (IJARET)
15. Vijayalakshmi, R., Prathibha, S.: A novel approach for task scheduling in cloud. In: Fourth International Conference on Computing, Communications and Networking Technologies, pp. 1–5. IEEE (2013)
16. Junwei, G., Yongsheng, Y.: Research of cloud computing task scheduling algorithm based on improved genetic algorithm. In: 2nd International Conference on Computer Science and Electronics Engineering, pp. 2134–2137. Atlantis Press (2013)
17. Sindhu, S., Mukherjee, S.: A genetic algorithm based scheduler for cloud environment. In: 4th International Conference on Computer and Communication Technology, pp. 23–27. IEEE (2013)
18. Devipriya, S., Ramesh, C.: Improved max-min heuristic model for task scheduling in cloud. In: International Conference on Green Computing, Communication and Conservation of Energy, pp. 883–888. IEEE (2013)

# PSO Based Tuning of a Integral and Proportional Integral Controller for a Closed Loop Stand Alone Multi Wind Energy System

**L.V. Suresh Kumar, G.V. Nagesh Kumar and D. Anusha**

**Abstract** The summary of the paper explains the optimal tuning of integral (I) and proportional integral (PI) controllers are applied to closed loop standalone integrated multi wind energy system by using particle swarm optimization. Tuning of I and PI controller gain values obtained from the optimization techniques to get the best possible operation of the system. For the optimal performance of the integrated wind energy system, the controller gains are tuned by using the PSO and genetic algorithms (GA). The system harmonics of voltage responses are observed with search heuristic algorithm that is nothing but a genetic algorithm. Similarly the system responses are observed and compared with PSO algorithm, and the PSO algorithm is proved better. The results establishes the proposed new stand alone multi wind energy system with I, PI controller gains are tuned by using PSO will gives less harmonic distortion and improves performance. The proposed system is developed in MATLAB/SIMULINK.

**Keywords** PSO · Wind energy system · Proportional integral controller

## 1 Introduction

The wind energy capacity generation has been increasing rapidly and became the fastest growing technology compared to the other non conventional energy sources. Most of the wind turbines capture the wind flow to generate kinetic energy to mechanical energy with the use of turbine blades [1–8].

L.V.S. Kumar (✉) · D. Anusha
Department of EEE, GMR Institute of Technology, Rajam, Andhra Pradesh, India
e-mail: lvenkatasureshkumar@gmail.com

D. Anusha
e-mail: daki.anusha92@gmail.com

G.V.N. Kumar
Department of EEE, GITAM University, Vishakapatnam, Andhra Pradesh, India
e-mail: drgvnk14@gmail.com

© Springer India 2016
H.S. Behera and D.P. Mohapatra (eds.), *Computational Intelligence in Data Mining—Volume 1*, Advances in Intelligent Systems and Computing 410, DOI 10.1007/978-81-322-2734-2_40

399

In modeling of closed loop stand alone integrated wind energy system it has controller block to control the voltage by using controllers. The controller is a feedback control technology to control the voltage and it gives better performance in renewable energy systems. The function of the controller is to operate with an algorithm and to maintain the output at a constant level for a given input so that there is no difference between the operating variable and the given input variable [2]. Due to the advantage of functional simplicity and reliability the popularity of Integral and PI controllers are increasing. These controllers provides reliable and robust operation performances for most of the systems and their gains are tuned to get a satisfactory closed loop performance of the system [2–5] and also improve the harmonic response of a system by controlling the voltage, and gives accurate performance of the system. For control Integral and PI parameters there is need to use optimization algorithms. To control the system operation properly then the Integral and PI control loop must be properly tuned. The optimization methods for tuning these controllers are PSO and GA techniques, and these are discussed in this paper [6, 7].

These methods are proposed for tuning the controllers. The Integral and PI controllers including in the system is necessary for the control the performance of minimum value to maximum value, and the parameters are must be tuned by using tuned algorithms and the controller performance are done with number of iterations. The process of tuning a controller is to set the proportional and integral gain values to get the best possible performance of a taken system [9–12].

In this paper, gain values are optimized by GA and PSO. First the initial gains are applied and are repeated by random, after that simulation has to be done. Using the PSO, GA to optimize the gain values for getting less level of harmonic distortion. The GA is operated by the predetermined iteration and each given iteration of a system. Gain of a controller was re-optimized by PSO algorithm is to reduce the search area of the GA algorithm and gives better optimization of gain values in each and every iteration [13]. The result shows for the proposed closed loop system and the system performance with PSO is better than the GA in terms of obtained output and number of iterations taken.

## 2 Modeling of Closed Loop Stand Alone Multi Wind Energy System

The proposed modeling of closed loop stand alone multi wind energy system is shown in Fig. 1. Here three wind energy systems taken as an input source. These wind energy systems produce mechanical energy and this is converted into electrical energy by connecting through induction generators. From the three systems nine phase electrical voltage is generated this voltage is converted to three phase voltage by using new proposed star/delta nine phase to three phase multi winding transformer [8, 13]. After this connection variable v/f coming from the transformer

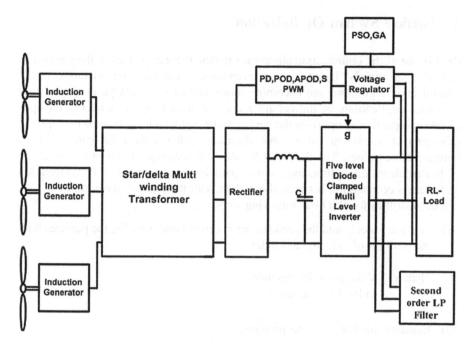

**Fig. 1** Configuration of closed loop stand alone multi wind energy system

is converted in fixed v/f by using back to back power electronic converter devices. For getting less harmonic distortion multi level inverter were introduced into the system. This voltage will be connected to the loads. Hence for controlling the voltage the feedback voltage will be taken from load side and this will be summed with generation side voltage and this will be controlled through integral and PI controllers and these will be generated sinusoidal voltage and it can be applied as a modulation index in pulse generation sections. Here the controllers are tuned to get optimum value by applying the particle swarm optimization algorithm, genetic algorithm. At optimum gain value of a controller then the system performance will get less harmonic distortion [8–10, 13].

$$V_a = L_{aa}\frac{di_a}{dt} + L_{ab}\frac{di_b}{dt} + \cdots\cdots\cdots + L_{an}\frac{di_n}{dt} \tag{1}$$

$$V_b = L_{ba}\frac{di_a}{dt} + L_{bb}\frac{di_b}{dt} + \cdots\cdots\cdots + L_{bn}\frac{di_n}{dt} \tag{2}$$

$$V_n = L_{na}\frac{di_a}{dt} + L_{nb}\frac{di_b}{dt} + \cdots\cdots\cdots + L_{nn}\frac{di_n}{dt} \tag{3}$$

## 3   Particle Swarm Optimization

PSO is one of the optimization algorithms to tune the gain values of the controllers and this is mainly based on an evolutionary computation methods [1–3]. Optimization algorithms are becoming more interest in recent years due to their advanced applications in the certain areas and industries. It is mainly an optimization algorithm to optimize the perfect value and which is obtained by applying minimum to maximum values, this algorithm will optimize the correct values between these values. The flowchart for the PSO technique is given in Fig. 2.

In Particle swarm optimization, it uses particles these are evolved by cooperation and make a competition between themselves with the pass of generations.

An algorithm for PSO has following steps:

(1)  Start the process and the particles are generated and initialize the particles with the position of velocity and index.

   Here $V_i$   Velocity in ith iteration
   $X_i$       Index in ith iteration

(2)  Evaluate the fitness of the problem.

   (a)   If the fitness of the index is greater than particle best fitness then execute the best particle fitness is equal to index.
   (b)   If the fitness of the index is greater than in all the particles then execute best particle fitness is equal to index.

Here $X_i = (x_{i1}, x_{i2}, ..., x_{iD})$ and $P_i = (p_{i1}, p_{i2}, ..., p_{iD})$, is named as $p_{best}$ and $g_{best}$ is the index of best particle among all the particles in the population area. The velocity of the particle $i$ is represented as $V_i = (v_{i1}, v_{i2}, ..., v_{iD})$. These particles are added to the system with the following Eqs. (4) and (5).

$$V_{id}^{n+1} = w \cdot v_{id}^{n} + c_1 \cdot \text{rand}\,(1...n) \cdot (p_{id}^{n} - x_{id}^{n}) \qquad (4)$$

$$X_{id}^{n+1} = x_{id}^{n} + v_{id}^{n+1} \qquad (5)$$

where $c_1$ and $c_2$ are two positive constant values.

In PSO algorithm it has the population with number of particles each particle is a dimensional valued vector, with number of optimized parameters. The PSO can be simply explained with the steps of (1) Initialization (2) Time updating (3) Weight updating (4) Velocity updating.

The tuned values of the controller parameters $K_p$, $K_i$ is obtained by using PSO algorithm are given in Table 1. For the best performance all the possible parameters are tested and apply to the system to maximize the system control. All the controller designs and parameter estimations are done for closed loop system performance to control the voltage.

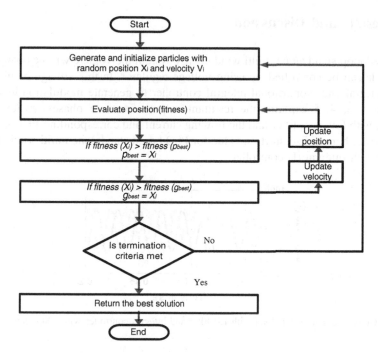

**Fig. 2** Flowchart for particle swarm optimization

# 4 Genetic Algorithm

The genetic algorithm is a one of the optimization technique to determine the fittest population for the given problem. It is a direct search method or algorithm. The GA-integral and PI controller consists of a controllers with its regarding parameter optimized by genetic algorithm [4–6].

By executing the following steps to execute Genetic algorithm, the gain values are optimized for controller to get the optimum solution for given problem [4–7].

(1) Start the program and for a given optimization problem, the initial population is generated.
(2) To achieve criterion for the termination, certain iterations have to be applied for the given population problem.
(3) Initialize generation counter and calculate the fitness value and evaluate the generations.
(4) Next generation value is created by using the operations for a given program.
    (a) Reproduction (b) Crossover (c) Mutation.
(5) If the termination criteria are achieved then print the result and stop the process and system to get the obtained optimized values in the system to get accurate performance with less harmonic content.

# 5  Results and Discussion

In closed loop stand alone multi wind energy system the voltage will be obtained at load side can be controlled by using voltage regulator. In this voltage regulator it have integral and proportional integral controller to generate modulation index.

Figures 3, 4, 5 explains the resultant outputs for three phase five level out voltage with low pass filter and the flowing current and corresponding FFT analysis to know the harmonic distortion value for PSO based standalone multi wind energy system with integral (I) controller.

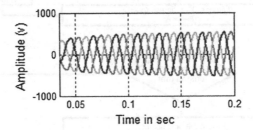

**Fig. 3** Output voltage of RL load with LP filter and integral controller with PSO technique

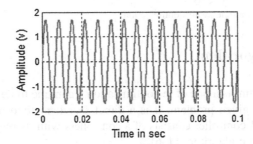

**Fig. 4** Output current of RL load with LP filter and integral controller with PSO technique

**Fig. 5** PSO based FFT analysis of the system with integral controller

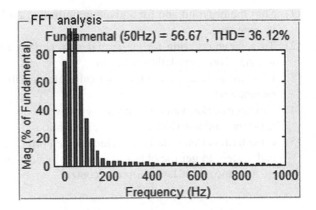

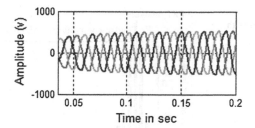

**Fig. 6** Output voltage of RL load with LP filter and integral controller with GA technique

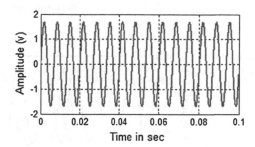

**Fig. 7** Output current of RL load with LP filter and integral controller with GA technique

Figures 6, 7, 8 explains the resultant outputs for three phase five level out voltage with low pass filter and the flowing current and corresponding harmonic distortion value for GA based standalone multi wind energy system with integral (I) controller.

Figures 9, 10, 11 explains the resultant outputs for three phase five level out voltage with low pass filter and the flowing current and corresponding harmonic distortion value for PSO based standalone multi wind energy system with proportional integral (PI) controller.

Figures 12, 13, 14 explains the resultant outputs for three phase five level out voltage with low pass filter and the flowing current and corresponding harmonic distortion value for GA based standalone multi wind energy system with proportional integral (PI) controller. And the system control will mainly depend on the gain values. This is applied as reference signal in pulse generation process. The obtained output voltage is five levels for RL load with less harmonic distortion. This voltage is controlled by integral and PI controllers with the help of particle swarm optimization algorithm, genetic algorithm techniques to get the optimized gain values and to reduce harmonic reduction and to improve the system performance. The following results are for output voltage and FFT analysis of the system with I and PI controllers with the techniques of PSO and GA. Compared to GA, PSO will give better optimum values for controllers and give the system performance with fewer harmonic.

**Fig. 8** GA based FFT
analysis of the system with
integral controller

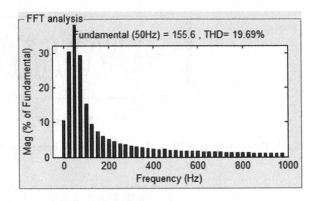

**Fig. 9** Output voltage of RL
load with LP filter and PI
controller with PSO technique

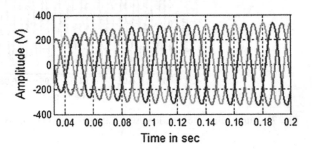

**Fig. 10** Output current of RL
load with LP filter and PI
controller with PSO technique

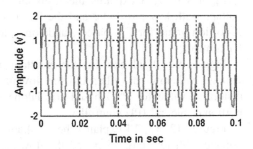

**Fig. 11** PSO based FFT
analysis of the system with PI
controller

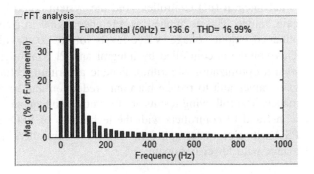

**Fig. 12** Output voltage of RL load with LP filter and PI controller with GA technique

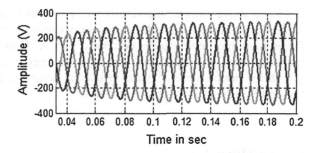

**Fig. 13** Output current of RL load with LP filter and PI controller with GA technique

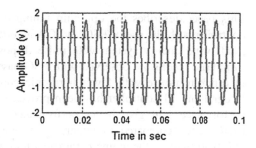

**Fig. 14** GA based FFT analysis of the system with PI controller

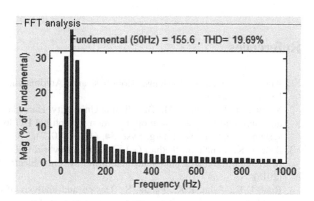

**Table 1** Comparision analysis of PSO and GA for different values of integral and PI controller gains

| S. no | For integral controller | | | | For PI controller | | |
|---|---|---|---|---|---|---|---|
| | $K_i$ | THD for PSO (%) | THD For GA (%) | $k_P$ | $K_i$ | THD for PSO (%) | THD for GA (%) |
| 1 | 0.8 | 62 | 83.03 | 0.7 | 0.8 | 23.32 | 31.05 |
| 2 | 0.9 | 47.87 | 57.22 | 0.8 | 0.9 | 20.13 | 31.06 |
| 3 | 1 | 36.12 | 51.44 | 0.9 | 1 | 16.99 | 19.69 |
| 4 | 1.1 | 43.56 | 71.78 | 1 | 1.1 | 19.59 | 29.05 |
| 5 | 1.2 | 43.88 | 74.87 | 1.1 | 1.2 | 17.50 | 31.06 |

Hence above showed the results will prove that the tuned parameters of PI controller for standalone multi wind energy system with PSO technique will give fewer harmonic performances compared to GA. Hence from the above tabular values the PSO base standalone multi wind energy system with PI controller will give better performance compare to integral controller and also compared to GA tuning PSO tuning is optimized technique.


# 6  Conclusion

This paper describes the optimum tuning of the gain values of Integral and PI controller with the use of PSO and GA algorithms for new closed loop stand alone multi wind energy system with pulse disposition multi carrier PWM technique. Simulated results are observed with PSO based integral and PI controller's results in quick response with fewer harmonic compared with GA. Hence PSO based Integral and PI controller has been proved to be an efficient method and System performance will give better performance and PI controller will give efficient performance compared to integral controller. Moreover, this method has good ability to adapt to the tuning parameters for control the load voltage and harmonics.


# References

1. Araki, M.: Control Systems, Robotics and Automation—vol II—PID Control. Kyoto University, Japan
2. Moradi, M.H., Abedini, M.: Bu Ali Sina University, Hamedan, Iran: A combination of genetic algorithm and particle swarm optimization for optimal DG location and sizing in distribution systems. Electr. Power Energ. Syst. **34**, 66–74 (2012)
3. Sahu, R.K., Panda, S., Sekhar, G.T.C.: A novel hybrid PSO-PS optimized fuzzy PI controller for AGC in multi area interconnected power systems. Electr. Power Energ. Syst. **64**, 880–893 (2015)
4. Rout, U.K., Sahu, R.K., Panda, S.: Design and analysis of differential evolution algorithm based automatic generation control for interconnected power system. Ain Shams Eng. J. **4**(3), 409–421 (2013)
5. Kim, D.H., Park J.I.: Intelligent PID controller tuning of AVR system using GA and PSO. ICIC 2005, Part II, LNCS 3645, pp. 366–375. Springer, Berlin (2005)
6. Yoshida, H., Kawata, K., Fukuyana, Y.: A particle swarm optimization for reactive power and voltage control considering voltage stability. In: IEEE International Conference on Intelligent System Applications to Power Systems, April 4–8 (1999)
7. Sivanandam, S.N., Deepa, S.N.: Introduction to Genetic Algorithms. Springer Publications Co. (2010)
8. Somashekar, B., Chandrasekhar, B., Livingston, D.: Modeling and simulation of three to nine phase using special transformer connection. Int. J. Emerg. Technol. Adv. Eng. **3**(6) (2013)
9. Hassanpoor, A., Norrga, S., Nee, H., Angquist, L.: Evaluation of different carrier-based PWM methods for modular multilevel converters for HVDC application. In: Proceedings Conference on IEEE Industrial Electronics Society, pp. 388–393 (2012)

10. Chen, Z., Senior Member, IEEE, Guerrero, J.M., Senior Member, IEEE, Blaabjerg, F., Fellow, IEEE: A review of the state of the art of power electronics for wind turbines. IEEE Trans. Power Electron. **24**(8) (2009)
11. Solihin, M.I., Tack, L.F., Kean, M.L.: Tuning of PID controller using particle swarm optimization (PSO). In: Proceeding of the International Conference on Advanced Science, Engineering and Information Technology (2011)
12. Malik, S., Dutta, P., Chakrabarti, S., Barman, A.: Parameter estimation of a PID controller using particle swarm optimization Algorithm. Int. J. Adv. Res. Comput. Commun. Eng. **3**(3) (2014)
13. Rodriguez, J., Lai, J.S., Peng, F.Z.: Multilevel inverters: a survey of topologies, controls, and applications. IEEE Trans. Ind. Electron. **4**(4), 724–738 (2002)

# Statistically Validating Intrusion Detection Framework Against Selected DoS Attacks in Ad Hoc Networks: An NS-2 Simulation Study

Sakharam Lokhande, Parag Bhalchandra, Santosh Khamitkar, Nilesh Deshmukh, Satish Mekewad and Pawan Wasnik

**Abstract** Network security is a weak link in wired and wireless network systems which if breeched then the functionality of the underlying network is impaired by malicious attacks and causes tremendous loss to the network. The DoS (Denial of Service) attacks are one of the harmful network security threats. Ad Hoc networks can be easily damaged by such attacks as they are infrastructure less and without any centralized authority. This paper narrates results of an undertaken research work to study the consequence of selected DoS attacks on the performance of Ad Hoc networks. The gained knowledge is used to design and develop a security framework to detect intrusion and perform response actions. The validations of proposed defense system are carried out using statistical approach. All Simulations were carried out in NS-2.

**Keywords** Ad hoc networks · Dos attacks · IDS · Mitigation

S. Lokhande (✉) · P. Bhalchandra · S. Khamitkar · N. Deshmukh · S. Mekewad · P. Wasnik
School of Computational Sciecnes, S.R.T.M. University, Nanded
MS 431606, India
e-mail: lokhande_sana@rediff.com

P. Bhalchandra
e-mail: srtmun.parag@gmail.com

S. Khamitkar
e-mail: s.khamitkar@gmail.com

N. Deshmukh
e-mail: nileshkad@yahoo.com

S. Mekewad
e-mail: satishmekewad@gmail.com

P. Wasnik
e-mail: pawan_wasnik@yahoo.com

© Springer India 2016
H.S. Behera and D.P. Mohapatra (eds.), *Computational Intelligence in Data Mining—Volume 1*, Advances in Intelligent Systems and Computing 410, DOI 10.1007/978-81-322-2734-2_41

# 1 Introduction

In recent years, the scenario of computing is drastically changed due to introduction of mobile computing devices. This lead to emerging ubiquitous computing [1] which merely does not dependent of capability provided by the personal computer and has been a research hot spot in last few years. The mobile technology has made it necessary to use wireless network for connection purpose. The wireless networks are cost economic and suitable for voice & data transmission as they use air as medium for transmission. Wireless networks broadly fall into two main categories, infrastructure based called wireless through wireless Access Point (AP) and Ad Hoc network, where the nodes communicate with each other without having a central access point [2]. Our study centers on the later on. These networks have a self-governing system of mobile nodes. Such networks are known as infrastructure less network because no any per-existing infrastructure is required for communications and transmission takes place via mobile nodes only [3, 4]. Wireless Ad Hoc networks have wide range of applications including military applications, disaster management, business environments for cooperative and collaborative computing and sensor networks. It is witnessed that the computing scenario is drastically changed due to applicability of wireless communications. Today we have number of application areas where wireless networks can play important role. However, the risk also increases of getting harmed by intrusions and threats as the network is open and without centralized coordinator. According to [5], any unauthorized attempt is called as threat. Threats are deliberate and are in terms of attacks. These attacks can be internal or external as well as active or passive [6, 7]. The Denial of Service (DoS) attacks can be internal and external. These attacks attempts partially or totally disallow access of intended services to user [8]. These exist in all protocol layers. These are very easy to generate but are very difficult to detect and hence they are popular choice of hackers [9]. These can easily be applied to wireless networks [10, 11]. The DoS attacks in wireless Ad Hoc networks are more harmful than wired network as they significantly downgrade performance causing congestion [12]. We have chosen two Dos attacks for our study, viz, data flooding attack and Black hole attack [10]. In data flooding attack, useless data packets will exhaust the communication bandwidth in the network and in black hole attack, a third party node behaves like routing strategy such that sender node will take it as shortest path [13–15].

# 2 Literature Review

It is evident from the literature reviewed in the context of research hypothesis that there are numbers of intrusion detection and prevention systems proposed for traditional wired networks as we need to monitor central monitoring devises only [16]. The same is not true with Ad Hoc network as they do not have physical ambiance or central monitoring devices. Case drastically changes in wireless Ad Hoc networks where the communication medium is widely open and shared among different users [2, 3]. This highlights chances of being accessed by both legitimate and malicious users by giving scope to intruders. In Ad Hoc network, nodes can move randomly in any direction and at any movement a node can leave or join a network. If a node is compromised, it will generate false routing information. Many solutions exist for this [17]. Numerous architectures have been evolved over the time and they have implicit considerations of intrusions [17, 18]. We have chosen Distributed and Cooperative Intrusion Detection Systems for the underlying study where all nodes participate in IDS. Preventing DoS attacks has got severe attention in past decade. The Sun et al. [19] proposed neighborhood based and route recovery scheme and the detection and prevention of black hole attack is carried out using neighboring node information. In another study Shurman et al. [20, 21] proposes redundant route and unique sequence number schemes for preventing black hole attack. The Tamilselvan et al. [22, 23] proposed a time based threshold detection method by enhancing the original AODV routing protocol and Glomosim simulator as proposed by Shurman et al. [14, 24]. The study of Djenouri et al. [14] proposed a Bayesian approach and Glomosim simulator for node allegation that enables node deliverance before decision. Other similar study is Bait DSR (BDSR) scheme [24, 25]. A hybrid routing protocols is proposed by combining proactive and reactive methods. On demand DSR routing protocol is used as a base routing protocol. All above surveyed literature is helpful in augmenting the defense mechanism for selected attacks in wireless Ad Hoc networks. After reviewing contemporary work on attacks and network security issues, it was found that the routing strategy was not been considered. If a fault is detected in the primary route, nodes can switch to an alternate route provided they are able to find multiple routes. We have come across some novel methods of intrusion detection which forced us to go for statistical analysis. We primarily plan for a detection algorithm as proposed by Huang et al. [26, 27] an IDS mechanisms that works similar to Tseng et al. [25, 27] and an IDS Model proposed by Brutch and Ko [25, 28] which has a statistical anomaly detection algorithm. All these cited literature review highlights novelties and contemporary research investigations in detection of DoS attacks. Since our research hypothesis deals with mitigation and prevention of DoS attacks, it is need of hour to address issues regarding responses related to specific attacks. Many studies, as cited above, have incorporated prevention and mitigation of such attacks. Such attempt needs understanding of Intrusion detection and intrusion response framework. The framework components are relevant and coexist with each other. It was our critical observation that while designing IDS, intrusion response framework has given less attention than intrusion detection framework [29]. After a rigorous analysis all above related issues, a common finding came in picture that the IDA for Ad Hoc network must work with localized and partial audit data. On the backdrop of all above discussions and research findings, the present research study investigates prevention and mitigation of selected DoS attacks in Ad Hoc networks.

# 3 Proposed Model

It was understood from earlier two sections of this paper that the data flooding and black hole attacks will affect the significant parameters as well as the performance of the Ad Hoc network. It was also understood that cooperative distributive framework is necessary for IDS designing. This fact is used as base for our intrusion detection and prevention framework. The proposed framework will monitor different significant parameters for the detection of intrusive activity in Ad Hoc network. If these parameters change rapidly, there are chances of getting an intruder. Later a remedial action is taken. The challenge here is the identification of sensitive parameters and their threshold values to predict the intrusion correctly. Many researchers' uses parameters and their threshold values for the intrusion detection in Ad Hoc network [30–32]. The parameter Threat (T) is used by detection framework to detect an attack or vulnerability. Threat is a number, which takes values between 0 and 3, when there is no attack, the network is in the normal state (NS) and this is indicated by the T range 0; when there is an attack, the network is in the vulnerable state (VS) and is indicated by the T range from 1 to 3; The threshold value of Threat (T) for the normal and vulnerable state are obtained by measuring the values of the significant parameters. When the network is in operating state the values of parameters are measures and the new value of Threat (TI). By comparing the computed TI with the T threshold, the node is classified as being in the normal or vulnerable state and this classification detects the attack. Each step of the threat detection framework, whose objective is to calculate the Threat to detect an attack, is explained below. Steps of the proposed framework are shown in Fig. 1.

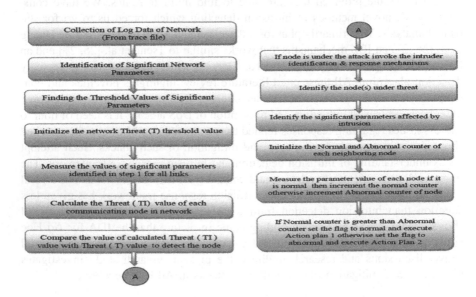

**Fig. 1** Steps in intrusion detection and prevention model

We refer to the terminology in [33] for the sketching of the response and prevention mechanism. The significant parameters identified and the thresholds for these significant parameters in the detection framework are used to defend the attack [33]. On the basis of threshold values, the node is reputed using the counters and flags on their behavior, which is achieved by monitoring significant parameters of node. Based on the reputation counter, the reputation flag is asserted as Normal, or Vulnerable. If the reputation flag is Normal, no action is necessary. If the reputation flag is Vulnerable then response action is executed. The response action may isolate, block or deny further connections to malicious nodes. This framework works on the basis of the network parameters and performing response action on the basis of assertion of reputation flag. Such a framework allows for response to flooding and black hole attacks. Those malicious and selfish nodes that generate abnormal parameter values are identified and isolated. The following action plans [33] are used in the response framework.

1. Plan 1: If a neighboring node to the node under threat is flagged as "normal", no action is needed.
2. Plan 2: If a neighboring node to a node under threat is flagged as "abnormal", action plan 2 is executed which may isolate or block the attacking node to protect the system.

# 4 Experimentations, Observations and Results

We have carried out two experimentations, one for each DoS attacks. They are narrated briefly herein. As the first objectives of this study is to measure the impact of data flooding DoS attack on server node. In the first stage, the simulated network runs the under the normal operation where the nodes are communicated by sending and receiving the data packets normally. A set of random and moderate Constant Bit Rate (CBR) traffic is generated over the simulated normal network. In the normal run the basic network performance parameters are measured and referred to as the baseline of the normal network behaviors. These include total packet sent, total packet received, total no. of packets forwarded, total number of packets dropped end to end delay. These parameters are then used to measure average packet delivery ratio, network throughput etc. of the network under normal operation using the AWK script. In the second stage, one node is launched and specified as an attacker node, which performs data flooding DoS attack over the service provider node in the simulated network during normal network operation. The data flooding DoS attacker node generates huge amount of data packets which are generally larger than the normal packet size. The impact of this attack and changes in networks/nodes performance parameters are measured. In the third stage, the proposed intrusion detection and prevention framework is launched in the network. The defense framework will identify the attacking node on the basis of nodes parameters by comparing it with the values of normal and under the data flooding attack. This will confront the flooding DoS attack going over the service provider node and will be identified the attacking node for isolation so as to stop all the further requests from this attacking node. The network performance parameters are recorded and compared to the values against the attacked phase. As second objective of this study, we try to observe the impact of multiple black hole nodes on the performance of Ad Hoc network. Firstly, the simulated network runs under the normal condition i.e. without the black hole nodes. A set of random and moderate Constant Bit Rate (CBR) traffic is generated through the simulated network. The basic networking performance parameters are measured and referred to as the baseline of the normal network behavior. These parameters are total packet sent, total packet received, total number of packets forwarded, packets dropped, packet delivery ration and throughput is measured. We are also curious to measure the percentage of packet loss between communicating nodes. In the second stage, Multiple Black hole attack is implemented, where multiple black hole nodes are injected individually into the network during normal network operation. The impact of this attack and changes in networks performance parameters are measured and recorded. Third stage, the proposed intrusion detection and prevention framework is implemented and launched in the network to which will defend against the multiple black hole node attack. The defense system launched in the network confronts this attack in the network. The parameters are recorded and will be compared to the values against the normal phase as well as attacked phase. Below figure Fig. 2a shows screen shot of Nam output of NS-2/simulation window for data flooding attack. The service provider is node 4 and the service requesters are node 1 and node 5 respectively. There is one attacker node 6, sending continues data packets towards service provider node 5. In the normal operation of network the service provider node is able to send data packets towards requester nodes 1 and 5; also both the nodes are receiving the data packets normally without any interruption as

**(a)**                                    **(b)**

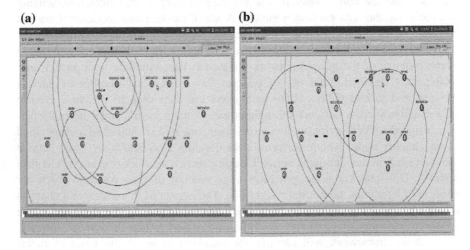

**Fig. 2** Simulated network for (**a**) data flooding attack and black hole attacks (**b**)

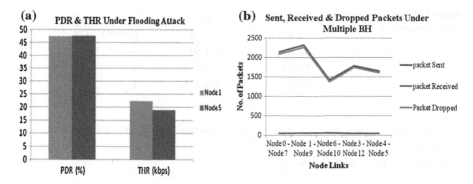

**Fig. 3** Graphical illustration of parameter values for data flooding (**a**) and black hole attacks (**b**)

shown in figure. In this scenario the effect of multiple black hole in the Fig. 2b. For this scenario we use the simulated network setup which has the 15 nodes. The multiple black hole nodes varies from 2 to 4.

Similarly we carried the same experiment by changing the number of nodes and recording values of parameters. The selected parameters include Packet Delivery Ratio (PDR) and Throughput (THR). The values are obtained by processing log (trace) files generated after the simulation. The graphical illustration is in Fig. 3a, b.

# 5 Statistical Validations in Terms of Chi Square Goodness of Fit Test

Usually, we understand any depict in the expected norm behavior of a subject by using Statistical tools [32]. For testing the fitness or effectiveness of our proposed framework for intrusion detection and prevention, we applied the Chi-Square test [32]. For calculations, we made a distribution for the whole network. The statistics data containing the packet delivery ratio (PDR), throughput (THR) and end to end delay (E2ED) in normal, under attack and with proposed response framework is collected in a table. The values of parameters over 900 s of simulation time for varying network size have been considered. The Chi-square test hypothesis with observed frequency (O) and expected frequency (E) is defined as,

$$\chi^2 = \Sigma(O - E)^2/E \tag{1}$$

We have two hypotheses. The first hypothesis is for specific distribution. The second hypothesis is for not following specific distribution. For testing our hypotheses, we verified the data of network under the normal operation first. The data distribution of network for node size 50 is considered. The hypothesis test is performed at five percentage significance level ($\alpha = 0.05$). There are 3 types of categories in the test data (k = 3). The degree of freedom is two (df = 3 − 1 = 2).

**Table 1** Normal network and relative frequency

| Category | Value | Relative frequency |
|----------|-------|-------------------|
| PDR | 87.28 | 0.2918917111 |
| THR | 211.50 | 0.7073223751 |
| E2ED | 0.235 | 7.85 |
| Total | 299.015 | 1 |

The observed values for parameters under normal operation and there relative frequency are shown in Tables 1 and 2.

From the above Table 2, we get

$$\chi^2 = \Sigma(O - E)2/E = 3.18037 \tag{2}$$

From above tables, we get Chi-square tabulated value as $\chi^2 = 5.991$ and Chi-square calculated value as $\chi^2 = 3.18037$.

We have selected another data set of same network size having 50 nodes and repeated all calculations. So the observed values for this test is

$$\chi^2 = \Sigma(O - E)^2/E = 8.13031 \tag{3}$$

For this observation, we get Chi-square tabulated value as $\chi^2\ 0.05 = 5.991$ and Chi-square Calculated Value as $\chi^2 = 8.13031$. Here the chi-square calculated value is larger than Chi-square tabulated value, so the first hypothesis is rejected and the second hypothesis is accepted. It shows presence of anomaly or intrusion (DoS) in the network. For testing the goodness test of our proposed framework against the DoS attack using the same Chi square test. We have calculated $\chi^2$ as

$$\chi^2 = \Sigma(O - E)^2/E = 0.440205538 \tag{4}$$

When same hypothesis is performed, we get Chi-square tabulated value as $\chi^2\ 0.05 = 5.991$ and the Chi-square Calculated Value as $\chi^2 = 0.440205538$. Here the Chi-square calculated value is smaller than Chi-square tabulated value, so the first hypothesis is accepted showing absence of intrusion and the second hypothesis is rejected.

**Table 2** Chi-square test statistic for normal

| Category | Relative frequency (f) | Observed frequency (O) | Expected frequency (E) = n * f | (O − E) | (O − E)² | (O − E)²/E |
|----------|------------------------|------------------------|-------------------------------|---------|----------|------------|
| PDR | 0.29189 | 87.28 | 87.28 | 5.43351 | 2.9523 | 3.38256 |
| THR | 0.70732 | 211.5 | 211.5 | 9.47352 | 8.9748 | 4.24339 |
| E2ED | 7.85 | 0.235 | 0.23472678 | 0.00027322 | 7.4652 | 3.18037 |
| Total | 1 | n = 299.015 | | | | 3.18037 |

# 6 Conclusion

This research paper narrates creation, deployment and cross validation of a cooperative distributive IDS model for Ad Hoc network. It is evaluated against selected DoS attacks namely, data flooding and black hole attack. Extensive simulations were performed to evaluate the performance of proposed framework in NS-2 simulator. The statistical validations are also carried out with the help of Chi-square—goodness of fit approach. The fitness of our proposed framework is tested under the various scenarios and the results obtained prove that our proposed framework is good for the intrusion detection and prevention in the Ad Hoc network.

# References

1. Conti, M.: Body, personal and local ad hoc wireless networks. In: The Handbook of Ad Hoc Wireless Networks, Chap. 1. CRC Press LLC (2003)
2. Ilyas, M.: The Handbook of Ad Hoc Wireless Networks. CRC Press (2003)
3. Stallings, W.: Wireless Communication and Networks. Pearson Education (2002)
4. Schiller, J.: Mobile Communication. Addison-Wesley (2000)
5. Anderson, J.P.: Computer Security Threat Monitoring and Surveillance. James P. Anderson Co., Fort Washington (1980)
6. Wu et al.: A survey of attacks and countermeasures in mobile ad hoc networks. In: Wireless/mobile network security, vol. 17. Springer (2006)
7. Virada, V.P.: Intrusion detection system (IDS) for secure MANETs: a study. Int. J. Comput. Eng. Res. (ijceronline.com) 2(6). ISSN 2250-3005(online) October 2012
8. Zhang, X., Sekiya, Y., Wakahara, Y.: Proposal of a method to detect black hole attack in MANET. In: International Symposium on Autonomous Decentralized Systems (ISADS '09), pp. 1–6 (2009)
9. Al-Shurman, M., Yoo, S.-M., Park, S.: Black hole attack in mobile ad hoc networks. ACM Southeast Regional Conference (2004)
10. Aad, I., Hubaux, J.-P., Knightly, E.W.: Denial of service resilience in ad hoc networks. In: Proceedings of the 10th Annual International Conference on Mobile Computing and Networking, pp. 202–215 (2004)
11. Bellardo, J., Savage, S.: 802.11 Denial-of-service attacks: real vulnerabilities and practical solutions. In: Proceedings of the 12th Conference on USENIX Security Symposium, vol. 12
12. Mistry, N., Jinwala, D.C., Zaveri, M.: Improving AODV protocol against blackhole attacks. In: Proceedings of the International Multi Conference of Engineer and Computer Science, vol. 2 (2010)
13. Soomro, S.A., Soomro, S.A., Memon, A.G., Baqi, A.: Denial of service attacks in wireless ad hoc networks. J. Inform. Commun. Technol. 4(2), 1–10
14. Bhardwaj, M., Singh, G.P.: Types of hacking attack and their countermeasure. Int. J. Educ. Plann. Admin. 1(1), 43–53
15. Atiya Begum, S.: Techniques for resilience of denial of service attacks in mobile ad hoc networks. Int. J. Sci. Eng. Res. 3(3) (2012)
16. Zhang, Y., Lee, W., Huang, Y.: Intrusion detection techniques for mobile wireless networks. ACM/Kluwer Wireless Networks J. (ACM WINET) 9(5) (2003)
17. Aad, I., Hubaux, J.P., Knightly, E.W.: Denial of service resilience in ad hoc networks. ACM, MOBICOM, Philadelphia, PA, USA (2004)

18. Mukherjee, B., Heberlein, L., Levitt, K.: Network intrusion detection. IEEE Network **8**(3), 26–41 (1994)
19. Chan, Y.K., et al.: IDR: an intrusion detection router for defending against distributed denial-of-service (DDoS) attacks. In: Proceedings of the 7th International Symposium on Parallel Architectures, Algorithms and Networks (ISPAN'04), pp. 581–586 (2004)
20. Buttyán, L., Hubaux, J.P.: Stimulating cooperation in self-organizing mobile ad hoc networks. ACM J. Mobile Networks (MONET) **8**(5), 579–592 (2003)
21. Luo, H., Lu, S.: Ubiquitous and robust authentication services for ad hoc wireless networks. Department of Computer Science, UCLA, Technical Report TR-200030 (2000)
22. Sun, B., Wu, K., Pooch, U.W.: Alert aggregation in mobile ad hoc networks. In: Proceedings of the 2003 ACM Workshop on Wireless Security (WiSe'03) in Conjunction with the 9th Annual International Conference on Mobile Computing and Networking (MobiCom'03), pp. 69–78 (2003)
23. Raj, P.N., Swadas, P.B.: DPRAODV: a dynamic learning system against blackhole attack in AODV based MANET. Int. J. Comput. Sci. **2**, 54–59 (2009)
24. Tseng, F.H., et al.: A survey of black hole attacks in wireless mobile ad hoc networks. Human-centric Comput. Inform. Sci. **1**, 4 (2011). doi:10.1186/2192-1962-1-4
25. Alampalayam, S.P.: Classification and review of security schemes in mobile computing, wireless sensor networks. **2**, 419–440 (2010)
26. Trost, B.W.: Authenticated ad hoc routing at the link layer for mobile systems. Wireless Networks **7**(2), 139–145 (2001)
27. Brutch, P., Ko, C.: Challenges in intrusion detection for wireless ad hoc networks. In: Proceedings of Symposium on Applications and the Internet Workshop, pp. 368–373 (2003)
28. Tseng, C., Balasubramanyam, P.: A specification-based intrusion detection system for AODV, In: Proceedings of ACM Workshop on Security of Ad Hoc and Sensor Networks, pp. 125–134 (2003)
29. Kachirski, O., Guha, R.: Effective intrusion detection using multiple sensors in wireless ad hoc networks. In: Proceedings of 36th International Conference on System Sciences, pp. 57–64 (2003)
30. Lou, K.H., Xu, K., Gu, D., Gerla, M., Lu, S.: Adaptive security for multilayer ad hoc networks. Special Issue of Wireless Communication Mobile Comput. **2**, 533–547 (2002)
31. Ye, N., Chen, Q.: An anomaly detection techniques based on a CHI-SQUARE statistics for detecting intrusion into information system. Quality Reliab. Eng. Int. (2001)
32. Ye, N., Li, X., Emran, M., Xu, M.: Probabilistic techniques for intrusion detection based on computer audit data. IEEE Trans. Syst. Man Cybern. (2001)
33. Alampalayam, S.P., et al.: Intruder identification and response framework for mobile ad hoc networks. In: Conference Proceedings, Conference: 22nd International Conference on Computers and Their Applications, CATA-2007, Honolulu, Hawaii, USA, March 28–30 (2007)

# Fourier Features for the Recognition of Ancient Kannada Text

A. Soumya and G. Hemantha Kumar

**Abstract** Optical Character Recognition (OCR) System for ancient epigraphs helps in understanding the past glory. The system designed here, takes a scanned image of Kannada epigraph as its input, which is preprocessed and segmented to obtain noise-free characters. Fourier features are extracted for the characters and used as the feature vectors for classification. The SVM, ANN, k-NN, Naive Bayes (NB) classifiers are trained with different instances of ancient Kannada characters of Ashoka and Hoysala period. Finally, OCR system is tested on epigraphical characters of 250 from Ashoka and 200 from Hoysala period. The prediction analysis of SVM, ANN, k-NN and NB classifiers is made using performance metrics such as Accuracy, Precision, Recall, and Specificity.

**Keywords** Fourier descriptors · Support Vector Machine (SVM) · Artificial Neural Network (ANN) · k-Nearest Neighbor (k-NN) · Naive Bayes (NB) classifier

## 1 Introduction

An automatic recognition system to decipher ancient documents without much human intervention is of much relevance to mankind, to know the history and culture of past. Works carried out on Indian and non-Indian modern scripts are very high compared to ancient Indian scripts. Also, the complexity of recognition increases in case of ancient Indian scripts, as the characters has evolved overtime.

A. Soumya (✉)
Department of Computer Science & Engineering, RV College of Engineering,
Bengalore, India
e-mail: soumyaa@rvce.edu.in

G.H. Kumar
Department of Studies in Computer Science, University of Mysore, Mysore, India
e-mail: ghk2007@yahoo.com

© Springer India 2016                                          421
H.S. Behera and D.P. Mohapatra (eds.), *Computational Intelligence
in Data Mining—Volume 1*, Advances in Intelligent Systems
and Computing 410, DOI 10.1007/978-81-322-2734-2_42

The objective of the work is to design and develop an automated system for reading ancient Kannada text documents of two different periods—Ashoka and Hoysala. The system enables the user to classify the characters in ancient Kannada script and gives the output in modern form. Thus, automating the task of reading ancient Kannada documents is of great significance to the departments of Ancient History and Archeology.

This paper is organized as follows: Sect. 2 details the related works in the area. The system architecture and methodology is discussed in Sect. 3. The mathematical background of feature extraction is highlighted in Sect. 4. Experimental results and performance analysis is covered in Sects. 5 and 6 provides conclusion.


# 2 Related Work

In this section, some of the works on Document Analysis and Recognition for Indian and non-Indian Scripts are described.

A system for recognition of ancient Tamil characters from temple wall inscriptions is developed, that uses Fourier and Wavelet features [1]. ANN is trained with scaled conjugate gradient back propagation algorithm and accuracy of 98 % is reported.

The features Normalized Central moments (NCM) and the Normalized Fourier Descriptors (NFD) of the character are invariant to character shifting, scaling, or rotation. These features are compared for Arabic OCR and a 100 % recognition rate with the NFD but in case of NCM the recognition rate reduces with change in image coordinates ratio is observed [2].

A method to recognize handwritten Tamil characters using a multilayer perceptron is proposed. Fourier Descriptors are extracted from the characters and taken as input to the Back Propagation Neural Network (BPNN) for recognition. Test results shows very high recognition accuracy of 97 % for handwritten Tamil characters [3].

A system for the discrimination between ancient document collections of Arabic and Latin scripts [4], provides 95.87 % recognition rate. The advantage of this method is that it can be easily implemented for the recognition of other ancient document collections and it can have better identification rates by providing relative features of each document base.

The concept proposed in paper [5] to perform classification of handwritten ancient Tamil scripts, utilizes Extreme Learning Machine. The performance of Extreme Learning Machine is compared with Probabilistic Neural Networks and observed that Extreme Learning Machine gives highest accuracy rate of 95 %.

Rotation invariant OCR System for Sinhala language is developed using Fourier Transform and ANN [6]. Sinhala characters of different fonts and font sizes are recognized with over 85 % recognition accuracy.

The performance of different classification techniques with different features for Kannada character recognition is studied [7]. BPN, k-NN and SVM classifiers with two feature extraction methods HU's invariant moments and Zernike moments are used.

## 3  System Architecture and Methodology

The OCR system shown in Fig. 1 comprises of five sub-components: Pre-processing, Segmentation, Feature Extraction, Classification and Post-processing. The preprocessing module reduces noise using spatial filtering and binarizes by applying Otsu's threshold method. This preprocessed image is given to the segmentation module, which applies the canny edge detector and extracts the edges of characters. Close character contours are detected from the edges, based on that characters are segmented. The Segmented characters are fed to the feature extraction module that computes the General Fourier features and based on that it also computes the rotation and scale invariant Fourier features. Output from all these computations is saved in data store. The classification module consist of four classifiers SVM, ANN, k-NN and NB which are trained using Fourier features and later used in testing. From the predicted class label ancient character is mapped to the modern Kannada character, during Post-processing.

## 4  Mathematical Background

In this work, Fourier features are used for training the classifier and later testing. Fourier features can be extracted from close character contours. $a_n$, $b_n$, $c_n$, and $d_n$. From these general Fourier features, scale and rotation invariant features are extracted.

- **General Fourier features**

Fourier features can be extracted from close character contours. $a_n$, $b_n$, $c_n$, and $d_n$ are the extracted feature and is given by Eqs. (1)–(4):

$$a_n = \frac{T}{2n^2\pi^2} \sum_{i=1}^{m} \frac{\Delta x_i}{\Delta t_i} [\cos \phi_i - \cos \phi_{i-1}] \tag{1}$$

$$b_n = \frac{T}{2n^2\pi^2} \sum_{i=1}^{m} \frac{\Delta x_i}{\Delta t_i} [\sin \phi_i - \sin \phi_{i-1}] \tag{2}$$

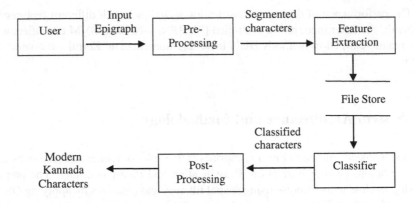

**Fig. 1** System architecture of OCR System

$$c_n = \frac{T}{2n^2\pi^2} \sum_{i=1}^{m} \frac{\Delta y_i}{\Delta t_i} [\cos \phi_i - \cos \phi_{i-1}] \tag{3}$$

$$d_n = \frac{T}{2n^2\pi^2} \sum_{i=1}^{m} \frac{\Delta y_i}{\Delta t_i} [\sin \phi_i - \sin \phi_{i-1}] \tag{4}$$

where $\phi_i = 2n\pi t_i/T$, $\Delta x_i = x_i - x_{i-1}$, $\Delta y_i = y_i - y_{i-1}$, $\Delta t_i = \sqrt{\Delta x^2 + \Delta y^2}$,
$T = t_m = \sum_{j=1}^{m} \Delta t_j$, $t_i = \sum_{j=1}^{i} \Delta t_j$ and m is the number of pixels along the boundary.

- **Rotation invariant Fourier features**

To obtain the features that are independent of the particular starting point, it requires calculating the phase shift from the first major axis as in Eq. (5):

$$\phi_1 = \frac{1}{2} \tan^{-1} \frac{2(a_1 b_1 + c_1 d)}{\sqrt{a_1^2 - b_1^2 + c_1^2 - d_1^2}} \tag{5}$$

Then, the coefficients can be rotated to achieve a zero phase shift as in Eq. (6):

$$\begin{bmatrix} a_n^* b_n^* \\ c_n^* d_n^* \end{bmatrix} = \begin{bmatrix} a_n & b_n \\ c_n & d_n \end{bmatrix} \begin{bmatrix} \cos n\phi_1 & -\sin n\phi_1 \\ \sin n\phi_1 & \cos n\phi_1 \end{bmatrix} \tag{6}$$

Now to obtain rotation invariant description, the rotation of the semi major axis can be found by Eq. (7):

$$\psi_1 = \tan^{-1} \frac{c_1^*}{a_1^*} \tag{7}$$

Now features can be obtained using Eq. (8):

$$\begin{bmatrix} a_n^{**} & b_n^{**} \\ c_n^{**} & d_n^{**} \end{bmatrix} = \begin{bmatrix} \cos\psi_1 & \sin\psi_1 \\ -\sin\psi_1 & \cos\psi_1 \end{bmatrix} \begin{bmatrix} a_n^{*} & b_n^{*} \\ c_n^{*} & d_n^{*} \end{bmatrix} \tag{8}$$

- **Scale invariant Fourier features**

To obtain scale invariant features, the coefficients can be divided by the magnitude, E, of the semi major axis, given by Eq. (9):

$$E = \sqrt{a_1^{*2} + c_1^{*2}} = a_1^{**} \tag{9}$$

## 5 Experimental Results and Analysis

### 5.1 Experimental Results

The OCR system for ancient Kannada script is trained with dataset of 258 characters from Ashoka period and 362 characters from Hoysala period with 6 different instances of each. The system is tested of 70 epigraphs from Ashoka and Hoysala period. Figure 2 shows the result of recognition for an ancient input epigraph.

### 5.2 Performance Analysis

In Predictive Analytics, the **confusion matrix** as shown in Table 1 reports the number of *false positives (FP)*, *false negatives (FN)*, *true positives (TP)*, and *true negatives (TN)*. The Performance Metrics of prediction such Accuracy, Precision, Recall, and Specificity are computed using Eqs. 10, 11, 12 and 13 respectively.

The performance metrics used to evaluate the developed OCR system are:

- **Classification rate**: This metric is used to determine the accuracy of Classifier and is given by Eq. (10):

**Fig. 2** Final result of recognition phase

**Table 1** Confusion matrix

|           |          | Actual   |          |
|-----------|----------|----------|----------|
|           |          | Positive | Negative |
| Predicted | Positive | TP       | FP       |
|           | Negative | FN       | TN       |

$$Accuracy = \frac{TP + TN}{TP + TN + FP + FN} \qquad (10)$$

- **Recall**: This metric is the proportion of positive cases that were correctly identified, as calculated using the Eq. (11):

$$recall = \frac{TP}{TP + FN} \qquad (11)$$

- **Precision**: This metric is the proportion of the predicted positive cases that were correct, as calculated using the Eq. (12):

$$precision = \frac{TP}{TP + FP} \qquad (12)$$

- **Specificity**: This metric measures the proportion of negatives which were correctly identified, as calculated using the Eq. (13):

$$specificity = \frac{TN}{TN + FP} \qquad (13)$$

These Performance Metrics (Accuracy, Precision, Recall, and Specificity) computed for the test data are shown in Tables 2 and 3 respectively, for characters of Ashoka and Hoysala period. The tabulations are shown in percentage, each column indicates the Classifier used and the rows indicate the Metric value.

**Table 2** Performance metrics in percentage for Ashoka period test data

|         |             | Ashoka |       |       |       |
|---------|-------------|--------|-------|-------|-------|
|         |             | SVM    | ANN   | KNN   | NB    |
| Metrics | Accuracy    | 83.60  | 76.80 | 49.20 | 64.80 |
|         | Precision   | 88.78  | 87.36 | 77.03 | 86.84 |
|         | Recall      | 91.00  | 83.00 | 52.00 | 66.00 |
|         | Specificity | 60.00  | 43.33 | 16.52 | 30.61 |

**Table 3** Performance metrics in percentage for Hoysala period test data

| Metrics | | Hoysala | | | |
| | | SVM | ANN | KNN | NB |
|---|---|---|---|---|---|
| Metrics | Accuracy | 80.50 | 71.50 | 48.50 | 62.50 |
| | Precision | 89.03 | 85.03 | 73.94 | 81.95 |
| | Recall | 86.25 | 78.12 | 55.00 | 68.12 |
| | Specificity | 51.11 | 33.96 | 11.11 | 23.88 |

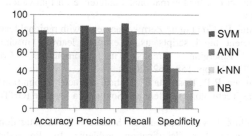

**Fig. 3** Performance metrics of classifiers for Ashoka characters

**Fig. 4** Performance metrics of classifiers for Hoysala characters

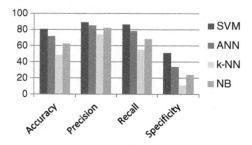

The performance characteristics for the OCR system of ancient Kannada script using SVM, ANN, KNN, NB Classifiers are graphed in Figs. 3 and 4 for Ashoka and Hoysala period respectively.

# 6 Conclusion

The developed system recognizes Kannada characters of two ancient eras Ashoka and Hoysala. The trained OCR system is tested on 70 epigraphs of ancient times. It is observed from the analysis that, all the classifiers have better performance for characters of Ashoka period compared to Hoysala period. SVM Classifier works better in classifying the test data. ANN Classifier's capability lies in between those SVM and NB Classifiers, k-NN have lowest classification rate.

# References

1. Raja Kumar, S., Subbiah Bharathi, V.: An off-line ancient tamil script recognition from temple wall inscription using fourier and wavelet features. Eur. J. Sci. Res. **80**(4), 457–464 (2012)
2. Dhabe, P.S., Bakki, S.R., Kulkarni, U.V., Sontakke, T.R.: Rotation invariant features: a method, problems and possible solutions. In: Proceedings of IEEEs (WCCI: IJCNN) Conference (2002)
3. Sutha, J., Ramaraj, N.: Neural network based offline tamil handwritten character recognition system. In: International Conference on Computer Intelligent and Multimedia Applications, pp. 446–450 (2007)
4. Zaghden, N., Mullot, R., Alimi, A.M.: Characterization of ancient document images composed by Arabic and Latin scripts. In: International Conference on Innovations in Info Technology, pp. 124–127 (2011)
5. Sridevi, N., Subashini, P.: Combining Zernike moments with regional features for classification of handwritten ancient tamil scripts using extreme learning machine. In: International Conference on Emerging Trends in Computing, Communication and Nanotechnology ICECCN, pp. 158–162 (2013)
6. Rimas, M., Thilakumara, R.P., Koswatta, P.: Optical character recognition for Sinhala language. Global Humanitarian Technology Conference: South Asia Satellite (GHTC-SAS), pp. 149–153 (2013)
7. Ramanathan, R., Rohini, P.A., Dharshana, G., Soman, K.P.: Investigation and development of methods to solve multi-class classification problems. In: Proceedings of International Conference on Advance in Recent Technology in Communication and Computing—ARTCom, pp. 27–28 (2009)

# A Spelling Mistake Correction (SMC) Model for Resolving Real-Word Error

Swadha Gupta and Sumit Sharma

**Abstract** Spelling correction has been haunting humans in various fields of society like while creating business proposals, contract tenders, students doing their assignments, in email communications, sending request for proposals, while writing content for website and so on. Already existing Dictionary based correction approaches have helped by providing solution to the problem when the written word doesn't even qualify to be called as a legal word. But is it the only challenge a writer faces while writing the desired documents! The words, which fall in the category of correct spelling words, may sometimes be the word which writer did not intend to write. The above illustrated genre of errors is called Real-Word error. This paper proposes a spelling correction system whose main focus is on automatic identification and correction of real word errors accurately and efficiently. The approach includes hybridization of Trigram and Bayesian approach and using Modified Brown corpus as a training set. A large set of commonly confused words is used in this case for evaluating the performance of the proposed approach.

**Keywords** Real-word errors · Spelling mistakes · Spelling corrector · Modified corpus · Supervised approach · Unsupervised approach

## 1 Introduction

Misspelled word correction refers to recognizing and rectifying the mistakes, which the user generally makes while writing text. The need of spell checker arose because mistakes are unavoidable part of writing as everyone cannot be skilful in representing language i.e. in writing process. So, Spelling correction has become very useful and widely used technology. When the spelling of the word is written

S. Gupta (✉) · S. Sharma
Department of Computer Science, Chandigarh University,
Gharuan, Mohali, India
e-mail: swadhagupta@rocketmail.com

S. Sharma
e-mail: sumit_sharma@mailingaddress.org

© Springer India 2016
H.S. Behera and D.P. Mohapatra (eds.), *Computational Intelligence in Data Mining—Volume 1*, Advances in Intelligent Systems and Computing 410, DOI 10.1007/978-81-322-2734-2_43

incorrectly, then that mistake is called non-word spelling error or mistake [1]. In others words, it can also be described as a word which does not exist in the list of words in the dictionary e.g. "I am fonde of music" rather than "I am fond of music". The word "fonde" is a non-word error. When the spelling of the word is written correctly but other word is written in place of actual word then that mistake is called Real-word spelling error [1, 2]. In other words, it can also be described, as a word user does not intend for e.g. "I am font of music" rather than "I am fond of music". The word "font" is a real-word error. Indeed, experimental analysis have highlighted that more than 40 % of errors are result of real-word error [1]. It is also known as context-based errors. Nearly all word processors have a built-in Spelling checker that flags the spelling mistakes i.e. non-word errors. The other mistakes such as real-word spelling mistakes go unrecognized by most of the current techniques. To identify the real-word spelling mistakes, there is a need to utilize the neighboring contextual information of the target word.

Let's take an example of a sentence, "I want to eat a piece of cake" and we have confused word set as (piece, peace). To identify that 'peace' cannot be used in this case, neighbouring contextual information like 'cake' for word 'piece' is to be used.

This paper is organized as follows: first, we describe previous work in the field of real-word correction. We then present proposed framework and results. This paper closes with a discussion of the choices made in formulating this methodology and plans for future work.

## 2  Advancements in Field

The need of real-word correction became prominent in the mid of 90's. This gained attention of researchers to usher in the field of real-word error correction. Peterson [3] discussed the errors which spelling checker computer program could not detect. Spell checker worked efficiently in identification and correction of non-word errors but failed to identify and correct context-sensitive errors. For removing non-word errors spell checker would check the word against the list of the given words. If the word exists in the list, then it is considered as correct otherwise flagged as incorrect. One possible to solution to this problem is an exhaustive list containing all the words existing in the vocabulary of the language chosen. In addition to it, there was also need of intelligent spell checker that detects and corrects both syntax and semantic errors in a sentence. Mays [4] introduced a statistical approach to deal with the problem of semantic errors. The probability of sentences was calculated by using the maximum likelihood estimation of probability and semantic errors are detected but it was limited to only few sentences. Yarowsky [5] introduced a learning algorithm to find out the sense of word that has more than one meaning so that it can be correctly used in a sentence. The learning method used in this case was unsupervised and the training set used did not include tagging. Words can as well as exist in more than one collocation, so the advantage of this feature of word was been used to sense disambiguity. Golding [6] introduced Tribayes, which was based on trigram, and Bayes to correct the context-sensitive spellings errors. Trigram was based on parts-of-speech of words and Bayes was based on features. Tribayes had used the best of both methods to deal with the problem of real-word errors but it worked only for a small dataset.

Golding [7] implemented Bayesian hybrid as another solution to the problem of context-sensitive spelling errors. In this, Bayesian classifier method put forward decision lists to make best use of both context texts and collocations. The target word with highest probability was substituted. To solve the problem of context-sensitive error, the collected evidences were transformed into a single piece of information. However, when it was applied to the real-word spelling errors correction, it performed much better than the component methods but the scope was still limited. Mangu [8] illustrated new way to correct the real-word spelling problem. Acquiring information in small set of rules was one of the important characteristic of this approach and was easily understandable. Rather than emphasis on large set of features and weight, it focused more on small set of rules. With the help of given technique, the machine could automatically understand and learn the rules, but the scope was very limited. Golding [9] proposed method called WinSpell to identify and correct context-sensitive spelling mistakes. The features during the training of Winspell were extracted and their weights were calculated and further assigned to them. It utilized information from multiple classifiers (features) rather than using single classifier to decide on the substitution of intended word, but it was not efficient for sentences consisting of more than three, four words. Davide Fossati [10] used mixed trigram model to correct the real-word errors. In this, the POS tagging was performed in order to tag the sentences using the Stanford tagger. The tagged sentences having the confusion words were compared with the HMM (hidden markov model) labeled tags. The difference was detected while comparing tags, which meant there was a misspelling in the sentence and the mixed trigram was applied to correct it. Therefore, the outcome of results exhibited increase in coverage of spell checker using the mixed trigram model.

Zhou [11] proposed a method known as RCW (real-word correction) for the real-word spelling errors based on tribayes. One of the major point of consideration in this work was that due to inadequate training set, there was exclusion of essential features. In this, Word Net was used to extract the pruned features and the problem of pruned features was solved to a certain degree but it became inefficient for long sentences. Context-based approaches, such as Tribayes [6], Bayesian hybrid method [7], Winnow [9], mixed trigram model [10], RCW (real-word correction with WordNet) based on Tribayes [11], correction with trigrams [12], have had certain degree of success for the problem of real-word correction. But the scope of correcting errors was limited to only predefine dataset because these all were corpus based methods.

Spelling correction had been a cumbersome task from long. The difficulty lied in identifying and correcting real-word spelling errors. The non-word spelling errors were not difficult to correct because of the availability of dictionary searching techniques. The foremost challenge for spelling correction in text was the development of real-word based correction techniques. Therefore, its correction was challenging as well as a very important task that needed to be achieved. It is being concluded after extensive analysis of literature that the identification and correction of real-word spelling errors or mistakes can be performed efficiently with trigram and Bayesian technique. Both techniques work well for real-word spelling correction. The limitation of using currently available corpus is that the contextual information is limited to only the text available in the corpus. Corpus contained contextual information of only limited text but does not cover all the contextual information of the language. Therefore, the scope of correcting mistakes remained limited to only the contextual information available in the corpus. Thus, the real-word error correction is limited to only small data set and its accuracy is also reduced.

## 3 Proposed Framework

Spelling correction is an application used to identify and correct the spelling mistakes in the text written by the user. Conventional spell checker fix only non-word errors and the real-word errors (words which are valid but are not intended by the user) goes undetected. Correcting this kind of problem requires a totally different approach from those used in the conventional spell checker.

Considering this problem, SMC method is proposed, which is based on trigram, and Bayesian approaches. This method would be able to solve the problem to a certain extent by adding the information in its sample, which is not available in the corpus so that knowledge of other contextual information can also be included in it. The addition of more information in the corpus will result in increasing the scalability of correcting large data set. Our approach also aims at retrieving the synonyms of the words, which is not available in the corpus by extracting synonyms from the dictionary of their corresponding words.

### 3.1 Training Feature

Brown corpus [13] is used as a training set in this proposed method but in the modified form to increase the scope of correcting real-word errors. The contextual information (sentences) is tagged with brown corpus [13] tagging and inserted in the samples of brown corpus [13]. The context information, which is added, is different from the information already available in the brown corpus [13].

### 3.2 Trigram Approach

Trigram approach takes full advantage of the data that is present in the surroundings of the target word i.e. collocation features. Trigram calculates the probability of collocation of three words in a sentence and adds all the calculated probabilities of a sentence. The probabilities of all the ambiguous words in the confusion set are calculated by substituting them one by one in a sentence. The target word having the highest probability is substituted in the final outcome and is considered as correct word.

$$p(w3|w1, w2) = \frac{f(w1, w2, w3)}{f(w1, w2)} \tag{1}$$

f(w1, w2, w3) $\rightarrow$ count of w3 is seen following w2 and w1 in brown corpus
f(w1, w2) $\rightarrow$ count of w2 is seen following w1.

## 3.3  Bayesian Approach

Bayesian approach optimally utilizes the data that is present in the surroundings of the target word i.e. context words. The extracted words surrounding the target word are named as features. Bayesian approach learns the contextual information surrounding the target word with the help of training corpus which further comprises of correct articles. The probabilities of context words are calculated by calculating the frequency of occurrences of features individually and the frequency of occurrences of features along with the target word. If the feature is not found in the corpus then synonym of that particular feature is extracted from the dictionary and its probability is calculated. The synonym having the highest probability is substituted in place of its corresponding feature. The ambiguous word having the highest score is substituted in the final outcome and is considered as correct word.

$$\textbf{Value}(\textbf{f})[\textbf{10}] = \log\left(\frac{p(c, a)}{p(c) * p(a)}\right) \tag{2}$$

p(w, a) $\rightarrow$ the joint probability between p(c) and p(a)
p(c) $\rightarrow$ probability of feature of the target word
p(a) $\rightarrow$ probability of target word

$$\textbf{Sum}(\textbf{w}_\textbf{a}) = \sum_{ci \in C} value(fi) + \sum_{sj \in S} \max(sj) \tag{3}$$

function $\max(s_j)$ $\rightarrow$ highest value of all synonyms of the feature $s_j$
function $value(f_i)$ is used to calculate the value of feature $c_i$
$c_a$ $\rightarrow$ ambiguous word

Bayesian approach is used when the POS tagging of ambiguous words are same else, in case of different tagging, its performance degrades.

The following procedure summarizes the algorithm:

**Input**: Sentence T =$w_1$, $w_2$, $w_3$..,$w_i$,...,$w_n$ ∈
*$w_j$ → input word*

$X → \{w_i, w_i{}^c\}$ *is confusion set*

**Output**: Corrected $w_i$ **If** $w_i$ ∈ X.
**If** $w_j$ ∈ C
    **then** tag the whole sentence
**Goto** Step 2
**Else**
        **print** " word not found in data set"
**For** i₌ 0,1,2....,n where i ∈ X do
**If** $w_j$ ∈ $w_i$, having different POS brown corpus tagging
    **then**
        trigram is applied
            Extract the collocation(T) ∈ A
                *A → training set*
            Find fr(col)
                *fr(col) → frequencies of the collocation of sentences*
            Combine corresponding collocations and frequencies.

            Calculate p($w_i$)
                *p($w_i$) → probability of $w_a$ in the C*

$$p(w3|w1, w2) = \frac{f(w1,w2,w3)}{f(w1,w2)}$$

                *f(w1,w2,w3) → count of w3 is seen following w2 and w1 in*
                        *brown corpus*
                *f(w1,w2) → count of w2 is seen following w1.*

        **Print** the word with highest probability
**Else If** $w_j$ ∈ $w_i$, having same POS brown corpus tagging
    **then**
        bayes is applied
            Extract the context words ∈ A
*A → training set*
            Find the fr(c)
            *fr(c) → frequencies of the context words*
            Combine corresponding features and frequencies.
**If** fr(c) = 0
    **then**
        extract S
*S → Synonyms of $w_i$*
        Calculate the sum of $w_i$.

$$\text{Value(f)} = log\left(\frac{p(c,a)}{p(c)\cdot p(a)}\right)$$
      *p(w, a) → the joint probability*
        *between p(c) and p(a)*
                           *p(c) → probability of feature o*
                              *the target word*
    *p(a) → probability of target word*

$$\text{Sum}(w_a) = \sum_{cl \in C} value(fi) + \sum_{sj \in S} max(sj)$$
                *function max($s_j$) → highest value of all*
                        *synonyms of the feature $s_j$*
                *function value($f_i$) is used to calculate the*
                      *value of feature $c_i$*
        *$c_a$ → ambiguous word*
      **Print** the word with highest probability

In above algorithm, different corpora are used for training set and testing set. Supervised learning approach is used for the training corpus, which is manually enhanced brown corpus, and unsupervised learning approach is used for testing and the test-set will be a collection of incorrect sentences which is manually created. We have supposed that the text in training and testing sets contains no spelling mistakes. Frequently occurring words in Brown corpus [13] is selected as the confusion sets. Test set is unsupervised as nobody indicates whether the spelling of the word it checks is correct or incorrect. SMC is better than RCW [11] to adapt because of the utilization of supervised and unsupervised strategy. We have found that, using this strategy, the performance of SMC is better on an unfamiliar test set. Two different corpora are used to enhance the performance of the system as well as supporting real life applicability.

## 4  Experimental Results

In this empirical study, two widely used and publicly available datasets i.e. brown [13] corpus and set of confused words are used to evaluate our proposed system. Brown corpus contains 1,014,312 words sampled from 15 text categories and set of 30 confused words are used. When the POS tagging for ambiguous words are different then trigram is used and when it is same, then bayes method is used. Trigram uses the context information in the form of collocation and bayes uses context information in the form of features. The probabilities for ambiguous words are calculated in both cases of trigram and bayes and word having highest probability is selected. The calculation of probabilities for set of 30 confused words is shown in Table 1.

1. Input: Everywhere capital and enterprise are lacking.

The experimental results of SMC is shown in Table 2.

The results are shown in above table. Number of test cases implies number of times confused words occurred in test corpus. Number of correct implies the number of cases SMC method corrected. The accuracy achieved for SMC is 92.17 % according to the results obtained.

Table 1  The information of the feature and value of ambiguous words—(capital, capitol)

| Collocation features | | | | Total |
|---|---|---|---|---|
| Everywhere capital and | Capital and enterprise | And enterprise are | Enterprise are lacking | |
| 0.156974 | 0.4385965 | 0.10 | 0.265897 | 0.9614675 |
| Everywhere capitol and | Capitol and enterprise | And enterprise are | Enterprise are lacking | |
| 0.235894 | 0.2698755 | 0.10 | 0.158975 | 0.7647445 |

**Table 2** The result of real-word error correction with the help of SMC model

| S. no. | Confusion items | No. of test cases | SMC | |
|--------|-----------------|-------------------|-----|--|
| | | | No. of correct | Accuracy (%) |
| 1. | Accept, except | 20 | 19 | 95 |
| 2. | Capital, capitol | 20 | 19 | 95 |
| 3. | Among, between | 20 | 18 | 90 |
| 4. | Brake, break | 20 | 19 | 95 |
| 5. | Desert, dessert | 20 | 19 | 95 |
| 6. | Device, devise | 20 | 18 | 90 |
| 7. | Farther, further | 20 | 18 | 90 |
| 8. | Formerly, formally | 20 | 19 | 95 |
| 9. | Hear, here | 20 | 19 | 95 |
| 10. | Instance, instant | 20 | 19 | 95 |
| 11. | Passed, past | 20 | 18 | 90 |
| 12. | Peace, piece | 20 | 18 | 90 |
| 13. | Principal, principle | 20 | 19 | 95 |
| 14. | Raise, rise | 20 | 19 | 95 |
| 15. | Sea, see | 20 | 17 | 85 |
| 16. | Stationary, stationery | 20 | 17 | 85 |
| 17. | Waist, waste | 20 | 19 | 95 |
| 18. | Weak, week | 20 | 18 | 90 |
| 19. | Than, then | 20 | 18 | 90 |
| 20. | Adverse, averse | 20 | 18 | 90 |
| 21. | Altar, alter | 20 | 19 | 95 |
| 22. | Appraise, apprise | 20 | 19 | 95 |
| 23. | Loose, lose | 20 | 19 | 95 |
| 24. | Pour, pore | 20 | 18 | 90 |
| 25. | Bare, bear | 20 | 18 | 90 |
| 26. | Censure, censor | 20 | 19 | 95 |
| 27. | Curb, kerb | 20 | 18 | 95 |
| 28. | Currant, current | 20 | 18 | 90 |
| 29. | Duel, dual | 20 | 17 | 85 |
| 30. | Storey, story | 20 | 19 | 95 |
| | Average | / | / | 92.17 |

# 5  Conclusion

The scope of correcting misspelled word is limited to contextual information of text used in corpus or contextual information of text written by author and becomes insufficient in correcting text apart from the contextual information of the corpus. To deal with the problem of real-word errors, we proposed an algorithm using trigram and bayes methods. Supervised and unsupervised learning strategy is used in this work. Supervised learning is used for training and unsupervised is used for testing. Unsupervised learning is used so that this work can be applied to any contextual information of the text. Brown corpus [13] is used as a training set and manually created sentences will be used as a test set because of unavailability of test sets. In our work, we have used supervised and unsupervised learning strategy so that evaluation is done on different contextual information of text from the training set. We have modified the brown corpus [13] by increasing the contextual information in the sample files of it to increase the scalability of correcting the real-word errors. The trigram method is used when POS tagging of ambiguous words in the confusion set are different and bayes method is used when POS tagging of ambiguous words in the confusion set are same. The algorithm also aim at taking an advantage of the synonyms of the context words, which is not found in the brown corpus [13]. The synonyms are extracted from the dictionary.

# 6 Future Scope

Although the SMC system developed in this research has gained some course of success in identifying and correcting the real-word spelling mistakes, it has also suggested several issues that needs to be addressed in the future development. Implementation and comparison with existing real-word spell checker such as MS Word 2007. Identification and correction of real-word errors is limited to small confusion sets. Therefore, the scalability of correction needs to be improved so that large set of words could be corrected. A large corpus is required that includes text of all possible contextual information. The number of real-word correction per sentence at a time is one; hence, algorithm has to be modified to do multiple corrections per sentence. Integration of proposed approach with the conventional spell checker needs to be done.

# References

1. Huang, Y., Murphey, Y.L., Ge, Y.: Automotive diagnosis typo correction using domain knowledge and machine learning. In: IEEE Symposium Series on Computational Intelligence, pp. 267–274 (2013)
2. Kukich, K.: Technique for automatically correcting words in text. ACM Comput. Surveys (CSUR) **24**, 377–439 (1992)
3. Peterson, J.L.: A note on undetected typing errors. Commun. ACM **29.7**, 633–637 (1986)
4. Mays, E., Damerau, F.J., Mercer, R.L.: Context based spelling correction. Inf. Process. Manage. **27**(5), 517–522 (1991)
5. Yarowsky, D.: Unsupervised word sense disambiguation rivaling supervised methods. In: Proceedings of the 33rd annual meeting on Association for Computational Linguistics. Association for Computational Linguistics, pp. 189–196 (1995)
6. Golding, A.R., Yves, S.: Combining trigram-based and feature-based methods for context-sensitive spelling correction. In: Proceedings of the 34th Annual Meeting on

    Association for Computational Linguistics. Association for Computational Linguistics,
    pp. 71–78 (1996)
 7. Golding, A.R.: A Bayesian hybrid method for context-sensitive spelling correction. arXiv
    preprint cmp-lg/9606001, pp. 1–15 (1996)
 8. Mangu, L., Eric, B.: Automatic rule acquisition for spelling correction. ICML **97**, 187–194
    (1997)
 9. Golding, A.R., Roth, D.: A winnow-based approach to context-sensitive spelling correction.
    Mach. Learn. **34**(1–3), 107–130 (1999)
10. Fossati, D., Eugenio, B.D.: I saw TREE trees in the park: how to correct real-word spelling
    mistakes. LREC, pp. 896–901 (2008)
11. Zhou, Y., et al.: A correcting model based on tribayes for real-word errors in english essays.
    In: 2012 Fifth International Symposium on Computational Intelligence and Design (ISCID),
    vol. 1, pp. 407–410. IEEE (2012)
12. Wilcox-O'Hearn, A., Graeme, H., Alexander, B.: Real-word spelling correction with trigrams:
    a reconsideration of the Mays, Damerau, and Mercer model. In: Computational Linguistics
    and Intelligent Text Processing, pp. 605–616. Springer, Berlin Heidelberg (2008)
13. Nelson, F.W., Kucera, H.: Brown Corpus Manual. Brown University (1979)

# Differential Evolution Based Tuning of Proportional Integral Controller for Modular Multilevel Converter STATCOM

L.V. Suresh Kumar, G.V. Nagesh Kumar and P.S. Prasanna

**Abstract** This paper discusses differential evolution algorithm for tuning of proportional integral controller in modular multilevel converter based STATCOM applications. Unlike conventional VSC based converters, an MMC is known for its distinctive features such as modularity, low harmonic content and flexibility in converter design. The MMC-STATCOM is capable of reactive power compensation, simultaneous load balancing and harmonic cancellation. Differential evolution algorithm is used to tune the proportional integral controller of STATCOM. The proposed model is verified in MATLAB/Simulink and the results are well in proximity with the theoretical analysis.

**Keywords** Modular multilevel converter · STATCOM · Differential evolution

## 1 Introduction

The rapid development and popularity of power electronics technology led to extensive use of industry loads which poses power quality problems. The power quality problems characterized with harmonic distortion, low power factor and phase imbalances create unexpected disturbances in the operation of electrical

L.V. Suresh Kumar (✉)
Department of EEE, GMR Institute of Technology,
Srikakulam, Andhra Pradesh, India
e-mail: lvenkatasureshkumar@gmail.com

G.V. Nagesh Kumar
Department of Electrical and Electronics Engineering, GITAM University,
Visakhapatnam, Andhra Pradesh, India
e-mail: drgvnk14@gmail.com

P.S. Prasanna
Department of EEE, GMR Institute of Technology, Rajam, Andhra Pradesh, India
e-mail: prasanna628@gmail.com

© Springer India 2016                                                                                                    439
H.S. Behera and D.P. Mohapatra (eds.), *Computational Intelligence
in Data Mining—Volume 1*, Advances in Intelligent Systems
and Computing 410, DOI 10.1007/978-81-322-2734-2_44

equipment. The compensation of non linear, poor power factor and unbalanced loads is an important issue in the modern power system. The static synchronous compensator (STATCOM) is one of the crucial FACTS controllers which can provide flexible control to mitigate power system disturbances. Converters presently employed in STATCOM are the voltage sourced type, but current sourced type converters may also be used. The major reasons for the preference of the voltage sourced converter are: (1) Current sourced converters require power semi conductor devices with bi directional blocking capability. (2) The losses of current source converters are higher as compared with the voltage source converters. (3) Automatic protection of power semiconductor devices against voltage transients is possible in voltage source converters. Current sourced converters may require additional over voltage protection or higher voltage rating for the semiconductors [1, 2].

Voltage sourced converters are broadly classified as: (1) Diode clamped multilevel converter (2) Flying capacitor multilevel converter (3) Cascaded multilevel converter. But for higher voltages, these converters fail to operate due to excessive usage of diodes, excessive use of flying capacitors and need for separate dc sources [3, 4]. Hence the industry has chosen MMC to deal with high power applications. They provide a reliable solution to build STATCOM eliminating the use of coupling transformer and enabling the power exchange with the power system. In addition, it can operate under unbalanced conditions such as during symmetrical and unsymmetrical faults thus avoiding the power system collapse [5, 6].

The modern power system network is dynamic in nature. Inherent disturbances which lead to parameter variations may result in degradation of the performance which eventually challenges the reliability of the system. In order to curb the menace of parameter variation, the controllers must be precisely designed. PI controllers are vastly used for practical applications due to their distinctive features such as faster response and reduction in the steady state error. Various methods such as ziegler-nichols method, linear quadratic regulation method, Genetic algorithm (GA) based tuning method, Particle swarm optimization (PSO) based tuning method etc., are used to tune the PI controller. The evolutionary algorithms (EA) and differential evolution (DE) share the benefit of simple structure as PSO. But DE is much simpler and it is easy to implement. It is known for its high computational efficiency and low space complexity [7, 8]. Hence DE is well suited to tune the PI controller for MMC based STATCOM.

## 2 MMC-STATCOM Configuration

Figure 1 shows the MMC based STATCOM configuration. A three phase MMC consists of three legs which inturn consists of two arms: upper arm and lower arm. Each arm consists of N number of series connected sub modules and an arm inductance $L_{arm}$. The sub modules are the building blocks of MMC. Each sub module consists of a half bridge configuration and a sub module capacitance. The

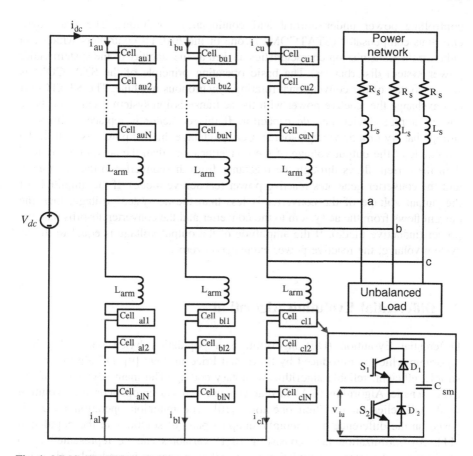

**Fig. 1** MMC-STATCOM configuration

half bridge cell consists of two switches which acts in complementary with each other. The output of the sub module is either equal to the capacitor voltage or zero depending upon the switching states. To synthesize an (N + 1) level waveform, 2N sub modules are required where N denotes the number of sub modules in an arm. The switching function of the sub modules is so controlled that at any instant, N sub modules out of 2N sub modules are inserted to generate the desired voltage level.

The increased development of power electronics technology and focus on renewable energy power generation led to significant changes in power system operation. The intermittent nature of renewable energy sources may cause frequency and voltage instability. The compensation of the non linear loads without affecting the normal operation of the power system is of major concern. Flexible AC transmission systems (FACTS) open up the new opportunities for controlling power under normal and contingency conditions. The static synchronous compensator (STATCOM) is one of the FACTS controllers which can enhance the transmission characteristics as well as the stability of the system under power system disturbances. The basic operating principle of the STATCOM is similar to that of the conventional rotating synchronous machine. The STATCOM can exchange the reactive power with the ac transmission system because of its dc energy storage device i.e., dc capacitor. If this capacitor is replaced with a dc storage battery or a dc voltage source it can exchange the real power as well. If the amplitude of the output voltage of the converter more than the ac system voltage, then the current flows through the reactance from the converter to the ac system, and the converter generates reactive power (capacitive mode). If the amplitude of the output voltage of the converter is less than the ac system voltage, then the current flows from the ac system to the converter and the converter absorbs reactive power (inductive mode). If the amplitude of the output voltage is equal to the ac system voltage, the reactive power exchange is zero.

# 3 Differential Evolution Algorithm

Differential Evolution (DE) algorithm is a population based search heuristic algorithm which is introduced by Storn and Price in 1997 [9]. DE algorithm is a simple, efficient, reliable algorithm with easy coding. The main advantage of DE over Genetic Algorithm (GA) is that GA uses crossover operator for evolution while DE relies on mutation operation [10]. The mutation operation in DE is based on the difference of randomly sampled pairs of solutions in the population [11]. An optimization task consisting of D variables can be represented by a D-dimensional vector. A population of $N_P$ solution vectors is randomly initialized within the parameter bounds at the beginning. The population is modified by applying mutation, crossover and selection operators. DE algorithm uses two generations; old generation and new generation of the same population size [12]. Individuals of the current population become target vectors for the next generation. The mutation operation produces a mutant vector for each target vector, by adding the weighted difference between two randomly chosen vectors to a third vector [13]. A trial vector is generated by the crossover operation by mixing the parameters of the mutant vector with those of the target vector. The trial vector substitutes the target vector in the next generation if it obtains a better fitness value than the target vector. The complete procedure and about its operators are clearly explained in [11–13].

# 4  Simulation Results

The proposed control algorithm is simulated in MATLAB/Simulink to verify the effectiveness of MMC based STATCOM. The circuit parameters are presented in Table 1.

The ac power network is maintained at 5 kV, 100 MVA. To examine the transient performance of the proposed system, a three phase rectifier is connected at the load terminals. Figure 2 represents the simulated waveforms of MMC-STATCOM at point of common coupling (PCC) with integral controller. Figure 3 represents the simulated waveforms of MMC-STATCOM at point of common coupling (PCC) with PI controller. It can be observed that the modular multilevel converter produces sinusoidal line voltages with low harmonic content. The low harmonic content totally eliminates the use of filters and the converter can be directly connected to the power line. Either integral controller or proportional integral controller can be used to generate the reference signal for MMC. However, the %THD of the STATCOM voltage with integral controller is 0.72 % and that of PI controller is 0.35 %. The %THD of the STATCOM current with integral controller is 3.30 % and that of PI controller is 2.64 %. The ripple content of active and reactive power with PI controller is less than that of the integral controller. The capacitor voltages are well balanced with the implementation of the PI controller rather than that of the integral controller. Hence MMC-STATCOM exhibits promising performance when the PI controller is employed.

DE algorithm is used to optimize the gains of integral controller and PI controller. The integral absolute error is used as the fitness function. The error is generated by considering the load currents of the ac network as reference values and the currents at PCC as the actual values. The integral gain of the integral controller is obtained as 2.21. The Proportional gain and the integral gain obtained by applying the DE algorithm are 0.12 and 1.5 respectively. DE-PI controller has achieved better optimal performance as compared with DE-I controller. However, a trade off exists between the converter efficiency and the computational speed.

**Table 1**  Circuit parameters

| | |
|---|---|
| Converter power capacity | 100 MVA |
| RMS line to line voltage | 5 kV |
| Source resistance | 25 mΩ |
| Source inductance | 200 μH |
| DC link voltage | 200 V |
| Arm inductance | 2.5 mH |
| Sub module capacitance | 3.6 mF |
| Fundamental frequency | 50 Hz |
| Carrier frequency | 0.5 kHz |
| Number of sub modules per arm | 2 |

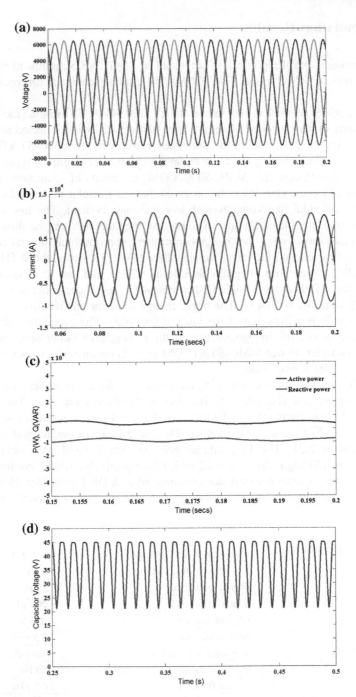

**Fig. 2** Simulated waveforms of MMC-STATCOM with I controller. **a** Voltage waveform at PCC, **b** current injected at PCC, **c** active and reactive power exchanges at PCC, **d** mean value of capacitor voltages

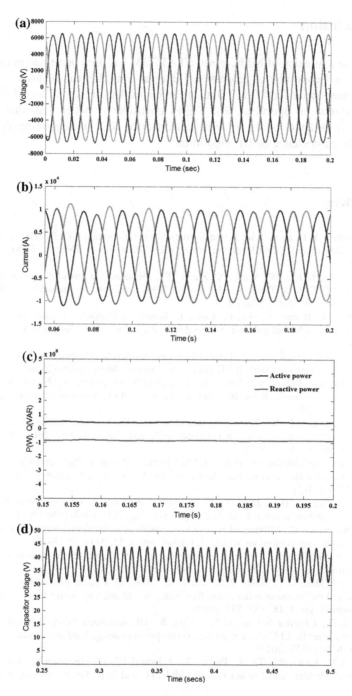

**Fig. 3** Simulated waveforms of MMC-STATCOM with PI controller. **a** Voltage waveform at PCC, **b** current injected at PCC, **c** active and reactive power exchanges at PCC, **d** mean value of capacitor voltages

# 5  Conclusion

The form and function of MMC based STATCOM has been presented in this paper. Either integral controller or proportional integral controller can be used to generate the reference signal for MMC. Both these controllers are tuned by adopting differential evolution algorithm. The DE-PI controller is better than that of DE-I controller in order to attain supreme performance. The simulation results clearly convey the competence of DE algorithm.

# References

1. Mohammadi, H.P., Bina, T.K.: A transformerless medium-voltage STATCOM topology based on extended modular multilevel converters. IEEE Trans. Power Electron. **26**(5), 1534–1545 (2011)
2. Yang, X., Li, J., Fan, W., Wang, X., Zheng, T.Q.: Research on modular multilevel converter based STATCOM. In: The Proceedings of 6th IEEE Conference on Industrial Electronics and Applications, pp. 2569–2574. (2011)
3. Rodriguez, J., Bernet, S., Wu, B., Pontt, J., Kouro, S.: Multilevel voltage-source-converter topologies for industrial medium-voltage drives. IEEE Trans. Ind. Electron. **54**(6), 2930–2945 (2007)
4. Franquelo, L.G., Rodriguez, J., Leon, J.I., Kouro, S., Portillo, R., Prats, M.A.M.: The age of multilevel converters arrives. IEEE Trans. Ind. Electron **30**(8), 28–39 (2008)
5. Marquardt, R.: Modular multilevel converter: an universal concept for HVDC networks and extended dc-bus-applications. In: Proceedings of the IEEE International Power Electronics Conference. (2010)
6. Nami, A., Liang, J., Dijkhuizen, F., Demetriades, G.D.: Modular multilevel converters for HVDC applications: review on converter cells and functionalities. IEEE Trans. Power Electron. **30**(1), 18–36 (2015)
7. Saad, M.S., Jamaluddin, H., Darus, I.Z.M.: Implementation of PID controller tuning using differential evolution and genetic algorithms. Int. J. Innovative Comput. Inf. Control **8**(11), 7761–7779 (2012)
8. Lara, L.D., Verma, H.K., Jain, Chesta: Differential evolution for optimization of PID gains in automatic generation control. Int. J. Comput. Sci. Eng. **3**(5), 1848–1856 (2011)
9. Stron, R., Prince, K.: Differential evolution—a simple and efficient adaptive scheme for global optimization over continuous spaces. J. Global Optim. **11**, 341–359 (1995)
10. Panda, S.: Differential evolution algorithm for SSSC-based damping controller design considering time delay. J. Franklin Inst. **348**, 1903–1926 (2011)
11. Chandra Sekhar, G.T., Sahu, R.K., Panda, S.: AGC of a multi-area power system under deregulated environment using redox flow batteries and interline power flow controller. Eng. Sci. Technol. Int. J. **18**, 555–578 (2015)
12. Sahu, R.K., Chandra Sekhar, G.T., Panda, S.: DE optimized fuzzy PID controller with derivative filter for LFC of multi source power system in deregulated environment. Ain Shams Eng. J. **6**, 511–530 (2015)
13. Sahu, R.K., Gorripotu, T.C.S., Panda, S.: A hybrid DE-PS algorithm for load frequency control under deregulated power system with UPFC and RFB. Ain Shams Eng. J. **6**, 893–911 (2015)

# Designing of DPA Resistant Circuit Using Secure Differential Logic Gates

Palavlasa Manoj and Datti Venkata Ramana

**Abstract** Crypto circuits can be attacked using the technique of differential power analysis by another/separate party, using power consumption dependence on secret message/information for extracting the critical data (information). To avoid DPA (Differential Power Analysis) and security bases, differential logic styles are basically used, because of constant power dissipation. This paper proposes a new design methodology using OR-NOR gates in 90 nm VLSI technology using SABL (Sense Amplifier Based Logic) for DPDN (Differential Pull down Logic) to secure/protect differential logic gates and to eliminate charge in the pull-down differential gate and to remove the memory effect.

**Keywords** Sense amplifier based logic · Differential Power Analysis · Differential Pull Down Network · Universal logic gates · 90 nm VLSI technology

## 1 Introduction

Now a days security is the most important concern. Encryption is usually based on exact protected algorithms, considered to make a secret message text from a basic text, which cannot be mathematically attacked [1]. However, the physical performance of the encryption algorithm—SCI (side-channel information) that preserves is used by an attacker to trace the important secret key. Side channel attacks (SCA's) [2] on the hiding policy used for extracting the information through different techniques and some defaults/weaknesses which give away the chance to find out the confidential key through power utilization, instant delay, and electromag-

P. Manoj (✉) · D. Venkata Ramana
Department of Electronics and Communication Engineering, GMR Institute of Technology,
Rajam, Andhra Pradesh, India
e-mail: slmh478@gmail.com

D. Venkata Ramana
e-mail: venkataramana.d@gmrit.org

© Springer India 2016
H.S. Behera and D.P. Mohapatra (eds.), *Computational Intelligence
in Data Mining—Volume 1*, Advances in Intelligent Systems
and Computing 410, DOI 10.1007/978-81-322-2734-2_45

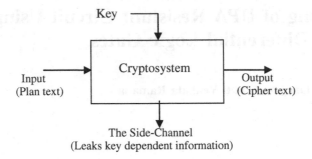

**Fig. 1** Basic side channel attack

netic emission, etc. DPA attack [3] is based on the well recognized information that active power utilization in a logic circuit. Thus, an attacker can acquire the top secret key by measuring the current from power supply of the cryptographic device while doing the encryption mechanism as power provide current of a hiding mechanism as it is the stage an encryption, and by analyzing the exact power traces (Fig. 1).

The output is dependent on the input signal when glitches are present. So, for these strict timing is required which is a drawback. They are depicted in given figure.

SABL (Sense Amplifier Based Logic) [4] is taken into consideration because in this style, the differential logic designs are provided with constant power supply. Some other logic styles are gate level masking, complementary circuits and DPA algorithms as depicted in the Fig. 2. The SABL logic style when it is used with, Low-Swing Current Mode Logic (LSCML) [5], Dynamic Current Mode Logic (DyCML) [6], Three-Phased Dual-Rail Pre-charge Logic (TDPL) [7] is less attacked by DPA. SABL is a differential logic technique that has the following

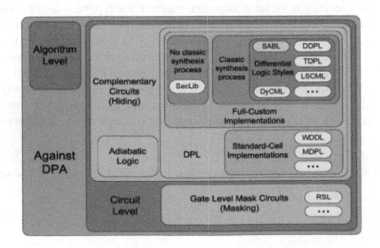

**Fig. 2** Various measures against DPA

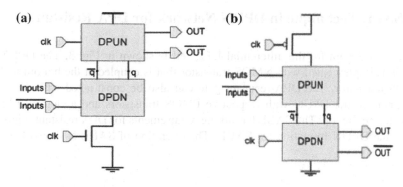

**Fig. 3** Dynamic and dual-rail gate logic style. **a** Using NMOS DPDN block logic. **b** Logic function with PMOS in (DPUN)

requirements apart from the other design styles: it has one charge event and uniform capacitance charges. SABL achieves better results because their internal structure suppresses the influence of internal capacitances better than the reduced output swing used by DyCML and LSCML.

Differential circuits utilize their inherent symmetry to have same consumptions for "0" and "1" evaluations, as both the true and complemented outputs are generated simultaneously. Figure 3 shows a typical scheme for a dynamic differential logic style. Such logic styles contain a differential pull-down network (DPDN) performing the logic function and a differential pull-up network (DPUN) working in alternate precharge and evaluation phases. They provide the complementary outputs at every clock cycle with a single clock event.

From Table 1, it shows that SABL gates provide the best trade-off in hardware resources, power and security, especially if balanced outputs are provided. It is inferred from Table 1 that Sense Amplifier Based Logic style is preferred because it is sensible to unbalanced outputs with constant power, low energy consumption, reduced delay and difficult to attack power traces by the effect of DPA. In this, the DPDN network is considered while pull down configuration energy consumption is reduced due the effect caused by sensing amplifier.

**Table 1** Features and performances of full-custom secure DPL techniques

| DPDN | No. of transistors DPDN | AND-NAND | | XOR-XNOR | | Drawbacks |
|------|------|----------|--------|----------|--------|-----------|
| | | $E_{avg}$ | $\sigma_e$ | $E_{avg}$ | $\sigma_e$ | |
| SABL [4] | 12 | 1 | 1 | 1 | 1 | Sensible to unbalanced outputs |
| LSCML [5] | 13 | 131.80 | 1.553 | 1.50 | $\approx$1e5 | Sensible to unbalanced outputs reduced output |
| DyCML [6] | 13 | 1.996 | 126.7 | 1.976 | $\approx$1e8 | Sensible to unbalanced outputs reduced output |
| TDPL [7] | 14 | 1.333 | 0.80 | 1.40 | 1.28 | Generation and routing of additional control signal |

## 2  SABL Technique in DPDN Network for DPA Resistance

The circuit design for the differential DPL cell is shown in Fig. 3. The DPDN in
Fig. 3a is implemented with NMOS transistor that is coupled at the bottom locked
NMOS transistor. The DPA-resistant gate can also be constructed with the logic
implement in the DPUN with respective PMOS transistors and a clocked PMOS
transistor on the top. The SABL has all the requirements for DPA resistant. Figure 3
shows the DPUN formation for SABL. The operation of SABL is shown below.

### *2.1  Operation*

SABL fulfills every necessity for differential power analysis. The DPUN formation
for SABL is shown in Fig. 4. The operation for SABL is given: the clock P-MOS
transistors (P6 and P8) here as DPUN in the pre charge phase are ON. In the
evolution stage (clock = 0), the sources of P7 and P10 NMOS transistors are
grounded because of discharge path in DPDN and switching action of transistor C
which is ON continuously and thus the logic function at the output depends on the
input values. The properties of SABL that makes it resistant to DPA are: (1) an
existence for clocked base transistor P11 (2) complete balance in DPUN, also
(3) Outputs of DPDN having no connection to the gate of output inverters (P1/P2
and P3/T4). Even using a proper common method for SABL in DPUN, the DPDN
must be balanced for implementing the logic regardless of input. For making an
effective DPDN to counter the DPA attacks, it should be completely balanced,

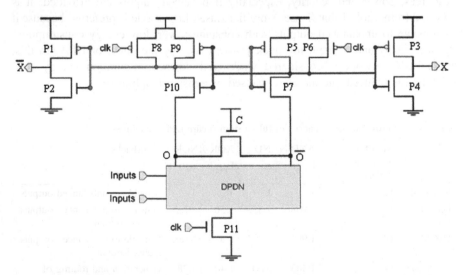

**Fig. 4** Sense Amplifier Based Logic Style (SABL)

equal number of NMOS series transistors with equal resistance in all paths. The gate operates with a constant delay regardless of input values. Figure 4 shows AND-NAND and XOR-XNOR for DPDN.

# 3  Realization of DPDN (Differential Pull Down Networks)

A technique for matching the charge in internal nodes during the pre-charge phase [8] is proposed in order to prevent the undesirable effects. This can be realized in two different ways: (1) by recycling the charge and distributing equally between the internal nodes and (2) All the internal nodes are by charged/discharged to the same value. In the two cases, it is enough to have specific transistors that are in the ON state only during precharge (Fig. 5).

## 3.1  Single Switch Solution (P)

The internal nodes in the similar level are attached together for a differential logic function for any DPDN execution through a switch which is ON during the pre-charge phase (clk = 0), by making voltages equal in the same level. There is only one switch for each transistor level in the DPDN except for the first one, which gives the complementary outputs. In the SABL structure, these are interconnected with the intermediate Vdd-gated NMOS transistor that is always ON. This measure confirms accurate distribution of charge during precharge without any leakage of information. In SABL, only a single phase clk is needed.

## 3.2  Dual Switch Solution (2P)

In the DPDN style the intermediate nodes are fixed to supply/ground terminals with independent switches while pre-charge, driving exactly the same voltage in all nodes. Exactly one pair of switches is required for each DPDN level except for first one, which generates the true and the complemented output. In the SABL structure,

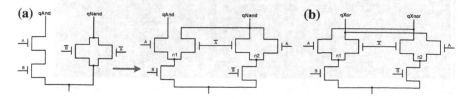

**Fig. 5** Realization for **a** N-MOS AND or NAND and **b** N-MOS XOR or XNOR DPDN

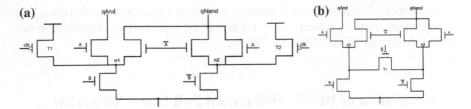

**Fig. 6** Single/dual switch solution for AND-NAND for DPDN gates

these are interconnected with the intermediate Vdd-gated NMOS transistor which
are ON continuously. Thus, for an N-depth DPDN, 2(N-1) switches are required. In
single-switch configuration, the only possible solution uses PMOS switches that are
ON during pre-charge, connected to Vdd (Fig. 6).

## 4 Simulation and Results

The proposed technology of SABL gates is performed using AND-NAND type and
XOR-XNOR type of design in sense amplifier based logic is simulated in Micro
Wind Tool and Digital Schematic Editor and Simulator (DSCH) using VLSI 90 nm
technology [9]. It is carried out using nominal conditions, i.e., for classic transistors,
Vdd and at temperature of 27 °C. The inputs and outputs of the entry individual
tested were passed through gates of same style having low clock frequency being

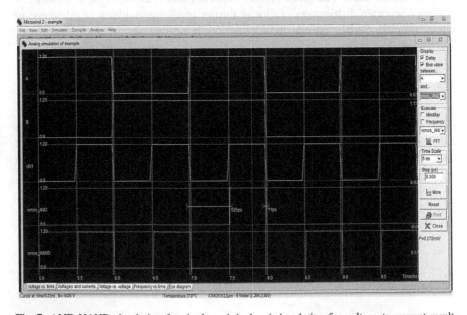

**Fig. 7** AND-NAND simulation for single and dual switch solution for voltage to current result

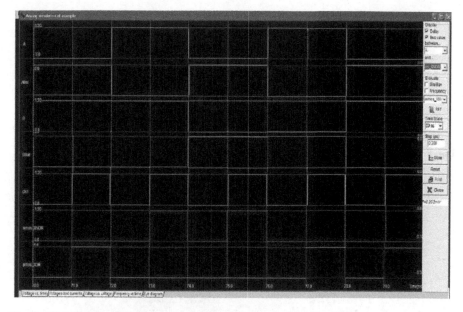

**Fig. 8** XOR-XNOR simulation for single and dual switch solution voltage to current result

0.500 GHz. The power consumption for all the possible combinations were measured. The Differential Pull down Logic (DPDN) in AND-NAND Gate for single and dual switch solution and in XOR-XNOR [10] Gate for single switch solution are presented respectively in Figs. 7 and 8.

Existing method was single (P) and dual (2P) switch solutions for AND-NAND and NOR-XNOR in power was changed for few cases like as input {i.e. 00 and 10} may be, and in the proposed technique are at cases as input {00, 01, 10, 11} for all conditions in single and dual switch solutions at constant power. So, constant power source to destination constant power such a case DPA is not attacked for finding out secret information/data.

## 5 Implementation

In the proposed technique, OR-NOR gates are used to secure/protect differential logic gates at 90 nm technology, to eliminate charge in the pull-down of a differential gate and remove the memory effect. In Sense Amplifier Based Logic (SABL) gate in Differential pull down network (DPDN) circuit using by OR-NOR gates for single switch solution and dual switch solution (Figs. 9 and 10).

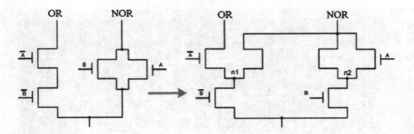

**Fig. 9** Single/dual switch solution for OR-NOR gates

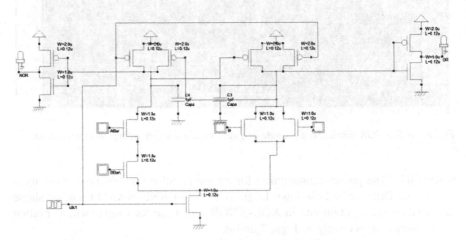

**Fig. 10** SABL style for DPDN for OR-NOR gates

## 5.1 Results

The OR-NOR gate are used in SABL style to secure differential logic gates for constant power supply for all conditions and calculated the power at all cases {00, 01, 10, 11}. So, all cases available in single and double switch solutions. Such a case constant power supply allows constant data/information for sending. The memory effect is due to effect of internal charge stored due power variations. Hence, in the proposed technique uses constant power and therefore memory effect is constant in data/information as flowing from source to destination (Figs. 11 and 12).

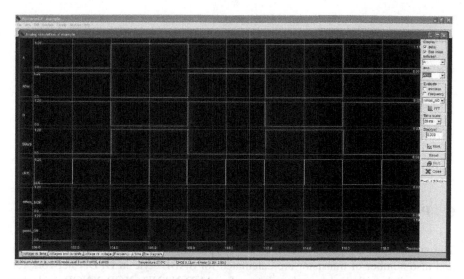

**Fig. 11** OR-NOR simulation for single and dual switch solution voltage to time result

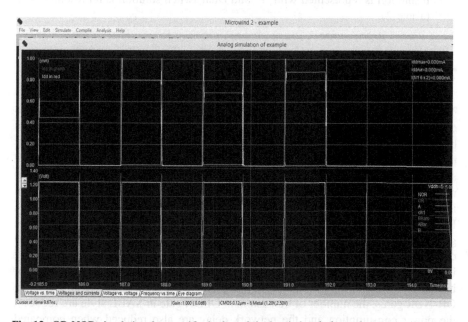

**Fig. 12** OR-NOR simulation for **a** and **b** single and dual switch solution voltage to current result

**Table 2** Performance comparison of DPDN gates

| DPDN gate in SABL style | No. of transistors in DPDN | Total power (mW) | Delay (ps) | Energy consumption (fj) |
|---|---|---|---|---|
| AND-NAND | 16 | P → 0.231 | 250 | 30.25 |
| | | 2P → 0.230 | | |
| XOR-XNOR | 15 | P → 0.203 | 190 | 25.03 |
| | | 2P → 0.200 | | |
| OR-NOR | 13 | P → 0.129 | 140 | 18.59 |
| | | 2P → 0.120 | | |

# 6 Performance Evaluations

The simulations are performed at 27 °C temperature. The delay, energy deviation and also number of transistor are calculated for SABL style in DPDN network for single switch and dual switch solution for AND-NAND, XOR-XNOR and also OR-NOR gates to protect security bases of view is given by in this table Single switch solution is represented with 'P' and Dual switch solution is represented with '2P' (Table 2).

# 7 Conclusion

This paper implemented to provide security data transmission through Crypto circuits. The implemented method uses OR-NOR gate in pull down configuration based on SABL (Sense Amplifier Based Logic) for DPA resistant circuits. This mechanism provide eliminates the memory effect with the help of single and dual switch solution in the use of differential pull down network. So the total power is reduced due to internal effect charge used. The memory effect and energy deviation is also successfully reduced to secure/protect differential networks against DPA Resistance Circuit. The proposed results for OR-NOR gates in Differential logic gates which are provided to eliminate the internal charge stored at the pull down configuration in SABL (Sense Amplifier Based Logic). The designed methodology is implemented using CMOS technology at 90 nm for complementary circuits for hiding the secret information/data. By implementing the design technique using OR-NOR gates, the number of transistors in DPDN gate in SABL style are reduced, the power consumption and the network delay are also reduced when compared with the existing methods.

# References

1. Tena-Sánchez, E., Castro, J., Acosta, A.J.: A methodology for optimized design of secure differential logic gates for DPA resistant circuits. IEEE J. Emerg. Sel. Top. Circuits Syst. **4**(2), 203–215 (2014)
2. Gutmann, P., Naccache, D., Palmer, C.C.: Side-channel attacks on cryptographic software. IEEE J. Co-published Comput. Reliab. Soc. **9**, 1540–7993 (2009)
3. Kocher, P., Jaffe, J., Jun, B.: Differential power analysis. In: Proceedings of the International Cryptal Conference, pp. 388–397. (1999)
4. Tiri, K., Akmal, M., Verbau Whede, I.: A dynamic and differential CMOS logic with signal independent power consumption to withstand differential power analysis on smart cards. In: Proceedings of the European Solid-State Circuits Conference, pp. 403–406. (2002)
5. Hassoune, I., Macé, F., Flandre, D., Legat, J.-D.: Low-swing current mode logic (LSCML): a new logic style for secure and robust smart cards against power analysis attacks. Micro Electron. J. **37**(9), 997–1006 (2006)
6. Allam, M.W., Elmasry, M.I.: Dynamic current mode logic (DyCML): a new low-power high-performance logic style. IEEE J. Solid-State Circuits **36**(3), 550–558 (2001)
7. Bucci, M., Giancane, L., Luzzi, R., Trifiletti, A.: Three-phase dual rail pre-charge logic. In: Proceedings of the International Workshop Cryptography. Hardware Embedded Systems, pp. 232–24. (2006)
8. Suzuki, D., Saeki, M.: Security evaluation of DPA countermeasures using dual-rail pre-charge logic style. In: Proceedings of the International Workshop Cryptography. Hardware Embedded Systems, pp. 255–269. (2006)
9. Lin, L., Burleson, W.: Leakage-based differential power analysis (LDPA) on sub-90 nm CMOS crypto systems. In: Proceedings of the IEEE International Symposium Circuits Systems, pp. 252–255. (2008)
10. Tiri, K., Verbau Whede, I.: Design method for constant power consumption of differential logic circuits. In: Proceedings of Design, Automat. Test European Conference on Exhibit, pp. 1530–1591. (2005)

# Efficient Algorithm for Web Search Query Reformulation Using Genetic Algorithm

Vikram Singh, Siddhant Garg and Pradeep Kaur

**Abstract** A typical web user imposed small and vague queries onto web based search engines, which requires higher time for query formulation. In this paper, a nature inspired optimization approach on term graph is employed in order to provide query suggestion by assessing the similarity. Term graph is simulated according to the pool of relevant documents of user query. The association among terms graphs is based on similarity and will be act as fitness values for genetic algorithm (GA) approach, which converges by deriving query reformulations and suggestions. Each user interactions with the search engine is a considered as an individual chromosome and larger pool help in convergence for significant reformulations. Proposed algorithmic solution select optimal path and extracts the most relevant keywords for an input search query's reformulation. The query user will select one the suggested reformulated query or query terms. The optimization performance of the proposed method is illustrated and compared with different optimization techniques, e.g. ACO, PSO, ABC.

**Keywords** Ant colony optimization · Genetic algorithm · Query reformulation · Particle swarm optimization · Query suggestion

## 1 Introduction

In recent decade's web based search engines have become more interactive and manipulative on user query, which leads to more automated knowledge acquiring and designing structures computation involved. Automated knowledge acquisition

V. Singh (✉) · S. Garg · P. Kaur
National Institute of Technology, Kurukshetra, Haryana, India
e-mail: viks@nitkkr.ac.in

S. Garg
e-mail: siddhantgarg@outlook.com

P. Kaur
e-mail: erpradeepkaur@outlook.com

© Springer India 2016
H.S. Behera and D.P. Mohapatra (eds.), *Computational Intelligence in Data Mining—Volume 1*, Advances in Intelligent Systems and Computing 410, DOI 10.1007/978-81-322-2734-2_46

is helpful to web users in effectively and efficiently navigating over a huge document collection. In the earlier days, query log analysis was a prominent approach for knowledge acquisition and thus a promising research area. Over the past decades, information retrieval (IR) emerges as a dominant research area along with web mining due to the comprehensive growth and evolution of data (documents) in the Internet [1, 2]. Subsequently, web users are contributing data into the web with almost exponential rate [2].

A typical search engine retrieve a large set of results (documents or URL's) on the submission of user query, upon which user may expect relevant documents in top pages of retrieved results. Generally, a web user frequently adjusts (modify) previously imposed query with an expectation of retrieving better results. These adjustments or modifications on input query are leads to query reformulations or query suggestion. Query reformulations also intend to resolve the expressions mismatch during query answering, keeping semantics of original query and reform it to new query that have best possibility to retrieves better and find better documents. Query reform includes various sub activities like; Query recommendation, Suggestion, Modification and Expansion [2].

The fundamental motivation of both a search engines and humans are to converge upon appropriate set of query reformulations. In the modern day information age, there are many web based search engines are available and each has its own ways of reformulation or suggestions. A typical search engine user reformulate queries based on the initial query search results and in some cases their information and knowledge of how particular search engines work. In current statistics, 28 % queries (approx.) are generalization of the previously submitted query of 2 billion daily (roughly) web searches on internet [3, 4]. For example, a search for 'team India', can be alternated to a query to 'cricket team India schedule' in case result are not satisfactory from initial query. Query Reformulations put up a large part of web based search commotion, in a study by Dogpile.com logs and Jansen et al. [5] showed that 37 % of queries are reformulated queries while Altavista logs [6] study shows that 52 % of web search users reformulate initial queries with anticipation of better results.

The reformulation is an interactive and iterative attempt amid users and search engines in order to achieve to satisfactory results. The retrieved results play an important role on effectiveness of reformulation strategy and accuracy of query suggestion phase. The query suggestion is entirely based on retrieved results, where results consists of documents or URL's, and key terms within results are future the possible suggestions for a user query. Each retrieved results can have multiple relevant (related to query) key terms and a term association graph is build so that best possible suggestion from graph can be made exhaustively. The nature based optimization algorithms are the appropriate techniques to solve an exhaustive search based scenario by generating the optimal solutions by keeping the best trade-off between design objectives. Our proposed query reformulation approach, primarily built upon genetic algorithm (GA) inspired suggestions. Each query or reformulated query submission to search engine is considered as individual and thus encoded as chromosome for GA. GA based approaches generate the optimal paths on term association graph, the reformulation is suggested based on terms in the optimal paths.

## 1.1 Related Work

The initial research contributions to design a query reformulation strategy for a typical search engine are based on query logs analysis. In [7] complete review on query reformulation based on query logs analysis is discussed, while in this paper our focus is to considered approach designed by analyzing query term and keyword retrieved from top–k ranked. In recent research works on query reformulation for web search engine are motivated and designed for generating query suggestions automatically to the user. There are various inherent sub activates are involves during the reformulations; query expansion [4], query substitution [8], and refinement techniques [4, 9]. User session based query reformulation are discussed in [5, 8, 10, 11], in which a user session is bounded as a series of communication by the user in intention to fulfill the information need. Thus, sessions are considered as single query, followed by multiple reformulated queries. An alternate approach for reformulation is according to the user click data analysis; searched results relevance is indicated by the pattern of click data by user. In [2, 12] such approaches are discussed and highlighted that click data is indicator of preferences of search results and proposed best methods of analyzing the same. In recent years, some potential milestones are achieved like of, approach in [13] classified a random sample sized 100 documents and then categorized into eleven classes to perform reformulations. In [10, 11, 14] similar reformulation categories are used on extracted term from linguistics. Query reformulation through query logs is principally based on re-finding behavior, in [15] an algorithms which to perceive a subset of the reformulations using query logs is demonstrated, similarly [7] constructed a small nomenclature and used a provisional random field model to forecast query refinements. More recent work [15] proposed conceptual reformulation according to the categories like content, format, resource of the result retrieved initially.

## 1.2 Contribution and Outline

The main contribution of the paper are mentioned below,

- Our key contribution is a GA based query reformulation strategy according to the similarity among document terms in association term graph.
- Another contribution, is detailed optimization performance analysis among the other nature inspired optimization techniques such, ACO (Ant Colony

Optimization), PSO (Particle Swarm Optimization) over the various performance parameter, Precision versus Recall, Optimization efficiency, distribution of query reformulation type etc.

In Sect. 2, query reformulation strategy is explained with other complementary activity. Proposed GA based query reformulation is explained in Sect. 3 and process is also explained. The result section illustrated the performance of the proposed approach with other optimization approach.

## 2 Query Reformulation

Query answering in a web based search engine has been considered critical and key issue in dynamic distributed computing environments [2]. A significant stride in this procedure is reformulating a query posed by a user and to target reformulated query while considering existing semantics correspondences between them. Traditional approaches usually aim at reformulating queries by means of equivalence correspondences. However, query terms from a initial query does not always have accurate or strict mapping on query terms in a reformulated query, which might have empty resultant reformulation and retrieve no answer to users [16]. In this case, if users highlight the relevance for them to receive semantically related answers, it may be better to produce an enriched query reformulation and, consequently, close answers than no answer at all.

The existence of multiple and heterogeneous data schemas, describing related data is a common phenomenon in those distributed settings, such as in the area of scientific research as biology, geography, health, education. In these scenarios, people have overlapping data, and they want to access other sources' additional information. In general, query answering in such environments is complex and hard to achieve. A typical query answering involves steps like: (i) query submission (ii) query analysis (iii) relevant data sources' identification (iii) query reformulation (iv) query execution (v) answers integration and finally (vi) query result presentation. Query reformulation has been considered one of the most critical, since it is concerned with the ability of translating the queries based heterogeneous data sources located at different locations and their inter-correspondence. Thus, when a user, at a given peer P, formulates a query posed over its schema, answers are computed at P, and the query is reformulated and forwarded to other peers through correspondence paths in the network. Since initial query usually do not enclosed with absolute information to answer a given query, any relevant refinement on query might add new or may be complementary answers. Furthermore, more paths of correspondences are explored which may poses different answers. In next section, a GA based query reformulation approach is discussed, which is focused on exploration of new path for reformulation.

# 3 GA Based Query Reformulation

On the arrival of user query to the web based engine, first step is to retrieve relevant documents (document, URL etc.) on relevance of importance, concurrently query keyword are loaded into a ternary search tree (TST) which is considered efficient for the search purpose over a search tree. In the next step, association among documents are define based on the similarity and association term graph is created, this term graph is attached at the end leaf node of TST constructed in previous step. Each graph traversal (based on the similarity value) is considered a genetic algorithm's (GA) chromosome and encoded. GA converges with top-k optimal paths, keyword present in the path are extracted and suggested for the user subsequently. The schematic diagram of the proposed query reformulation is shown in Fig. 1.

Next, the important processes involved in query reformulations are discussed:

*Formation of Ternary Search Trees*: based on the terms retrieved from top documents, terms are inserted into the ternary search structures (TST), The TST has preferred over other type of structures, as its effectiveness on correcting frequent spelling errors from user queries and also efficient for the purpose, in terms of computing storage space during strings to be stored share a common prefix. As

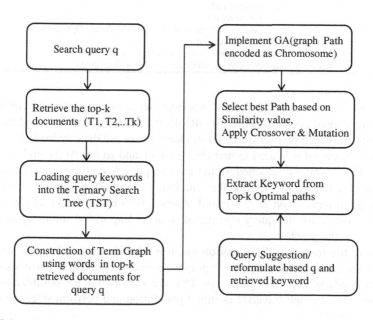

**Fig. 1** Schematic diagram of proposed GA based query reformulation

shown in Fig. 1, a TST for key terms is shown along with associated term graph. At end of each keyword in ternary tree, a term graph is linked. Term graph is a graph in which terms ($t_{1 \ldots n}$) present in top-k documents retrieved for the keyword query are interlinked based on its similarity with other document. Set of query Q ($q_1$, $q_2$, ..., $q_n$) are fed into ternary tree and query node will act as a root node in the graph to be constructed in next phase.

*Construction of Associated Term Graph*: For every keyword loaded into the ternary search tree, an associated term graph is constructed where the node of the graph is the document which contains the word. The following algorithm is used for the construction of the term graph:

---

***Term_Graph_Construction ($q_i D_n$) and Associated_term_graph Creation ($D_i D_j$, $W_{ij}$)***

**Input:** query terms $q_i$, keywords retrieved from documents $d_t$
**Output:** Graph structure $G_t$
    For each retrieved document $d_i \in D$ retrieved for the each query q
        Extract terms from $d_i$ and create node for each term and create edges between
        these nodes to be complete graph. Based on **Similarity value**
        Computing Similarity (Di, $D_j$)
        **Input**: Keywords ($k_i$,$k_j$) of two documents $D_i$ and $D_j$
        **Output**: Similarity value between $D_i$ and $D_j$
        For each keyword in document $D_i$
            Load the $k_i$ a map M.
            For each keyword in document $D_j$
            If $k_j$ present in M
                Increment count
                Assign the weight on edge between $D_i$ and $D_j$ as count
                For each term $d_t$
            If a term $d_t$ present in more than on document then
    Create an edge connecting $d_t$ with terms present in documents

---

In Fig. 2, constructed TST and association term graph is shown. The ternary search tree consists of three terms 'WORMHOLE', 'BLACKHOLE' and 'SPACE' is constructed based on user query term. Query terms are directly loaded into TST, first query keyword is loaded as first child of TST and so on. At the end of each of child of TST a graph is attached. In the term graph, relevant documents are loaded on nodes and edge represents the similarity among the pair of documents. A association term graph is drawn and attached with each of TST key word. The graph is traverse for the query reformulation according to the similarity values on the edges of the graph.

Each of the traversal of association graph is treated as input for next step.

*Applying Genetic Algorithm*: each graph traversal is encoded as chromosome for GA, traversals of all term graphs in TST are considered for optimization. GA considers, set of graph traversal as initial population and perform selection of best path according to threshold similarity values (Thr$_j$) and apply genetic operator subsequently. GA used for selection of top-k paths is described below,

---

*Genetic Algorithm*

---

**Step 1: Generating all Paths,**
    **Input:** Associated Term Graph G$_i$
    **Output:** Path sequences $\pi_i$
        For each graph G$_i$
            Insert all the nodes, except the root node, in an array and    Compute all the permutations of the array and each permutation with root node in the beginning is a new path.
**Step 2: Selecting_appropriate_paths($\pi_i$,Th$_j$),**
    **Input:** Paths($\pi_i$)
    **Output:** Selected paths.
        For each path $\pi_i$
           if sum of weighrs of each edges> threshold of pass j
           Add the path to selected set
**Step 3: Genetic Operation (Crossover, Mutation)**
    **Crossover:**
    **Input:** Path $\pi_i$ and Path $\pi_j$ from the selected list of paths, P$_c$
    **Output:** New path formed after crossover
        Interchange the first half of $\pi_i$ with $\pi_j$ and $\pi_j$ with of $\pi_i$.
        if sum of weights of new path formed >threshold of pass I,
        Add the newly generated path to list of selected paths.
    **Mutation:**
    **Input:** Remaining Path $\pi_j$ from the selected list of paths, P$_m$
    **Output:** New path formed after mutation
**Repeat until** Stopping criteria met (top-k paths)

---

*Query Reformulation*: GA converges with a set of optimal (relevant and good) paths; each optimal path consists of number of nodes and edges of associated term graph. Each nodes in a optimal path is considered relevant to the query keyword and query. In proposed solution of query reformulation, a path with higher similarity values is preferred as it is assumed that for a query these documents will suggest more important and relevant query terms. Once path is selected, next step is to identify the distinct terms occurs in this path(on documents), algorithm (Query reformulation and Suggestion) extracts all the distinct terms appearing on documents falling in the path. The extracted words are then suggested for the query reformulation and send to the user, these set of extracted words are suggested to the web based search users as part of query reformulation. If any incomplete or irrelevant words exist in the query then a relevant word is suggested to the user depending on the stored documents. And hence this system is independent of query

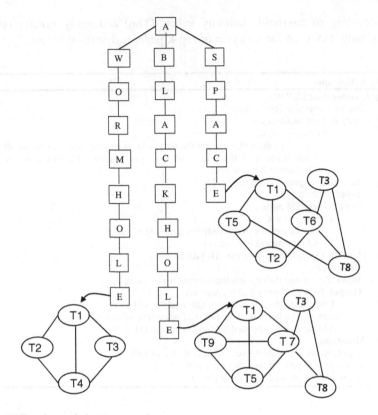

**Fig. 2** TST and association term graph

logs and depends only on the existing documents for suggestion of the queries. The
process for suggestion and reformulated is discussed below,

---

**Query_Reformulation (keywords K$_i$, Path Set π) and Suggestions (Q$_i$)**

**Input:** Set of keywords K$_i$ and Path Set π.
**Output:** Reformulated Query.

        For each Keyword K$_i$ and paths π$_j$
            Store the list of keywords of documents present in π$_j$.
            Find the top occurring words in the map and
        **Suggest ( );** return the reformulated query.
        **Input:** Query Keyword (Ki)
        **Output:** final set of query (Q$_i$)
            For each query word,
                Select the similar word from the documents
                Suggest those similar along with the word obtained

---

**Example** For given set of user queries, below given documents are retrieved based on which the queries are reformulated. Reformulation queries are also shown in the below given table.

Queries:

$Q_1$: *What is a wormhole?*
$Q_2$: *Wormhole and physicist working on phenomena*
$Q_3$: *Year of identification of wormhole*
$Q_4$: *Relation between wormhole, black hole and spacetime*
$Q_5$: *Theory of relativity and wormhole*

| Doc no | Input text | Terms |
|--------|-----------|-------|
| Doc 1 | A wormhole is a hypothetical topological feature that would fundamentally be a shortcut through space time. A wormhole is much like a tunnel with two ends, each in separate points in space time. For a simplified notion of a wormhole, visualize space as a two-dimensional (2D) surface. In this case, a wormhole can be pictured as a hole in that surface that leads into a 3D tube (the inside surface of a cylinder). This tube then re-emerges at another location on the 2D surface with a similar hole as the entrance. An actual wormhole would be analogous to this, but with the spatial dimensions raised by one. For example, instead of circular holes on a 2D plane, the entry and exit points could be visualized as spheres in 3D space. Researchers have no observational evidence for wormholes, but the equations of the theory of general relativity have valid solutions that contain wormholes. Because of its robust theoretical strength, a wormhole is one of the great physics metaphors for teaching general relativity. The first type of wormhole solution discovered was the Schwarzschild wormhole, which would be present in the Schwarzschild metric describing an eternal black hole, but it was found that it would collapse too quickly for anything to cross from one end to the other. Wormholes that could be crossed in both directions, known as traversable wormholes, would only be possible if exotic matter with negative to suggest that tiny wormholes might appear and disappear spontaneously at the Planck scale, [*wikipedia*] | General great hole hypothesis hypothetical inflated inflation inside instead leads location macroscopic matter metaphors metric natural naturally negative negative-mass notion observational one ordinary other others physicists physics pictured plane point. points possible predicted present process proposed quantum quickly raised re-emerges regions relative relativity relativity relativity. |

# 4   Result Analysis

Experiments are carried out by supplying user queries (Q.1 to Q.5) on the 500 documents (similar to above mentioned documents) on the GA based query reformulation approach along with similar ACO and PSO based approach.

(a).  **Precision versus recall of query reformulation techniques**

(b).  **Distribution of reformulation type**: The figure shows that most popular reformulations types are those where users move to a more specific intent or express the same intent in a different way. Reformulations with spelling suggestions and query generalizations are less popular.

(c).  **Reformulation Efficiency**: Reformulation efficiency is inferred to indicate the quality of search results, in our experiments using GA, PSO and ACO based reformulations both reformulated and initial queries are used to for comparative analysis of approaches. Some PSO and ACO based reformulations are misidentified as initial queries, which is overcome in GA based reformulation, due to which identification of reformulated query has no consequence on our study of reformatted queries. GA shows the better Precision/recall ratio and reformulation efficiency are different for each of the strategy, which is significant as evaluation is done over a huge dataset (Fig. 3).

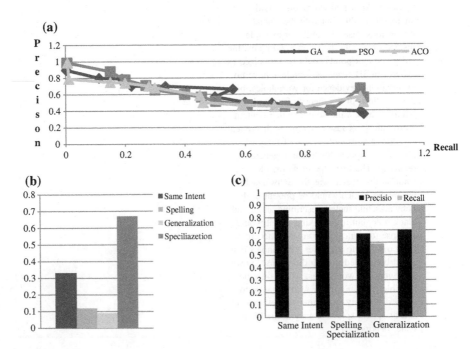

**Fig. 3**  Graphs for **a** precision versus recall for optimization techniques, **b** reformulation type and **c** precision and recall for reformulation types

# 5 Conclusion

In this paper, GA based effective and efficient approach for web based query reformulation is proposed. The query reformulation approach is build upon ternary search tree and term graph extracted from relevant documents retrieved for initial user query. Query reformulation based on the relevant terms will ensured the semantically similar query suggestion from the initial query suggestion. There are multiple equivalent query suggestion possible for user query, optimal plans is selected by genetic algorithm based approach according to the similarity values (act as fitness function). In the paper optimization performance of various approaches for the query reformulation is also discussed, such as ACO and PSO on various parameters. GA emerges as competitively effective on the query reformulation.

# References

1. Manning, C.D., Raghavan, P., Schutze, H.: Introduction to Information Retrieval. Cambridge University press, Cambridge (2008)
2. Fonseca, B.M., Golgher, P.B., de Moura, E.S., Possas, B., Ziviani, N.: Discovering search engine related queries using association rules. J. Web Eng. 2(4), 215–227 (2003)
3. Chirita, P.A., Firan, C.S., Nejdl, W.: Personalized query expansion for the web, In: Proceedings of the 30th International ACM SIGIR Conference on Research and Development in Information Retrieval, pp. 07–14, (2007)
4. Ozmutlu, S.: Automatic new topic identification using multiple linear regressions. Information Processing and Management 42(4), 934–950 (2006)
5. Dignum, S., Kruschwitz, U., Fasli, M., Kim, Y., Song, D.: In-corporating seasonality into search suggestions derived from intranet query logs. In: Proceedings of the IEEE/ACM International Conference on Web Intelligence and Intelligent Agent Technology, pp. 425–430 (2010)
6. Jones, R., Rey, B., Madani, O., Greiner, W.: Generating query substitutions. In: WWW '06, pp. 387–396, (2006)
7. Jones, R., Rey, B., Madani, O., Greiner, W.: Generating query substitutions. In: Proceedings of the 15th ACM International Conference on World Wide Web, pp. 387–396 (2006)
8. Nm Rieh, S.Y., Xie, H.: Analysis of multiple query reformulations on the web: the interactive information retrieval context. Inf. Process. Manage. 42(3), 751–768 (2006)
9. Jansen, B.J., Spink, A.: Real life, real users, and real needs: a study and analysis of user queries on the web, In: Information Processing and Management, pp. 207–227, (2000)
10. Wang, X., Zhai, C.: Learn from web search logs to organize search results. In: Proceedings of the 30th ACM SIGIR Intl. Conf. Research and Development in Information Retrieval, pp. 87–94 (2007)
11. Mitra, M., Singhal, A., and Buckley, C.: Improving automatic query expansion. In: SIGIR '98, pp. 206-214, (1998)
12. Yin, Z., Shokouhi, M., Craswell, N.: Query Expansion using external evidence. In: Advances in Information Retrieval. Springer, Heidelberg, pp. 362–374, (2009)
13. Jeh, G., Widom, J.: Simrank: a measure of structural-context similarity. In: Proceedings of the 8th ACM SIGKDD International Conference on Knowledge Discovery and Data Mining, pp. 538–543, (2002)

14. Agichtein, E., Brill, E., Dumais, S.: Improving web search ranking by in-corporating user behaviour information. In: Proceedings of the 29th ACM SIGIR International Conference Research and Development in Information Retrieval, pp. 19–26 (2006)
15. Whittle, M., Eaglestone, B., Ford, N., Gillet, V.J., Madden, A.: Data mining of search engine logs. In: Journal of the American Society for Information Science and Technology, pp. 2382–2400, (2007)
16. Cui, H., Wen, J.R., Nie, J.Y., Ma, W.-Y.: Query expansion by mining user logs, In: IEEE Transactions on Knowledge and Data Engineering, pp. 829–839, (2003)

# Throughput of a Network Shared by TCP Reno and TCP Vegas in Static Multi-hop Wireless Network

Sukant Kishoro Bisoy and Prasant Kumar Pattnaik

**Abstract** Previous study has shown the superiority of TCP Vegas over TCP Reno in wired and wireless network. However, when both Reno and Vegas co-exist on a wired link, protocol Vegas dominated by Reno due to its conservative nature. However, previous study does not include all the issues which might be arising during communication in the network. This paper finds the performance (throughput and fairness) of a network shared by Reno and Vegas in static multi hop wireless ad hoc network focusing on four issues. Here we introduced a new parameter called PST (Percentage Share of Throughput) which measures fairness among different connections shared by a common link. The issues are: (1) Both Reno and Vegas flows in forward direction, (2) Reno flows in forward and Vegas flows in reverse direction, (3) Both Reno and Vegas flows in forward direction along with bursty UDP traffic in background, (4) Both Reno and Vegas flows in forward direction where receiver is enabled with delayed acknowledgement (DelAck) schemes. Our result shows that Reno and Vegas are more compatible (or fairer) when DSDV protocol is used and their PST is much closer to each other (around 50 %). For increased hop length, DSDV is better routing protocol for all issues we have considered. However, use of DelAck techniques improves the throughput than others.

**Keywords** Reno · Vegas · Forward · Reverse · Background traffic · DelAck and PST

S.K. Bisoy (✉)
Department of Computer Science and Engineering, C. V. Raman College
of Engineering, Bhubaneswar, India
e-mail: sukantabisoyi@yahoo.com

P.K. Pattnaik
School of Computer Engineering, KIIT University, Bhubaneswar, India
e-mail: patnaikprasantfcs@kiit.ac.in

© Springer India 2016
H.S. Behera and D.P. Mohapatra (eds.), *Computational Intelligence
in Data Mining—Volume 1*, Advances in Intelligent Systems
and Computing 410, DOI 10.1007/978-81-322-2734-2_47

# 1   Introduction

MANET forms a network randomly using mobile nodes. In MANET, many routing protocols are available to route the packets from one node to another. Among them, AODV [1] and OLSR [2] are standardized by IETF MANET working groups [3]. AODV and DSR [4] are two reactive protocols and OLSR and DSDV [5] are proactive routing protocols. All of these routing protocols work at the network layer. TCP [6] is reliable protocol of the transport layer which provides in-order delivery of data to the TCP receiver. Previous study has shown that Reno and Vegas is not compatible when they share the same link in wired and wireless. In this work, we measure the performance in terms of aggregate throughput and fairness of a network shared by Reno and Vegas in wireless multi hop network. In order to analyze fairness, we introduce PST parameter which can judge the amount of throughput shared by each connection out of the aggregate (total) throughput.

The rest of the paper is structured as follows. Section 2 presents related work and Sect. 3 will explain TCP variants. Section 4 will present the simulation set up and Sect. 5 will explain results and analysis. Finally, we conclude our work in Sect. 6.

# 2   Related Work

Author [7] analyzed the interaction between TCP variants and routing protocols in MANET. Result shows that OLSR is best routing protocol with respect to throughput and packet loss ratio irrespective of TCP variants and provides a lower packet loss rate (15–20 %) than others in most situations. Author [8] formulated a practical framework to optimize the TCP flows, and proposed a distributed algorithm to improve the end-to-end throughput, and obtained per-flow fairness by exploiting cross-layer information. Author in [9] improves the fairness between Reno and Vegas by two methods. First method modifies TCP Vegas to compete with Reno and second method modifies random early detection (RED) to detect misbehaving flow so as to improve them. TCP fairness in wireless ad hoc wireless network is enhanced in [10] through a new scheme called Neighborhood RED (NRED). Author has shown that individual queue may not run with RED scheme in wireless network. Author in [11] shown that Reno and Vegas is more compatible through a careful analysis of router buffer space and proper configuration of Vegas parameter. Also proved the default configuration of Vegas is the main cause of incompatible and overall performance of a network can be increased with proper setting of Vegas parameter. Achieving fair sharing of network resource among TCP variants is possible through a novel approach mentioned in [12] by isolating delay-based and loss-based flows from each other via different queues. Author experimented and analyzed this approach using ns-2 simulator. Author in [13] analyzed the bandwidth sharing of responsive (FTP) and unresponsive flow (CBR) when they share a common link in the router. It is found that an unresponsive flow gets more bandwidth than responsive flow. So author has proposed an algorithm named RED with dynamic thresholds to achieve fair bandwidth of link capacity shared by them.

## 3 TCP Variants

Basic objective of TCP is to provide reliable communication over an unreliable network layer and to deliver the data from sender to receiver correctly and in order. In fact TCP has its variants of protocols like Reno and Vegas.

### 3.1 Reno

TCP Reno estimates the available bandwidth through packet loss in the network. For each round trip time the window size of Reno is increased by one until no packet loss occurs. Up on a packet loss Reno reduces its window size to half of the current window size. This process is called additive increase and multiplicative decrease (AIMD) which is based on the results given in [14]. This algorithm allocates fair amount of bandwidth between different flows. However Reno fails to achieve such fair bandwidth allocation because of not implementing rate based synchronized control techniques.

### 3.2 Vegas

TCP Vegas [15] uses different congestion control techniques which are quietly different from the previous one. It uses accurate estimation schemes to measure the bandwidth. It finds a difference amount of actual and expected rates to estimate the available bandwidth on the network and updates the window accordingly.

## 4 Simulation Set Up

In this, we analyze the performance in terms of aggregate throughput and fairness of a network shared by Reno and Vegas in wireless multi hop network. In order to analyze fairness properties we introduced PST value which can judge the amount of throughput of each connection out of the aggregate (total) throughput. The PST for each connection (%) is calculated by using the formula given below

$$PST_i = T_i/AT * 100.$$

where $i$ is the Reno or Vegas, $T_i$ is the throughput of each connection and AT ($= \sum T_i$) is the aggregate throughput of Reno and Vegas. In theory two connections are fair or compatible to each other if their $PST_i$ value is 50 %. In other ways, we say both connections use equal bandwidth of a link shared between them. Using NS2 simulator [16] we examined this with focus on four issues: (1) Both Reno and Vegas flows in forward direction, (2) Reno flows in forward and Vegas flows in reverse direction, (3) Both Reno and Vegas flows in forward direction along with background traffic (UDP), (4) Both Reno and Vegas flows in forward direction where receiver is enabled with DelAck schemes.

**Network Model and Parameter**

The basic network model used is shown in Fig. 1 consist of two sources (S1 and S2) and two destinations (D1and D2) where intermediate nodes are R1 to Rn. Connection 1 (TCP Reno) starts from S1 to D1 and connection 2 (TCP Vegas) starts from S2 to D2. Intermediate nodes R1 to Rn are organized in static chain topology. In our work with the number of nodes in the chain (N) varies from N = 2 to N = 10. Each connection stays for 1200 s long. Each node is separated by 200 m distances whereas transmission range of each node is 250 m. Both connections start their FTP traffic with packet size 512 bytes simultaneously at 0 s. The aggregate throughput of each connection is observed after the simulation started for 5 s.

Initially both Reno and Vegas connection flows from left to right (forward direction (issue 1) as shown in Fig. 1a. Then Reno flows in the forward direction and Vegas flows in the reverse direction (issue 2) as shown in Fig. 1b. Next issue 1 is used while running UDP as background (competing) traffic from S3 to D3 which flows simultaneously with Reno and Vegas as shown in Fig. 2. At last influence of delayed acknowledgement scheme is examined with issue 1 for topology shown in Fig. 1a.

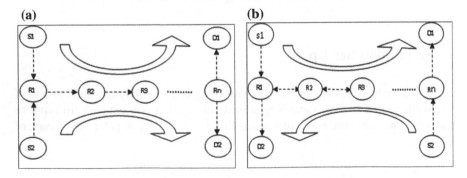

**(a)**                                        **(b)**

**Fig. 1** Topology with **a** forward traffic, **b** reverse traffic

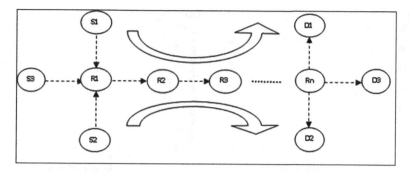

**Fig. 2** Topology with background traffic

# 5 Result and Analysis

In this section, we present the results gathered from the number of simulation experiments for different hop length considering all four issues mentioned earlier. Here we analyze two different properties of wireless network such as aggregate throughput and fairness (PST).

## 5.1 Issues with Forward Traffic

Initially we measure the aggregate throughput of Reno and Vegas for AODV, DSR and DSDV when they flow in the same direction (forward) using intermediate node R1 to R10 depends on the number of hops. In forward traffic, aggregate throughput of AODV is high up to chain length N = 3, then DSDV protocol takes lead slowly as shown in Fig. 3a. However, irrespective of ad hoc routing protocol used throughput decreases with increase of hop length. Generally when hop length increases the collision between a data packet and an ACK packet occurs more frequently and interference may cause more packet loss and decreases throughput. In a static environment, the maximum achievable throughput is limited by the interaction (at the MAC level) between neighboring nodes [17]. In addition, Throughput of Reno protocol decreases due to increase of window size aggressively and hidden node problem [18].

Then PST of each connection is measured to find fairness among Reno and Vegas using AODV, DSR and DSDV protocol. In theory, we say fairness between two connections is maintained, if PST value of each connection is 50 %. As Table 1 shows when AODV and DSR are used PST in Vegas is much higher than Reno after hop length reaches 3 (N > 3). In other ways, for increased number of hops, Vegas steals more bandwidth than Reno. However, Reno and Vegas maintain better fairness while using DSDV protocol as compared to AODV and DSR (see Table 1).

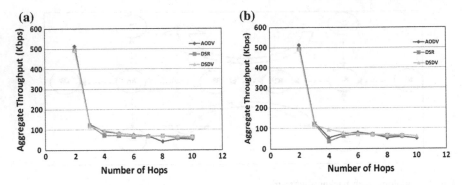

**Fig. 3** Aggregate throughput versus number of hops in chain (N). **a** Forward. **b** Reverse

**Table 1** Percentage share of throughput (PST) for forward traffic

| Number of hops in chain (N) | AODV | | DSDV | | DSR | |
|---|---|---|---|---|---|---|
| | Reno | Vegas | Reno | Vegas | Reno | Vegas |
| 2 hop | 50.38 | 49.62 | 50.28 | 49.72 | 51.18 | 48.82 |
| 3 hop | 83.87 | 16.13 | 67.15 | 32.85 | 68.78 | 31.22 |
| 4 hop | 34.00 | 66.00 | 52.68 | 47.32 | 16.40 | 83.60 |
| 5 hop | 22.85 | 77.15 | 54.96 | 45.04 | 15.66 | 84.34 |
| 6 hop | 23.47 | 76.53 | 55.42 | 44.58 | 15.52 | 84.48 |
| 7 hop | 39.10 | 60.90 | 60.30 | 39.70 | 16.34 | 83.66 |
| 8 hop | 19.37 | 80.63 | 58.02 | 41.98 | 12.00 | 88.00 |
| 9 hop | 15.30 | 84.70 | 59.80 | 40.20 | 20.00 | 80.00 |
| 10 hop | 12.90 | 87.10 | 64.83 | 35.17 | 16.00 | 84.00 |

## 5.2 Issues with Reverse Traffic

It has been shown that reverse traffic significantly impacts the performance of TCP, in wired and wireless network [19, 20] because acknowledgement of forward traffic and reverse traffic contend for same channel and pass in the same direction. In order to find influence of revere traffic we examined the performance of wireless network shared by Reno and Vegas. Here we measured the aggregate throughput and PST using AODV, DSR and DSDV protocol for topology shown in Fig. 1b. In this, Reno connection flows in the forward path and Vegas flows in the reverse path of Reno. As like forward traffic, AODV takes lead (aggregate throughput) up to hop length three, then DSDV protocol achieves higher throughput for increased hop length (N > 3) as shown in Fig. 3b. However, throughput of reverse traffic is slightly less as compared to forward traffic. From Table 2 we summarize that Reno and Vegas maintains better fairness keeping PST as close as 50 % using DSDV protocol. However for increased hop length (n > 3), Vegas dominates Reno by PST while using ADOV and DSR are used is routing protocol.

**Table 2** Percentage share of throughput (PST) for reverse traffic

| Number of hops in chain (N) | AODV | | DSDV | | DSR | |
|---|---|---|---|---|---|---|
| | Reno | Vegas | Reno | Vegas | Reno | Vegas |
| 2 hop | 51.10 | 48.90 | 51.23 | 48.77 | 51.00 | 49.00 |
| 3 hop | 70.20 | 29.80 | 69.40 | 30.60 | 73.10 | 26.90 |
| 4 hop | 44.47 | 55.53 | 52.68 | 47.32 | 39.66 | 60.34 |
| 5 hop | 50.70 | 49.30 | 56.85 | 43.15 | 22.46 | 77.54 |
| 6 hop | 18.55 | 81.45 | 55.51 | 44.49 | 11.64 | 88.36 |
| 7 hop | 33.43 | 66.57 | 56.84 | 43.16 | 13.90 | 86.10 |
| 8 hop | 17.26 | 82.74 | 61.60 | 38.40 | 12.30 | 87.70 |
| 9 hop | 12.50 | 87.50 | 57.48 | 42.52 | 19.20 | 80.80 |
| 10 hop | 17.82 | 82.18 | 62.47 | 37.53 | 17.73 | 82.27 |

## 5.3 Issues with Background Traffic

Next we measured the throughput and fairness (PST) of a network shared by Reno and Vegas in the presence of UDP as background traffic which flows from S3 to D3 as shown in Fig. 2. In background traffic, AODV protocol performs better irrespective of the number of hops as compared to DSDV and DSR. However, throughput of DSR and DSDV is below 20 kbps for the hop length N > 3. In presence of background traffic aggregate throughput of a network is lower as compared to forward and reverse traffic. However, Vegas dominates Reno in PST while using AODV and DSR as routing protocol. Among three routing protocols PST (fairness) of Reno and Vegas is better while using DSDV as routing protocol as shown in Table 3.

**Table 3** Percentage share of throughput (PST) for background traffic

| Number of hops in chain (N) | AODV | | DSDV | | DSR | |
|---|---|---|---|---|---|---|
| | Reno | Vegas | Reno | Vegas | Reno | Vegas |
| 2 hop | 58.00 | 42.00 | 57.26 | 42.74 | 57.00 | 43.00 |
| 3 hop | 62.00 | 38.00 | 79.65 | 20.35 | 69.47 | 30.53 |
| 4 hop | 46.20 | 53.80 | 40.00 | 60.00 | 34.26 | 65.74 |
| 5 hop | 30.50 | 69.50 | 53.70 | 46.30 | 28.80 | 71.20 |
| 6 hop | 20.90 | 79.10 | 40.30 | 59.70 | 24.65 | 75.35 |
| 7 hop | 38.60 | 61.40 | 34.70 | 65.30 | 20.55 | 79.45 |
| 8 hop | 17.23 | 82.77 | 42.42 | 57.58 | 49.48 | 50.52 |
| 9 hop | 24.53 | 75.47 | 82.70 | 17.29 | 28.37 | 71.63 |
| 10 hop | 19.27 | 80.73 | 54.60 | 45.40 | 42.46 | 57.54 |

## 5.4 Issues with Delayed Acknowledgement

As we know TCP receiver generates one acknowledgment denoted as ACK for each data packet that has received. If the data packet and ACK use the same path, then there may be a collision between them due to channel contention. Finally TCP throughput is reduced due to huge ACK packet generation. In order to find influence of DelAck we measured throughput of a network using DelAck schemes following Fig. 1. As shown in Fig. 4b throughput of AODV is slightly higher up to hop length 6, then DSR achieves high. From Table 4, it is found that Reno is dominated by Vegas with PSC value for increased hop length (N > 3) while using AODV and DSR protocol. However, DSDV maintains better fairness (PST close to 50 %) between Reno and Vegas as compared to AODV and DSR.

In order to find the best issues, we measure the performance of the network (aggregate throughput) for all issues like forward, backward, background and DelAck for specific routing protocol. From Figs. 5 and 6 we summarize that

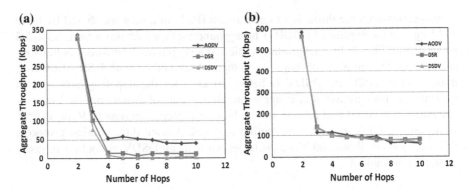

**Fig. 4** Aggregate throughput versus number of hops in chain (N). **a** Background traffic. **b** Delayed ack

**Table 4** Percentage share of throughput (PST) for delayed ACK schemes

| Number of hops in chain (N) | AODV | | DSDV | | DSR | |
|---|---|---|---|---|---|---|
| | Reno | Vegas | Reno | Vegas | Reno | Vegas |
| 2 hop | 51.33 | 48.67 | 51.40 | 48.60 | 51.87 | 48.13 |
| 3 hop | 75.72 | 24.28 | 78.37 | 21.63 | 71.00 | 29.00 |
| 4 hop | 23.70 | 76.30 | 51.00 | 49.00 | 21.78 | 78.22 |
| 5 hop | 29.87 | 70.13 | 59.00 | 41.00 | 22.76 | 77.24 |
| 6 hop | 43.78 | 56.22 | 55.13 | 44.87 | 15.80 | 84.20 |
| 7 hop | 16.00 | 84.00 | 57.82 | 42.18 | 25.00 | 75.00 |
| 8 hop | 14.87 | 85.13 | 57.18 | 42.82 | 27.20 | 72.00 |
| 9 hop | 21.00 | 79.00 | 58.46 | 41.54 | 20.80 | 79.00 |
| 10 hop | 20.50 | 79.50 | 55.42 | 44.58 | 11.00 | 89.00 |

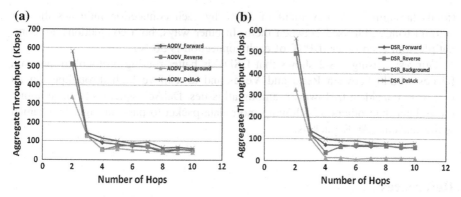

**Fig. 5** Aggregate throughput versus number of hops in chain (N). **a** AODV. **b** DSR

**Fig. 6** Aggregate throughput versus number of hops in chain (N)

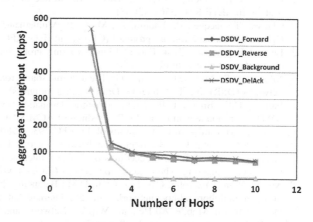

DelAck schemes achieve higher throughput than others irrespective of routing protocol used and hop length increased. DelAck schemes avoid collision in the channel by significantly reducing the number of ACK and thereby increases overall throughput. However, background traffic gets lower throughput as compared to others because in the presence of unresponsive traffic (UDP), responsive traffic like FTP cannot steal more bandwidth from the channel. In other ways they both are in compatible to each other while sharing the same channel.

# 6 Conclusions

We analyze the performance (aggregate throughput and fairness) of a network shared by Reno and Vegas in wireless multi hop network considering all four issues using three ad hoc routing protocols such as DSDV, AODV and DSR. In order to analyze fairness of a flow, we introduced PST parameter of each connection which can judge the amount of throughput used by each connection out of the aggregate (total) throughput. Achievement of 50 % by each connection indicates the fair amount bandwidth used between them. In other ways, both connections are compatible with each other if PST of each connection is 50 %.

Our result using NS-2 shows that DSDV is better routing protocol to maintain fairness (PST) between Reno and Vegas and to achieve high throughput on all issues we considered. However, among all issues, DelAck schemes achieve higher throughput than others by allowing more data packet to pass through the channel while reducing ACK packets.

# References

1. Perkins, C.E., Belding-Royer, E., Das, S.R.: Ad-hoc on-demand distance vector (AODV) routing. In: IETF RFC 3561. (2003)
2. Clausen, T., Jacquet, P., Laouiti, A., Minet, P., Muhlethaler, P., Qayyum, A., Viennot, L.: Optimized link state routing protocol (OLSR). In: IETF RFC 3626
3. Internet Engineering Task Force.: Manet working group charter. http://www.ietf.org/html. charters/manet-charter.html
4. Johnson, D.B., Maltz, D.A., Hu, Y.C.: The dynamic source routing protocol for mobile ad-hoc networks (DSR). In: IETF Internet Draft (work in progress). (2004)
5. Parkins, C.E., Bhagwat, P.: Highly dynamic destination sequence distance vector routing (DSDV) for mobile computers. In: Proceedings of ACM SIGCOMM'94, London, UK. (1994)
6. Postel, J.: Transmission control protocol. In: RFC 793. (1980)
7. Bisoy, S.K., Pattnaik, P.K.: Interaction between internet based TCP variants and routing protocols in MANET. In: Proceedings of Springer International Conference on Frontiers of Intelligent Computing: Theory and Applications (FICTA), vol. 247, pp. 423–433. (2013)
8. Xu, Y., Wang, Y., John, C.S. Lui, Chiu, D.M.: Balancing throughput and fairness for TCP flows in multihop ad-hoc networks. In: 5th International Symposium on Modeling and Optimization in Mobile, Ad Hoc and Wireless Networks and Workshops (WIOPT'07), April 16–20, Limassol, Cyprus, pp. 1–10. (2007)
9. Hasegawa, G., Kurata, K., Murata, M.: Analysis and improvement of fairness between TCP reno and vegas for deployment of TCP vegas to the internet. In: Proceeding of IEEE International Conference on Network Protocols, pp. 177–186. (2000)
10. Xu, K., Gerla, M., Qi, L., Shu, Y.: Enhancing TCP fairness in ad hoc wireless networks using neighborhood RED. In: Proceedings of the 9th Annual International Conference on Mobile Computing and Networking, Sept 14–19, San Diego, CA, USA. (2003)
11. Feng W., Vanichpun, S.: Enabling compatibility between TCP reno and TCP vegas. In: Proceeding of Symposium on Applications and the Internet (SAINT'03), Jan 2003, Florida, USA, pp. 301–308. (2003)
12. Podlesny, M., Williamson, C.: Providing fairness between TCP new reno and TCP vegas with RD network services. In: 18th International Workshop on Quality of Service (IWQoS), Beijing, Chaina, pp. 1–9. (2010)
13. Vukadinovic, V., Trajkovic, L.: RED with dynamic thresholds for improved fairness. In: Proceedings of the ACM Symposium on Applied Computing (SAC), Nicosia, Cyprus, March 14–17. (2004)
14. Chiu, D., Jain, R.: Analysis of the increase and decrease algorithms for congestion avoidance in computer networks. Comput. Netw. ISDN Syst. **17**, 1–14 (1989)
15. Brakmo, L., O'Malley, S., Peterson, L.: TCP vegas: new techniques for congestion detection and avoidance. In: Proceedings of ACM SIGCOMM, pp. 24–35. (1994)

16. Information Sciences Institute: The network simulator Ns-2. University of Southern California. http://www.isi.edu/nanam/ns/
17. Li, J., Blake, C., De C.D., Lee, H., Morris, R.: Capacity of ad hoc wireless networks. In: Proceedings of ACM/IEEE International Conference in Mobile Computing and Networking (MobiCom 2001), pp. 61–69. (2001)
18. Freeney, L.M.: An energy consumption model for performance analysis of routing protocols for mobile ad hoc networks. Mob. Netw. Appl. **6**(3), 239–249 (2001)
19. Balakrishnan, H., Padmanabhan, V.N., Fairhurst, G., Sooriyabandara, M.: TCP performance implications of network path asymmetry. In: IETF RFC 3449. (2002)
20. Balakrishnan, H., Padmanabhan, V.: How network asymmetry affects TCP. IEEE Commun. Mag. **39**(4), 60–67 (2001)

# A Pronunciation Rule-Based Speech Synthesis Technique for Odia Numerals

Soumya Priyadarsini Panda and Ajit Kumar Nayak

**Abstract** This paper introduces a model that uses a rule based technique to determine the pronunciation of numerals in different context and is being integrated with the waveform concatenation technique to produce speech from the input text in Odia language. To analyze the performance of the proposed technique, a set of numerals are considered in different context and a comparison of the proposed technique with an existing numeral reading algorithm is also presented to show the effectiveness of the proposed technique in producing intelligible speech out of the entered text.

**Keywords** Speech synthesis · Text to speech system · Rule based concatenative technique · Numerals · Odia language

## 1 Introduction

Recent research in the area of Speech and Language Processing enables machines to speak naturally like humans [1]. A Text-to-Speech (TTS) system in this aspect converts natural language text into its corresponding speech [2]. The intelligible speech synthesis systems have a widespread area of application in Human–computer interactive system like, talking computer systems, talking toys, etc. Text analysis is the front end language processor of the TTS system, which accepts input text, analyzes it and organizes into manageable list of words [3]. The input text may contain symbols (double quote, comma, report, etc.), numbers, abbreviations or special symbols.

Text normalization is an important problem in TTS systems which involve transformation of the raw input text into the equivalent of pronounceable units [4].

S.P. Panda (✉)
Department of CSE, Institute of Technical Education and Research,
Siksha 'O' Anusandhan University, Bhubaneswar, India
e-mail: sppanda.cse@gmail.com

A.K. Nayak
Department of CS&IT, Institute of Technical Education and Research,
Siksha 'O' Anusandhan University, Bhubaneswar, India
e-mail: ajitnayak@soauniversity.ac.in

© Springer India 2016
H.S. Behera and D.P. Mohapatra (eds.), *Computational Intelligence in Data Mining—Volume 1*, Advances in Intelligent Systems and Computing 410, DOI 10.1007/978-81-322-2734-2_48

It also involves converting all letters of lowercase or upper case, removing punctuations, accent marks, stop words or too common words (like "Don't" vs. "Do not", "I'm" vs. "I am", "Can't" vs. "cannot", etc.). Sentences are a group of word segments and these segments may be an acronym, a single word or a numeral. While for the abbreviations or acronyms, one time normalized pronunciations may be maintained, the pronunciation of numerals varies depending on the context of its use in the sentence or word [5]. A number need to be pronounced differently in different situations (e.g.: 2015-two thousand and fifteen (as quantifier) two zero one five (phone number)) and needs to be converted into their appropriate pronounceable forms to produce the desired speech outputs.

In this aspect most of the foreign languages are well researched [3], where the pronunciation rules are simpler due to the occurrence of pronunciation repetitions. For example in English language the numerals after 20 are pronounced as "Twenty one", "twenty two", …, "Thirty one", "Thirty two",…etc. However, the Indian language TTS techniques sill presents gap for its acceptance by the users due to the unavailability of appropriate pronunciation rules. Hardly, we can find any published work for the numeral to speech conversion. As the probability of repetition of pronunciation is relatively very less in Indian languages, the simple digit reading model that maintains the phonetic representation of the digits for producing the desired output speech is adopted by the researchers.

There are only fewer models documented for speech synthesis in Indian languages [6]. The dhvani TTS system for Indian language [7], maintains the pronunciations of numerals up to hundred as the phonetic representation and use the position pronunciations for 'hundred', 'thousand', etc. positions attached to the up to hundred pronunciation for reading the numerals. However, the context dependent numeral reading aspect is not considered in speech production. In this paper, we present a pronunciation rule based approach for the up to hundred pronunciations and incorporate it with the rule based concatenative technique to produce output speech for Odia numerals.

In the next section, we discussed about the rule based concatenative technique. Section 3 describes about the proposed model, Sect. 4 discusses about the result analysis of the proposed technique and Sect. 5 concludes the discussion explaining the findings of our experiments and the future directions of this work, where further work can be undertaken.

## 2    Rule-Based Concatenative Technique (RCT)

As compared to English, most of the Indian languages have approximately twice as many vowels and consonants along with a number of possible conjunct characters formed by combination of two or more characters [8]. Instead of storing hundreds of recorded speech units in the database to achieve highly natural speech segments, RCT [9] uses only 35 basic speech units of the consonant (C) and vowel (V) sounds and derive all other sound units from these basic units using a rule based waveform

Portion from "\ra"      Portion from "\ae"
sound                   sound

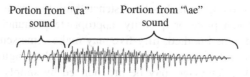

**Fig. 1** Wave pattern of *"\re"* (C-M) sound after concatenating portions from \ra and \ae sound

concatenation technique. Figure 1 shows the portion concatenation process to produce the sound "\re" from *"\ra"* and *"\ae"* in RCT.

# 3  Proposed Model

The input given to the TTS system is segmented text (words) from a sentence. For any Indian language, a word may contain the characters $\{ch_1, ch_2, ..., ch_L\}$ from the character set Ch or numerals $\{n_1, n_2, n_3, ..., n_M\}$ from the numerals set N, where L and M are the length of the word and numeral respectively or the number of Unicode units in the input text. Depending on the Unicode code point ranges for character set and numeral sets, the category of the word is determined. The rules for obtaining the pronounceable units are different for characters and numerals. While the characters forming a word in any Indian script may be pronounced as they are written, the numerals are needed to be parsed to readable units consisting of character units of pronunciation. i.e. the input numerals are normalized to a readable format before identifying the sound units involved. The context dependent pronunciation is also an important issue for the numerals. Figure 2 shows the overview of the proposed numeral reading module and the details of the phases are discussed next.

## 3.1  Context Dependent Numeral Pronunciation

As a number may be pronounced differently in different situations, simple digit reading may not provide the desired understandability in all aspects. For example, while reading a larger quantity or price say 1,45,955 by simply reading the digits as

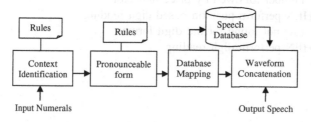

**Fig. 2** Text-to-speech conversion process

"one-four-five-nine-five-five" makes the listener think to rearrange the numbers to understand the spoken price or quantity, appropriate pronunciation as one-lakh, forty five-thousand, nine-hundred, fifty-five may make some sense to the listeners. The similar variation of pronunciation occurs in the Odia language.

A number in Odia language may be pronounced by simply reading the digits while mean for a quantity, phone number or credit card number, etc.; the number may be read by the relationship with its positions while meant for a price indicator or year. In case of a fraction value the left part before the decimal point is read based on the relationship between the position of the character and the numbers after the period are read as single digits. While reading a date people always read as "aek–saat–dui–hajaar–padara" for the date "01-07-2015" in triplet format (dd/mm/yyyy). In Odia language. For reading a time interval separated by a colon the format is different for the number before the colon and after the colon. To incorporate all the considered variation of a numeral pronunciation a set of manually coded rules are prepared. The rules for identifying the context dependent pronunciations are given below where $n$ is the number of digits in the number and $d_i$ is the ith digit in the number.

*Context dependent pronunciation rules*

Rule 1:  IF n >= 10
            AND no separation in between
            THEN perform digit reading
Rule 2:  IF n >= 10
            AND $d_i$ separated by ","
            THEN perform position based digit reading
Rule 3:  IF digits separated by "-"or "/" in a triplet format
            THEN perform date format digit reading
Rule 4:  IF digits separated by "-"
            THEN perform digit reading
Rule 5:  IF digits separated by ":"
            THEN perform time format digit reading (digit reading for digits before ":"
            and position based digits reading for digits after ":"
Rule 6:  IF digits separated by "."
            THEN Perform position based digit reading for digits before"." and digit reading for digits after "."
Rule 7:  IF number followed by price indicator
            THEN perform position based digit reading
Rule 8:  IF rule not found for the digit format
            THEN perform digit reading

## 3.2 Up to Hundred Pronunciation Rules

As we are considering a rule based approach, where new pronunciations may be derived based on the pronunciation units involved instead of available speech units in the database, first the pronounceable units are needed to be identified to produce the desired speech output. There is no generalized rule available for the pronunciation of numbers up to 100, due the pronunciation variation of numerals in different Indian languages. While, the probability of repetition of pronunciation is relatively very less in Indian languages (e.g.: pronunciations of the numbers in Odia: *21-"ae-ko-is", 22-"baa-is", 23-"te-is"*, etc.), for numbers greater than 100, a repetition of pronunciation occurs (e.g.: 122-"*ae-ka-saha-baa-is*", 123-"*ae-ka-saha-te-is*", 2100-"*dui-hajara-ae-ka-saha*", etc.). Therefore, the numerals after 100 may be formed by concatenating the 100th 1000th,...etc. place pronunciations with their respective up to 100 pronunciations.

As in our approach the main focus is to reduce the memory requirement, keeping the speech quality into consideration, storing all the desired units may not be the best solution. Therefore, we try to find out the pronunciation similarities among the numbers. For example in the Odia scripts, the pronunciation similarities for a multi digit number with a 2 at the tenth place have the pronunciation "*\is*" at the end. Similarly digit 2 at the unit position, the pronunciation may have a "*\ba*" at the beginning of the pronunciation. Table 1 lists the pronunciation and the repetitions in the pronunciations for the Odia numerals up to 100 starting from 11. The pronunciation of 1–9 are needed to be maintained as {"aeka", "dui", "tin", "chari", "pancha", "chha", "saata", "aatha", "na"} along with the unit place pronunceations as {"sun", "dasa", "saha", "hajara", "lakhya", "koti"}. However there are certain variations in the pronunciations needed to be handled separately where dissimilarity in the pronunciation exists as highlighted in the Table 1.

To derive all the pronunciations for the numerals, we have prepared different groups considering the above discussed similarities. We have classified the pronounceable units to be into three state groups as: Begin state (B), Middle state (M) and End state (E). Depending on the position of the number i.e. unit or tenth, the states are determined and the pronunciation is derived. For example for obtaining the pronunciation of a number N having length L, as $\{n_1, n_2, ..., n_L\}$, There exist 3 states representatives of the pronunciations, {B, M, E} for the unit and tenth positions, $n_L$ and $n_{L-1}$ respectively and the units from $n_1$ to $n_{L-1}$ may be derived using the common pronunciation rules by concatenating hundredth, thousandth, etc. position's pronunciation with the up to 100 pronunciation. For example, in producing the pronunciation of the numeral 11 as "*aek-gaa-ra*" in Odia language the units involved in the pronunciation are "\aek" from "B set", "\gaa" from "M set" and "\ra" from "E set" as shown in Fig. 3. i.e. B(L)-M(L−1)-E(L−1). However the above repetitive pronunciation is not same for numbers having a 9 or 0 at the unit place. To overcome this we separate the numbers of these category from the groups and produce their pronunciation by maintaining special cases of pronunciation as {9: "*na*", 19: "*une-is*", 29: "*ana-tiris*", etc.}, {20: "*ko-die*", 30: "*tiris*", etc.}

**Table 1** Pronunciation of Odia numerals with starting and ending similarities

| Pronunciation of numerals | starting unit as "aek" | starting unit as "baa" | starting unit as "te" | starting unit as "cha" | starting unit as "pan" | starting unit as "chau" | starting unit as "sat" | starting unit as "ath" | starting unit as "ana" |
|---|---|---|---|---|---|---|---|---|---|
| Ending unit as "ra" | aek-gaa-ra | Baa-ra | Te-ra | **Cha-uda** | Pan-da-ra | **So-hala** | Sa-ta-ra | Ath-a-ra | **Une-is** |
| Ending unit as "is" | aek-ko-is | Baa-is | Te-is | **Cha-b-is** | Pan-ch-is | Chha-b-is | Sa-te-is | Ath-e-is | Ana-tir-is |
| Ending unit as "is" | aek-tir-is | Ba-t-is | Te-t-is | Cha-u-tir-is | **Pain-tir-is** | Chha-t-is | Sa-in-tir-is | Ath-tir-is | Ana-chaa-lis |
| Ending unit as "lis" | aek-chaa-lis | Ba-ya-lis | Te-yaa-lis | Cha-u-raa-lis | **Pain-chaa-lis** | Chha-a-yaa-lis | Sa-t-chaa-lis | Ath-chaa-lis | Ana-chaas |
| Ending unit as "ban" | aek-aa-ban | Baa-ban | **Te-pan** | Cha-u-ban | Pan-chaa-ban | **Chha-a-pan** | Sa-taa-ban | Ath-aa-ban | Ana-sathi |
| Ending unit as "sathi" | aek-a-sathi | Baa-sathi | Te-sathi | Cha-u-sathi | Pan-cha-sathi | Chha-a-sathi | Sa-t-sathi | Ath-a-sathi | Ana-saturi |
| Ending unit as "stari" | aek-a-stari | Baa-stari | Te-stari | Cho-u-stari | Pan-cha-stari | Chha-a-stari | Sa-ta-stari | Ath-a-stari | Ana a-asi |
| Ending unit as "asi" | aek-aa-asi | Ba-yaa-asi | Te-yaa-asi | Cha-u-raa-asi | Pan-chaa-asi | Chha-a-yaa-asi | Sa-taa-asi | Ath-aa-asi | Ana a-nabe |
| Ending unit as "nabe" | aek-aa-nabe | Ba-yaa-nabe | Te-yaa-nabe | Cha-u-raa-nabe | Pan-cha-nabe | Chha-a-yaa-nabe | Sa-taa-nabe | Ath-aa-nabe | Ana-sat |

However, for some units, the pronunciation rules does not include the middle state, for example for the numeral two at the unit place, the mapping may be "\baa" from the B state and the next one is from the E state as "\ra" to form the pronunciation "baa-ra" for the numeral "12".

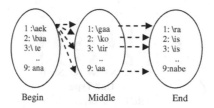

**Fig. 3** Possible states of a numeral in Odia language

The up to 100 pronunciation may fail for certain numerals, e.g.: consider the numeral 14 pronunced as "cha-uda". This does not follow the pronunciation similarities. An obvious (brute force) workaround is to have a small dictionary of such dis-similar units, and check whether a given number matches any of them at the beginning of text analysis phase. If so, break it up into the corresponding pronounceable units separately and parse them to the next phase separately. This works satisfactorily, and we've implemented this with a few numerals (14-"cha-uda", 16-"so-ha-la", 35-"pain-tir-is", 53-"te-pan", 56-"chha-pan", etc.).

## 3.3 Speech Database Mapping and Waveform Concatenation

As the model uses the RCT technique to produce the desired output speech, the respective base sound units in the speech database are needed to be obtained for performing waveform concatenation to produce the output speech. The speech database mapping phase identifies the respective speech database units to perform rule based waveform concatenation. The RCT technique is then used to produce the desired output speech for the Odia numeral. Figure 4 shows the portion concatenation process for the numeral "1" pronounced as "ae-ka" in Odia language from the two speech database units "\ae" and "\ka".

Portion from
"\ae" sound

Portion from
"\ka" sound

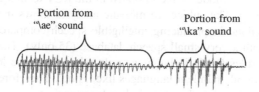

**Fig. 4** Wave pattern of numeral "1" (one) in Odia ('ଏ') pronounced as "\ae-ka"

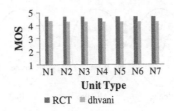

**Fig. 5** MOS test results for understandability

# 4 Result Analysis

The proposed numeral reading technique and the RCT technique is implemented in C/C++ and is being tested for producing different types of numerals in different context in Odia language. To analyze the quality of the produced speech, we use the most common approach adopted by most of the researchers, the Mean Opinion Score (MOS) test [8]. For performing the MOS tests, a set of random numerals, $N_1$, $N_2$, ..., $N_7$ are selected representing different category of pronunciations for the specified rules. The output speech is generated by the proposed technique as well as by the existing numeral reading technique used in dhvani TTS system. A group of 15 native Odia speakers are selected to perform the listeners test and are asked to give their feedback on the basis of ease of understandability on output speech produced by different techniques in a 5 point scale (1–5: very low to high). All the tests were performed with a headphone set. Figure 5 shows the average MOS test results by all listeners for different numerals by different techniques. The results of all our experiments shows the effectiveness of the proposed technique in producing better quality speech compared to the existing technique.

# 5 Conclusions

In this paper, we describe a pronunciation rule based method for speech synthesis for Odia numerals. The model is incorporated in the RCT technique to produce the desired output speech. The subjective measure analysis shows the effectiveness of the proposed technique in producing intelligible speech compared to the existing technique, even with a very small speech database (35 only). The model provides the pronunciation rules for Odia language only while the same level of similarity may be observed in other Indian languages like, Hindi. Therefore, the model may further be enhanced to work for other Indian languages.

# References

1. Feng, J., Ramabhadran, B., Hansel, J., Williams, J.D.: Trends in speech and language processing. IEEE Signal Process. Mag. (Jan, 2012)
2. Nukaga, N., Kamoshida, R., Nagamatsu, K., Kitahara, Y.: Scalable implementation of unit selection based text-to-speech system for embedded solutions. In: ICASSP, IEEE (2006)
3. Raj, A.K., Sarkar, T., Pammi, S.C., Yuvaraj, S., Bansal, M., Prahallad, K., Black, A.W.: Text processing for text-to-speech systems in Indian languages. In: 6th ISCA Workshop on Speech Synthesis (2007)
4. Alías, F., Sevillano, X., Socoró, J.C., Gonzalvo, X.: Towards high-quality next-generation text-to-speech synthesis: a multidomain approach by automatic domain classification. IEEE Trans. Audio Speech Lang. Process. **16**(7) (2008)
5. Tiomkin, S., Malah, D., Shechtman, S., kons, Z.: A hybrid text-to-speech system that combines concatenative and statistical synthesis units. IEEE Trans. Audio Speech Lang. Process. **19**(5) (2011)
6. Narendra, N.P., Rao, K.S., Ghosh, K., Vempada, R.R., Maity, S.: Development of syllable-based text to speech synthesis system in Bengali. Int. J. Speech Technol. **14**(3), 167–181 (2011)
7. http://dhvani.sourceforge.net/ 10th Feb 2015
8. Panda, S.P., Nayak, A.K.: An efficient model for text-to-speech synthesis in Indian languages. Int. J. Speech Technol. **18**(3), 305–315 (2015)
9. Panda, S.P., Nayak, A.K.: A rule-based concatenative approach to speech synthesis in Indian language text-to-speech systems. In: Proceedings of ICCD, pp. 523–531 (2014)

# References

# Erratum to: A New Approach on Color Image Encryption Using Arnold 4D Cat Map

**Bidyut Jyoti Saha, Kunal Kumar Kabi and Chittaranjan Pradhan**

**Erratum to:**
**Chapter 'A New Approach on Color Image Encryption**
**Using Arnold 4D Cat Map' in: H.S. Behera**
**and D.P. Mohapatra (eds.),** *Computational Intelligence*
*in Data Mining—Volume 1,* **Advances in Intelligent Systems**
**and Computing 410, DOI 10.1007/978-81-322-2734-2_14**

The original version of the book's chapter 'A New Approach on Color Image Encryption Using Arnold 4D Cat Map' was inadvertently published with an incorrect affiliation as 'Industrial and Financial Systems, Capgemini, Bhubaneshwar, India' whereas the correct affiliation is 'Industrial and Financial Systems, Capgemini, Mumbai, India'. The erratum chapter and the book has been updated with the correction.

---

The updated original online version for this chapter can be found at
DOI 10.1007/978-81-322-2734-2_14

---

B.J. Saha (✉)
Industrial and Financial Systems, Capgemini, Mumbai, India
e-mail: bidyutjyotisaha@gmail.com

K.K. Kabi
Guidewire Practice, CGI, Bhubaneswar, India
e-mail: kunal.kabi90@gmail.com

C. Pradhan
School of Computer Engineering, KIIT University, Bhubaneshwar, India
e-mail: chitaprakash@gmail.com

© Springer India 2016                                                                          E1
H.S. Behera and D.P. Mohapatra (eds.), *Computational Intelligence*
*in Data Mining—Volume 1,* Advances in Intelligent Systems
and Computing 410, DOI 10.1007/978-81-322-2734-2_49

# Author Index

**A**

Abhishek, Kumar, 73
Abirami, S., 325
Abraham, Siby, 347
Ahmed, Shoiab, 171
Ali, Kadampur, 367
Anusha, D., 399

**B**

Bebarta, D.K., 377
Behera, H.S., 57
Bhalchandra, Parag, 149, 411
Bisoy, Sukant Kishoro, 471
Boggaram, Achyut Sarma, 1
Bormane, D.S., 49

**C**

Chahal, Rajanpreet Kaur, 337
Chidambaram, S., 103
Chopra, Vinay, 159

**D**

Dam, Bivas, 273
Danti, Ajit, 171
Darkunde, Nitin, 149
Das, Asit Kumar, 227
Das, P.K., 181
Das, Sunanda, 227
De, Soumitra, 139
Delhibabu, Radhakrishnan, 95
Deolekar, Subodh, 347
Dash, Judhisthir, 273
Deshmukh, Nilesh, 411
Divya, T.L., 121
Dubey, Sanjay Kumar, 39
Dwivedi, Kalpana, 39

**E**

Enugala, Raju, 367

**G**

Garg, Siddhant, 459
Gowri, R., 191, 199
Gupta, Swadha, 429

**H**

Haritha, K., 113
Hemantha Kumar, G., 421
Hema, P.H., 217

**J**

Jawaharlal, M., 113
Joshi, Mahesh, 149

**K**

Kabi, Kunal Kumar, 131
Kalyan Chakravathi, P., 113
Kameswara Rao, T., 293
Kaur, Pradeep, 459
Kedar, Seema, 49
Khamitkar, Santosh, 149, 411
Krishnan, Rumesh, 325
Kurapati, Sravanthi, 367

**L**

Lokhande, Sakharam, 149, 411

**M**

Mahata, Dulal, 85
Mallampalli, Pujitha Raj, 1
Manjusha, R., 1
Manoj, Palavlasa, 447
Mekewad, Satish, 411

© Springer India 2016
H.S. Behera and D.P. Mohapatra (eds.), *Computational Intelligence in Data Mining—Volume 1*, Advances in Intelligent Systems and Computing 410, DOI 10.1007/978-81-322-2734-2

Mishra, Jaydev, 139
Misra, M., 181
Misra, U.K., 181
Muley, Aniket, 149
Murali Krishna, R.V.V., 261
Mustafi, A., 237
Mustafi, D., 237
Muthyala, Chandrasekhar Reddy, 1

**N**
Nagesh Kumar, G.V., 399, 439
Naik, Bighnaraj, 57
Nair, Vaishnavi, 49
Nayak, Ajit Kumar, 483
Nayak, Ameeya Kumar, 13
Nayak, Janmenjoy, 57

**O**
Om, Hari, 29

**P**
Padhy, B.P., 181
Pal, Souvik, 387
Panda, Soumya Priyadarsini, 483
Patra, Chanchal, 85
Pattanayak, Hadibandhu, 13
Pattnaik, Prasant Kumar, 387, 471
Paul, Animesh, 249
Poovammal, E., 283
Pradhan, Chittaranjan, 131
Prasad, P.M.K., 113
Prasad, T.V., 293
Prasanna, P.S., 439

**R**
Rajalakshmy, K.R., 357
Rajamani, Lakshmi, 367
Raja, M. Arun Manicka, 307
Rama Sree, S., 209
Ramesh, S.N.S.V.S.C., 209
Rathipriya, R., 191, 199
Remya, P.C., 357
Reshma, T.J., 317

**S**
Saha, Bidyut Jyoti, 131
Saha, Sagnika, 387
Sahoo, G., 237
Samanta, P., 181
Sastry, G.V.K.R., 21
Satyananda Reddy, Ch., 261
Satyanarayana Raju, K., 21
Sethy, Nirakar Niranjan, 13
Sharma, Kavita, 159
Sharma, Sumit, 429
Singh, Avadhesh, 73
Singh, M.P., 73
Singh, Rishav, 29
Singh, Sarbjeet, 337
Singh, Trailokyanath, 13
Singh, Vikram, 459
Sivabalan, S., 191, 199
Sivarathinabala, M., 325
Soumya, A., 421
Srinivasagan, K.G., 103
Sunitha, C., 217
Suresh Kumar, L.V., 399, 439
Surya Kalyan, G., 21
Surya Prasath, V.B., 95
Swain, Rajkishore, 273
Swamynathan, S., 307

**T**
Tejeswar Rao, K., 21

**U**
Upadhyay, Shubhnkar, 73

**V**
Vareed, Jucy, 317
Venkata Ramana, Datti, 447
Venkatesh, G., 377
Vidhyalakshmi, M.K., 283
Vijayalakshmi, M.N., 121

**W**
Wasnik, Pawan, 149, 411

Printed in the United States
By Bookmasters